Geographical
Information Handling
– Research and Applications

Geographical Information Handling – Research and Applications

Edited by
Paul M. Mather
Department of Geography, University of Nottingham, UK

JOHN WILEY & SONS
Chichester • New York • Brisbane • Toronto • Singapore

Copyright © 1993 by John Wiley & Sons Ltd,
Baffins Lane, Chichester,
West Sussex PO19 1UD, England

All rights reserved.

No part of this book may be reproduced by any means,
or transmitted, or translated into a machine language
without the written permission of the publisher.

Other Wiley Editorial Offices

John Wiley & Sons, Inc., 605 Third Avenue,
New York, NY 10158-0012, USA

Jacaranda Wiley Ltd, G.P.O. Box 859, Brisbane,
Queensland 4001, Australia

John Wiley & Sons (Canada) Ltd, 22 Worcester Road,
Rexdale, Ontario M9W 1L1, Canada

John Wiley & Sons (SEA) Pte Ltd, 37 Jalan Pemimpin #05-04,
Block B, Union Industrial Building, Singapore 2057

Library of Congress Cataloging-in-Publication Data
Mather, Paul M.
 Geographical Information Handling : research and applications /
edited by Paul M. Mather.
  p. cm.
 Includes bibliographical references and index.
 ISBN 0-471-94060-7
 1. Geographical information systems. I. Title.
G70.2.M38 1994
025.06'91—dc20   93-8835
                                                         CIP

British Library Cataloguing in Publication Data

A catalogue record for this book is available from the British Library

ISBN 0-471-94060-7

Typeset in 10/12pt Times from author's disks by Text Processing Department,
John Wiley & Sons Ltd, Chichester
Printed and bound in Great Britain by Bookcraft (Bath) Avon

Contents

List of contributors	vii
Acknowledgements	xi

1	Introduction: the ESRC/NERC Joint Programme on Geographical Information Handling P. M. Mather	1
2	Geographical Information Systems: the environmental view J. Knill	7
3	Geographical Information Systems: an ESRC perspective H. Newby	17
4	Commercial impacts of the ESRC/NERC Joint Programme on Geographical Information Handling M. J. Jackson	19
5	GIS research and development: the view from Southampton D. W. Rhind and I. T. Logan	23

SECTION I: BASIC RESEARCH — 31

6	Meeting expectations: a review of GIS performance issues B. M. Gittings, T. M. Sloan, R. G. Healey, S. Dowers and T. C. Waugh	33
7	Simulating the effects of error in GIS C. Brunsdon and S. Openshaw	47
8	Towards a generalization machine to minimize generalization effects within a GIS E. João, G. Herbert, D. W. Rhind, S. Openshaw and J. Raper	63
9	Datum transformations and data integration in a GIS environment A. H. Dodson and R. H. Haines-Young	79
10	The object-based paradigm for a geographical database system: modelling, design and implementation issues M. F. Worboys, K. T. Mason and B. R. P. Dawson	91

11	Spatiotemporal GIS techniques for environmental modelling M. A. O'Conaill, D. C. Mason and S. B. M. Bell	103
12	Development of a generic spatial language interface for GIS J. Raper and M. Bundock	113

SECTION II: ENVIRONMENTAL APPLICATIONS OF GIS — 145

13	Analytical tools to monitor urban areas M. J. Barnsley, S. L. Barr, A. Hamid, J.-P. A. L. Muller, G. J. Sadler and J. W. Shepherd	147
14	The application of GIS to the monitoring and modelling of land cover and use in the United Kingdom R. W. Gooding, D. C. Mason, J. J. Settle, N. Veitch and B. K. Wyatt	185
15	GIS and distributed hydrological models R. Romanowicz, K. Beven and R. Moore	197
16	Automated derivation of stream-channel networks and selected catchment characteristics from digital elevation models J. Hogg, J. E. McCormack, S. A. Roberts, M. N. Gahegan and B. S. Hoyle	207

SECTION III: ECONOMIC AND PLANNING APPLICATIONS OF GIS — 237

17	Mapping natural hazards with spatial modelling systems G. Wadge, A. Wislocki, E. J. Pearson and J. B. Whittow	239
18	An evaluation of GIS as an aid to the planning of proposed developments in rural areas D. A. Davidson, A. I. Watson and P. H. Selman	251
19	Property-based GIS: data supply and conflict D. J. Fairbairn	261
20	The impact of GIS on local government in Great Britain I. Masser and H. Campbell	273
21	An information system for the Eastern Thames Corridor A. Church, S. John, J. W. Shepherd, M. Frost and A. Macmillan	287
22	Intelligent, interactive and analysis-based GIS: principles and applications M. Clarke, J. Chesworth, J. Harmer, A. McDonald, Y. L. Sui and A. Wilson	325

Index — 339

List of contributors

M. J. Barnsley, *Remote Sensing Unit, Department of Geography, University College London, 26 Bedford Way, London WC1H 0AP, UK*

S. L. Barr, *Remote Sensing Unit, Department of Geography, University College London, 26 Bedford Way, London WC1H 0AP, UK*

S. B. M. Bell, *NERC Unit for Thematic Information Systems (NUTIS), Department of Geography, University of Reading, Whiteknights, Reading RG6 2AB, UK*

K. Beven, *Centre for Research on Environmental Systems and Statistics, Institute of Environmental and Biological Sciences, Lancaster University, Lancaster LA1 4YQ, UK*

C. Brunsdon, *Department of Geography, University of Newcastle upon Tyne, Newcastle upon Tyne NE1 7RU, UK*

M. Bundock, *Department of Geography, Birkbeck College, 7–15 Gresse Street, London W1P 1PA, UK*

H. Campbell, *Department of Town and Regional Planning, University of Sheffield, Sheffield S10 2TN, UK*

J. Chesworth, *School of Geography, University of Leeds, Leeds LS2 9JT, UK*

A. Church, *South-East Regional Research Laboratory, Department of Geography, Birkbeck College, 7–15 Gresse Street, London W1P 1PA, UK*

M. Clarke, *Managing Director, GMap Ltd, GMap House, Cromer Terrace, Leeds, LS2 9JU, UK*

D. A. Davidson, *Department of Environmental Science, University of Stirling, Stirling, FK9 4LA, UK*

B. R. P. Dawson, *Centre for Land Use and Environmental Sciences, University of Aberdeen, Aberdeen, UK*

A. H. Dodson, *Institute of Engineering Surveying and Space Geodesy, University of Nottingham, University Park, Nottingham NG7 2RD, UK*

S. Dowers, *Parallel Architectures Laboratory for GIS, Department of Geography, University of Edinburgh, Drummond Street, Edinburgh EH8 9XP, UK*

D. J. Fairbairn, *Department of Surveying, University of Newcastle upon Tyne, Newcastle upon Tyne NE1 7RU, UK*

M. Frost, *Department of Geography, King's College, University of London, UK*

M. N. Gahegan, *School of Computer Studies, University of Leeds, Leeds LS2 9JT, UK*

B. M. Gittings, *Parallel Architectures Laboratory for GIS, Department of Geography, University of Edinburgh, Drummond Street, Edinburgh EH8 9XP, UK*

R. W. Gooding, *NERC Unit for Thematic Information Systems (NUTIS), Department of Geography, University of Reading, Whiteknights, Reading RG6 2AB, UK*

R. H. Haines-Young, *Department of Geography, University of Nottingham, University Park, Nottingham NG7 2RD, UK*

A. Hamid, *Department of Photogrammetry and Surveying, University College London, Gower Street, London WC1E 6BT, UK*

J. Harmer, *School of Geography, University of Leeds, Leeds LS2 9JT, UK*

R. G. Healey, *Parallel Architectures Laboratory for GIS, Department of Geography, University of Edinburgh, Drummond Street, Edinburgh EH8 9XP, UK*

G. Herbert, *Smallword Systems Ltd, Burleigh House, 13–15 Newmarket Road, Cambridge CB5 8EG, UK*

J. Hogg, *School of Geography, University of Leeds, Leeds LS2 9JT, UK*

B. S. Hoyle, *Department of Electronic and Electrical Engineering, University of Leeds, Leeds LS2 9JT, UK*

M. J. Jackson, *LaserScan Ltd, Cambridge Science Park, Milton Road, Cambridge, CB4 4FY, UK*

E. João, *Department of Geography, London School of Economics, Houghton St., London WC2A 2AE, UK*

S. John, *South-East Regional Research Laboratory, Department of Geography, Birkbeck College, 7–15 Gresse Street, London W1P 1PA, UK*

J. Knill, *Chairman, Natural Environment Research Council, Polaris House, North Star Avenue, Swindon, SN2 1EU, UK*

I. T. Logan, *Ordnance Survey, Romsey Road, Maybush, Southampton SO9 4DH, UK*

List of Contributors

A. Macmillan, *SERPLAN, 14, Buckingham Gate, London SW1E 6LB, UK*

D. C. Mason, *NERC Unit for Thematic Information Systems (NUTIS), Department of Geography, University of Reading, Whiteknights, Reading RG6 2AB, UK*

K. T. Mason, *Department of Geography, Keele University, Keele ST5 5BG, UK*

I. Masser, *Department of Town and Regional Planning, University of Sheffield, Sheffield S10 2TN, UK*

P. M. Mather, *Department of Geography, University of Nottingham, University Park, Nottingham NG7 2RD, UK*

J. E. McCormack, *School of Computer Studies, University of Leeds, Leeds LS2 9JT, UK*

A. McDonald, *School of Geography, University of Leeds, Leeds LS2 9JT, UK*

R. Moore, *Institute of Hydrology, Crowmarsh Gifford, Wallingford, Oxfordshire OX10 8BB, UK*

J.-P. A. L. Muller, *Department of Photogrammetry and Surveying, University College London, Gower Street, London WC1E 6BT, UK*

H. Newby, *Chairman, Economic and Social Research Council, Polaris House, North Star Avenue, Swindon, SN2 1UJ, UK*

M. A. O'Conaill, *NERC Unit for Thematic Information Systems (NUTIS), Department of Geography, University of Reading, Whiteknights, Reading RG6 2AB, UK*

S. Openshaw, *School of Geography, University of Leeds, Leeds LS2 9JT, UK*

E. J. Pearson, *NERC Unit for Thematic Information Systems (NUTIS), Department of Geography, University of Reading, Whiteknights, Reading RG6 2AB, UK*

J. Raper, *Department of Geography, Birkbeck College, 7–15 Gresse Street, London W1P 1PA, UK*

D. W. Rhind, *Director-General, Ordnance Survey, Romsey Road, Maybush, Southampton SO9 4DH, UK*

S. A. Roberts, *School of Computer Studies, University of Leeds, Leeds LS2 9JT, UK*

R. Romanowicz, *Centre for Research on Environmental Systems and Statistics, Institute of Environmental and Biological Sciences, Lancaster University, Lancaster LA1 4YQ, UK*

G. J. Sadler, *South-East Regional Research Laboratory, Department of Geography, Birkbeck College, 7–15 Gresse Street, London W1P 1PA, UK*

P. H. Selman, *Cheltenham and Gloucester College of Higher Education, Cheltenham GL50 4AZ, UK*

J. J. Settle, *NERC Unit for Thematic Information Systems (NUTIS), Department of Geography, University of Reading, Whiteknights, Reading RG6 2AB, UK*

J. W. Shepherd, *South-East Regional Research Laboratory, Department of Geography, Birkbeck College, 7–15 Gresse Street, London W1P 1PA, UK*

T. M. Sloan, *Parallel Architectures Laboratory for GIS, Department of Geography, University of Edinburgh, Drummond Street, Edinburgh EH8 9XP, UK*

Y. L. Sui, *School of Geography, University of Leeds, Leeds LS2 9JT, UK*

N. Veitch, *The Environmental Information Centre, Institute of Terrestrial Ecology, Monks Wood, Abbots Ripton, Peterborough, Cambridgeshire PE17 2LS, UK*

G. Wadge, *NERC Unit for Thematic Information Systems (NUTIS), Department of Geography, University of Reading, Whiteknights, Reading RG6 2AB, UK*

A. I. Watson, *Department of Environmental Science, University of Stirling, Stirling, FK9 4LA, UK*

T. C. Waugh, *Parallel Architectures Laboratory for GIS, Department of Geography, University of Edinburgh, Drummond Street, Edinburgh EH8 9XP, UK*

J. B. Whittow, *Department of Geography, University of Reading, Whiteknights, Reading RG6 2AB, UK*

A. Wilson, *Vice-Chancellor, University of Leeds, Leeds LS2 9JT, UK*

A. Wislocki, *NERC Unit for Thematic Information Systems (NUTIS), Department of Geography, University of Reading, Whiteknights, Reading RG6 2AB, UK*

M. F. Worboys, *Department of Computer Science, Keele University, Keele ST5 5BG, UK*

B. K. Wyatt, *The Environmental Information Centre, Institute of Terrestrial Ecology, Monks Wood, Abbots Ripton, Peterborough, Cambridgeshire PE17 2LS, UK*

Acknowledgements

My thanks are due to all those who participated in the Joint Programme, especially the principal investigators and their research assistants. Their patience and forbearance during my occasional visits were appreciated. It is good to see so many of the latter group now working in the GIS field, both within and outside the academic arena. I also thank the members of the Steering Committee, and its chairman, Lord Chorley, for their guidance and their patience. The secretarial support from both ESRC and NERC was invaluable. Amanda Hewes of John Wiley & Sons Ltd is a model editor – helpful, undemanding and, apparently, confident in the ultimate ability of an editor to deliver! In this last task I have been aided by the authors, who have been cooperative and efficient. It is no small feat on their part to have a manuscript such as this assembled within a dozen weeks of the end-of-programme conference. Lastly, the valuable assistance of Alison Glazebrook and her colleagues at the Royal Geographical Society helped considerably in the organizing of that conference. The logistical arrangements required for a two-day conference with 250+ delegates were handled smoothly and effectively.

Chapter One

Introduction: the ESRC/NERC Joint Programme on Geographical Information Handling

P. M. MATHER
Department of Geography, University of Nottingham

In 1989 the UK Economic and Social Research Council (ESRC) and the Natural Environment Research Council (NERC) instituted a three-year Joint Programme of research and research training in Geographical Information Handling, following the recommendations of the Department of the Environment's Committee of Enquiry into Handling Geographic Information, chaired by Lord Chorley (Department of the Environment, 1987). The purpose of the programme was to increase British research and development effort in the area of geographical information handling and to encourage greater collaboration between the environmental and the social sciences. This volume represents a summary of the research findings of the Joint Programme, which were presented at a two-day conference held at the Royal Geographical Society, London, on 4/5 February 1993.

The principal objectives of the Joint Programme were:

1. To carry out basic research into the handling of geographic information using newly developed geographical information systems (GIS) technology.
2. To develop and apply GIS methods to selected applications areas and to demonstrate their value.
3. To encourage the transfer of results from information technology (IT) programmes elsewhere.
4. To generate additional training opportunities in GIS.

The programme had a budget of approximately £1 million, and its activities were supervised by a Steering Committee, under the chairmanship of Lord Chorley, which included representatives from ESRC and NERC, local and central government, and industry. A total of 17 research projects were selected, by the normal peer review process, from a wide range of submissions, and in addition 12 studentships were allocated to the Universities of

---

Geographical Information Handling – Research and Applications. Edited by P. M. Mather
© 1993 John Wiley & Sons Ltd

Edinburgh, Leicester, London and Nottingham to fund students wishing to take Masters courses in GIS.

Readers of this volume will judge the quality of the research findings reported here, and the value for money obtained by the Research Councils on their investment. As Evaluation Consultant to the Joint Programme, my experience indicates that the benefits accruing from the Programme include rather more than the published findings of the individual research groups. Participation in a 'community programme' should enhance communications between different specialist subgroups and increase awareness of both the problems faced by these various subgroups and solutions to these problems. GIS have developed through the conjunction of interests of several specialist disciplines, each with its own view of the world. Cartography, remote sensing and photogrammetry, architecture and planning, surveying, geodesy and computer science have all had a role in the development of GIS. The Joint Programme brought together experts in these fields, as well as environmental and social scientists, and over a period of three years these experts have had the opportunity to present and discuss research findings at annual seminars and via the circulation of progress reports. In a small way, then, the Joint Programme has helped in increasing the cohesiveness and self-awareness of the GIS research community in the United Kingdom. This coming together of researchers with different and wide-ranging interests has helped to reduce a suspicion which results, perhaps, from a historic and imperialist view of academic research, and which revolves around the 'G' in GIS. There is a view, held perhaps by a minority, which considers that GIS must be forcibly removed from the geographical community in order that it might develop more vigorously and be used more widely in the social, economic and environmental spheres. The involvement in the Joint Programme of researchers from different backgrounds demonstrates that geographers have not been inward-looking by attempting to restrict or restrain the development and use of GIS. Since that technology is essentially concerned with the description and analysis of features of the earth's surface, both natural and cultural, it is inevitable and understandable that geography has been the parent discipline to GIS. Like any good parent in a multicultural society, geography allowed its offspring to mix with and talk to others and to learn from them. GIS does not have to be wrested from the geographers – it is open to all.

The papers making up this volume represent different, though interrelated, strands of research. The first section, *Basic Research*, includes contributions that illustrate the linkages between GIS and computer science, statistics and geodesy. In the opening chapter, Gittings and fellow researchers at Edinburgh University review GIS performance issues, especially those related to input/output and machine architectures, using a critical path analysis approach. In the face of the growing sophistication of GIS software and increasing user expectations, parallelization is one of the more promising hardware architecture options. The second chapter in this section, contributed by Brunsdon and Openshaw, is a discussion of the effects of error and uncertainty associated with spatial information, concentrating on the effects of error propagation through GIS operations, and the estimation of uncertainty in vector data. João and co-workers examine another aspect of error in Chapter Eight, which deals with the question of generalization effects within GIS. These authors examine the effects on data accuracy of scale changes or other transformations. The authors propose a 'generalization machine' capable of minimizing generalization error through the use of artificial intelligence techniques. Dodson and Haines-Young concentrate on yet another source of error in Chapter Nine; in this case, the authors concentrate on what might loosely be called locational error, that is, error resulting from the discrepancies that exist between recordings of positions that are based upon differing geodetic estimates. This chapter concentrates on the problem as it applies to Great Britain, examining in particular the difficulties that are involved in the integration of locational data measured from Ordnance Survey maps and positions that are produced from the Global Positioning System satellites.

No volume on GIS would be complete without an extended discussion of databases. Chapter Ten of this volume, contributed by Worboys and colleagues at Keele University, examines the object-based paradigm for a geographical database system. These authors consider an underlying model for two-dimensional spatial objects which is developed to provide generic models of temporal GIS. An experimental implementation of the formal model using an enhanced relational database management system is described. O'Conaill, Mason and Bell of the NERC Unit for Thematic Information Systems (NUTIS) at the University of Reading concentrate on higher-dimensional data representations, in this case those underlying a spatiotemporal GIS for environmental applications. A four-dimensional raster data structure forms the basis of their model.

The final chapter of Section I covers the important topic of the human–computer interface. Raper and Bundock, of Birkbeck College, London, review research and development in the field, and present their solution, called UGIX, which is a user-customizable interface built on top of a spatially-extended SQL3. The importance of this, and related, work is considerable, since in many instances people wishing to gain access to spatial information held in a GIS are frustrated or deterred because of the apparent difficulties of interacting with the GIS. To say that the 'look and feel' aspects of GIS are those that provide the initial user response to GIS is not, perhaps, an overstatement.

Section II contains a set of papers describing *Environmental Applications of GIS*. Barnsley and co-workers provide an overview of photogrammetric and remote sensing tools for the extraction of spatial information from remotely-sensed images. Methods of inferring land use and land-use change in the urban context, and of deriving the height and volume of individual buildings for the derivation of three-dimensional CAD models, are described. Remotely-sensed images also provide the environmental data input to the GIS developed by research workers at NUTIS and the Institute for Terrestrial Ecology at Monks Wood. This work is described by Gooding *et al.*; it is important because it represents an attempt to integrate ecological modelling and GIS. Another modelling application is presented by Romanowicz and Beven, of the University of Lancaster, in collaboration with Moore, of the Institute of Hydrology. They show how a specific hydrological model, TOPMODEL, can be interfaced to a GIS to improve predictive performance. An interdisciplinary group from the University of Leeds consisting of members of the departments of geography, computer studies, and electronic and electrical engineering concentrate on a specific problem, namely the derivation of channel networks and drainage basin characteristics from elevation models, in Chapter Sixteen. This work also shows how the improved performance of desktop computers enables their application to solve problems which, a few years ago, required the use of mainframe computers.

The third section of this volume contains papers that can be conveniently grouped under the heading of *Economic and Planning Applications of GIS*. The first two contributions relate to environmental/landscape planning. The mapping and modelling of natural hazards with reference to planning is considered by Wadge and colleagues from the University of Reading. This chapter could equally well have been placed in Section II, but it is placed here as a good example of the way in which the academic research funded under the Joint Programme has formed links with operational uses of GIS, including local authority planning. A similar and equally effective linkage underlies the work reported by Davidson and his fellow researchers in Chapter Eighteen. This project, though of limited duration (one year), shows how the needs of environmental planning can be met in an efficient manner, provided that suitable data exist in an accessible and affordable form. A second short project, on property management, is described by Fairbairn, of the University of Newcastle upon Tyne, in Chapter Nineteen. The work of his group has led to the development of a PC-based land information system that has attracted wide interest from the surveying profession.

One of the major markets for GIS in the United Kingdom is represented by local authorities. The needs, perceptions and behaviour of this group of actual and potential users represent significant controls on the development of commercial GIS. Masser and Campbell, from the University of Sheffield, report on an in-depth survey of local authority use of GIS which provides strategic insight into the behaviour and aspirations of this diverse market. A large-scale example of the use of GIS in socio-economic planning is contributed by a group from the University of London led by Shepherd of Birkbeck College. They have chosen the eastern Thames corridor as their study region, and – like Davidson *et al.* and Wadge *et al.* – show that links with local planning authorities and groups can benefit both the researcher and the user. The eastern Thames corridor is the unseen side of the Greater London conurbation. Recent planning decisions, including the route of the Channel Tunnel railway, have given this region a role of some importance in the economic development of the South-East of England; the GIS developed under the Joint Programme should prove to be of some significance as economic development takes place. The final contribution, from Clarke, of Gmap Ltd, and a team from the University of Leeds led by McDonald, takes us back to the issue of modelling, a topic which is addressed by other contributors to this volume. Clarke and his co-authors are concerned by the archive/overlay aspects of many commercial GIS – they see simulations and models as providing both a greater intellectual challenge and a greater degree of utility to users of GIS.

These three substantive sections are preceded by short introductions from Professor John Knill, Chairman of NERC, Professor Howard Newby, Chairman of ESRC, Dr Michael Jackson, Chief Executive of LaserScan Ltd, and Professor David Rhind and Mr Ian Logan, Ordnance Survey. These introductions set the scene; the two chairmen provide insights into the reasoning that led up to the Joint Programme in 1989, and also indicate the way in which GIS is viewed within the research communities that they represent. Perhaps Professor Knill has the easier task, in that the earth and environmental sciences require, and are developing, models that describe and help to predict changes in the atmosphere, hydrosphere and cryosphere, as well as on the land surfaces of the earth. Raster data sets, often derived from earth observation systems, are appropriate to many of the applications described by Professor Knill, though the complexities of modelling (especially in more than three dimensions) are considerable. Professor Newby speaks for a more diverse community which covers economics, urban and rural planning, human geography, demographics, and – an area of growing importance – the political, social and economic consequences of environmental change. ESRC has been instrumental in taking the lead in several key areas, for example in the development of the Global Environmental Network for Information Exchange (GENIE), and in the stimulation of interest in GIS via the Regional Research Laboratories programme. Professor David Rhind and Ian Logan draw attention to the importance of the role of the Ordnance Survey of Great Britain as a research and development organization, a supporter of University research and as a supplier of data. The authors provide an interesting vision of the future, and also specify the Ordnance Survey's needs from research and development, which is relevant to the present government's policy on research funding for science as set out in the recent White Paper ("Realising our Potential – A Strategy for Science, Engineering and Technology", Cm 2250. HMSO (1993)). Dr Jackson provides a critical view of the Joint Programme from a position outside academia; his observations are a timely reminder that success is not necessarily counted only in numbers of publications and the sizes of research grants but should ultimately be reflected in the Programme's impact on the GIS industry in the UK. His views are not necessarily accepted by the Research Councils whose involvement in "near-market" research is limited by their Charters. Nevertheless, Dr. Jackson poses some pertinent questions which many would like to see answered.

The value of the Joint Programme can be summarized by pointing to the stimulus it has given to academic research in GIS in the United Kingdom. From another perspective, its value may lie in the increased contacts between academic researchers and industry. A third view may be that its greatest value is in bringing together research workers from a variety of disciplines with a common interest in spatial data handling and associated problems, while a fourth opinion might place greater stress on the short-term increase in the training opportunities made available to British students to study GIS at postgraduate level. I believe that it is important to see all of these strands together; I hope that this volume will stand as a record of a significant inter-Research Council initiative that has been drawn to a successful conclusion.

REFERENCES

Department of the Environment (1987) *Handling Geographic Information. Report of the Committee of Enquiry chaired by Lord Chorley*. HMSO, London.

HMSO (1993) *Realising our Potential: A Strategy for Science, Engineering and Technology.* Cmnd 2250. HMSO, London.

Chapter Two

Geographical Information Systems: the environmental view

J. KNILL
Natural Environment Research Council, Swindon

1. INTRODUCTION

At the present time I suspect that there are few environmental scientists working on multiple data sets collected in the natural environment who do not use GIS, or at least covet access to it. Forty years ago I suspect that most such scientists would simply not have known how to apply such a technique, or how to exploit the potential it offered. Working as a young geologist in the field at that time, my geological data was by definition geographically referenced on the traditional six-inch map sheet. Satellites did not exist until a few months after I finished my PhD research. Aerial photography, publicly available for the first time only a matter of 20 or 30 years earlier, offered a synoptic view of a mapping area, providing for both location, on Scottish maps which commonly had little more than a couple of spot heights and a stream or two, as well as a new vision of the relationship between landform and geology. Airborne magnetometry, developed during World War II, offered an entirely different perspective to the maps acquired so painstakingly in the field. Regional magnetic maps not only provided an interpretation of the distribution of rock types at the surface but also enabled a penetration below the surface to interpret rock structure at depth. This was possibly my first realization that environmental data collected in the field by different methods, when integrated together on the same geographic base, provided a spectacular added value.

Geological uses of GIS have now exploited this approach beyond measure. The ability to bring together the results of conventional field mapping, satellite-based interpretations of geological structure, multi-sourced geophysical data and the results of regional geochemical surveys, and then to vary the weighting given to the different parameters can provide a unique insight into, for example, the location of potential mineral deposits. The addition of topography, together with basic hydrological and climatic information, then permits a first pass at an engineering evaluation of mine planning, disposal of waste solids and fluids, and the tracking of airborne dust plumes. The further addition of population distribution, ecological factors, sites of particular importance, and access takes one to the stage of planning and environmental impact. The power of such a tool is that variations in weighting can explore human choice in how the natural environment is best used, exploited and protected. We also recognize the increased importance of public opinion, commonly articulated by

---

Geographical Information Handling – Research and Applications. Edited by P. M. Mather
© 1993 John Wiley & Sons Ltd

well-informed public interest groups. In the case of the selection of the site for a necessary but unwelcome intrusion into the local environment of an incinerator or waste disposal unit, GIS can offer a wholly transparent means of weighting the relevant factors. For example, GIS has been applied in Australia to assess the relative suitability of potential sites for the location of a federal low level radioactive waste disposal facility. Changes in the weighting given to the different criteria, such as minimum rainfall, a low water table, a simple geological environment amenable to modelling, low population density and so on can be input in a matter of seconds, enabling freedom of public access to a system that can fully explore the site selection process.

2. NERC'S INVOLVEMENT IN DATA COLLECTION AND ANALYSIS

I make no apologies for turning now to the work of the Natural Environment Research Council as an illustration of the way in which GIS is applied in environmental research and applications. No other single organization offers quite such a comprehensive overview of the range of processes and situations that occur within the natural environment. We work in both polar regions, on and below the land surface, in mountains, deserts and forests, in and below the shallow coastal seas and deep ocean, and within the atmosphere. We are concerned with both the natural and the managed environment.

In addition, NERC has a long history of research into the application of GIS, both through its own institutes and within universities. It is worth commenting that it was in 1967 that NERC established, within a few yards of the apartments of the Royal Geographical Society in which we meet, in the Royal College of Art, what was to become the Experimental Cartography Unit. The grant was for £264 000 over five years; a not inconsiderable sum even at the present time. The work of the Unit was to underpin much of NERC's future developments, notably in the British Geological Survey (BGS).

Finally, NERC works in this field both through its own institutes and with universities, ensuring that the very best minds and technology are brought to bear together, as was so clearly illustrated by this end-of-programme conference.

NERC is heavily involved in both national and international programmes, including many concerned with global environmental research. These programmes are often multidisciplinary and are inevitably multi-sited, being drawn from all areas of NERC interest, involving very large volumes of geographically referenced data and increasingly requiring the development of computer models. Examples of these programmes include:

1. The European Geotraverse (EGT). This programme produced a coordinated geological and geophysical transect of Europe from northern Norway to Tunisia, spanning all major tectonic provinces. The experimental phase was completed in 1990, and a huge volume of data is now being synthesized by NERC-supported scientists at universities and at the BGS. The UK holds, through the NERC and industry-funded British Institutions Reflection Profiling Syndicate (BIRPS), one-third of the world's deep seismic reflection data, critical to an understanding of the deeper structure of the crust and the underlying mantle.

2. Northern Hemisphere Ozone Depletion. During this programme, NERC-supported scientists participated in analysis of data from the 1992 European Arctic Stratospheric Ozone Experiment, as well as preliminary data from the Upper Atmosphere Research Satellite (UARS).

3. Global Energy and Water Cycle Experiment (GEWEX). The World Climate Research Programme's GEWEX will quantify and model the global movements of energy, which are strongly determined by water in all of its forms. Understanding of this activity is crucial because water balance processes lead to uncertainty in predicting climate change, and because it dominates the impacts of climate change. A UK National Plan for GEWEX proposes a coherent programme of research into the observation and modelling of clouds and rain, atmospheric water vapour, and the exchange of water and energy with the ground.

4. The North Sea Project. This programme, now completed, has produced three-dimensional hydrodynamical and transport models for predicting water quality and to aid the future environmental understanding and management of the North Sea.

5. The World Ocean Circulation Experiment (WOCE). WOCE aims to understand and assess the role of the ocean in modifying the earth's climate, in scales from weeks to decades.

6. Fine Resolution Antarctic Model (FRAM). FRAM sets out to model the Antarctic circumpolar current, which links all three major oceans. The ultimate aim is to construct a model of world ocean circulation. This programme has now been completed and has been followed by the Ocean Climate and Circulation Model.

3. THE ENVIRONMENTAL DATA EXPLOSION AND THE NEED FOR GIS

Through programmes of the type listed above, environmental science has generated enormous amounts of data of many different types, most of which are, or can be, referenced spacially. These data volumes are likely to increase at an accelerating rate in the future, due principally to the anticipated growth in remotely-sensed data sets. GIS have, therefore, an obvious relevance to environmental science, not only for storage and display of the data, but also because of their ability to bring together data sets for improved data extraction and integration.

Though GIS are currently used for many environmental science applications, they do tend to be rather under-utilized in environmental research. There are a number of reasons for this. In some cases, there may be a lack of awareness of the potential for GIS to act as anything more than automated map editing systems. In other cases, a large effort may be needed to convert existing analogue data into digital form, or different digital data types into a common format. There are also limitations in GIS technology. While current GIS are most effective when dealing with static two-dimensional digital map data, many environmental data sets are inherently three-dimensional, as in solid geology, or even four-dimensional, as with marine and atmospheric circulation data sets. As a result, environmental scientists have tended to develop their own very sophisticated systems for simulation process modelling, which often include data capture, storage and display facilities. This is particularly true in the field of meteorology, where the need for continuous updating of forecasts has led to the development of advanced four-dimensional data handling systems.

A substantial proportion of GIS use in environmental science revolves around fairly straightforward, though extremely effective, GIS operations, such as efficient data input, storage, and selective retrieval and display. Effective pursuit of NERC's goals therefore requires that data derived from such research are handled and processed efficiently.

Recognizing the importance of NERC's data holdings, the Council has taken the lead among the Research Councils in development and determination of its Data Policy. This policy, established in November 1991, creates the framework to make environmental data more widely available to the research community, and sets out principles for exchanging and charging for data while ensuring that they remain freely available for research purposes.

The implementation of this policy includes establishment of five Designated Data Centres (DDCs) and development of an on-line catalogue of all NERC's corporate data. The Data Centres provide a focus for NERC's data holdings: they have responsibility for seeking out and managing NERC data and each acts as a referral centre to the others.

The British Oceanographic Data Centre (BODC) is operated by NERC's Proudman Oceanographic Laboratory, and plays a central role in processing, quality control, assembly and dissemination of data collected in research projects supported by NERC's Marine and Atmospheric Sciences Directorate. The Data Centre also maintains the UK's national oceanographic database and is part of an international network of data centres operating under the auspices of the Intergovernmental Oceanographic Commission.

The Environmental Information Centre (EIC) at the Institute of Terrestrial Ecology (ITE), Monks Wood, is concerned primarily with information on ecology and land use. It is a unique source of terrestrial ecological data, and holds extensive databases for the UK environment. The EIC has responsibility for the management of major data holdings throughout the six regional stations of the ITE.

The National Geosciences Information Service is provided by the BGS, which has accumulated a substantial inventory of geological data and samples over many decades. These take the form of maps, field records, borehole logs and cores, chemical analyses, geophysical logs and rock samples. There is, in fact, a statutory obligation for information from boreholes deeper than a few tens of metres to be supplied to BGS. BGS has embarked on a major programme of digitization to establish databases incorporating the results of its own and others' work. This programme is associated with the Government's requirement that BGS provide a service tailored to meet customer needs and timetables.

The National Water Archive (NWA) is located at the Institute of Hydrology at Wallingford, which has long been the focus for the acquisition and exploitation of major hydrological databases. Archives of national river flows and groundwater levels form the core of the NWA, but a broad range of hydrological and related data sets are being assimilated into the coordinated management provided by the NWA.

The Antarctic Environmental Data Centre is located at the British Antarctic Survey. It controls a vast amount of data, extending from the ocean depths to the outer limits of the atmosphere. Key examples are the digital map of Antarctica, for which several nations have provided digital data and maps; and the ice thickness database, providing ice thickness and surface elevations for glaciologists and geologists and forming the world's largest single collection of Antarctic ice thickness data.

In so far as they help environmental scientists to answer substantive scientific issues, the technologies being developed to handle geographic information maintained in such centres are necessarily viewed with close interest within NERC.

NERC scientists have long depended on computer-based tools for data visualization in order to increase user understanding of data, to enable environmental questions to be answered. As I have indicated, NERC was a pioneer of automated cartography, producing the first prototype geological map to be generated entirely by computer-based techniques in

1973. During the 1970s and 1980s many new scientific opportunities were afforded by the advent of remotely-sensed data from satellites and also from NERC's own aircraft, both in Europe and in Antarctica. Substantial effort was invested into the extraction of meaningful information from the large quantities of raster data produced by this new technology. Thus, the twin technologies of digital cartography and remote sensing, and their integration, remain very important to NERC because of their unique capabilities for addressing environmental science issues.

In 1987, the handling of geographic information in the UK was reviewed by the Chorley Committee (Department of the Environment, 1987), and in the Government's response the need for national coordination was recognized, with NERC being identified as one of the principal agencies to carry this out. As a result, NERC convened a Working Group to recommend the part that NERC should play in the development of GIS and their application within the environmental science community. One of the recommendations of this Group was that a GIS research programme should be set up in collaboration with the universities through a Special Topic and other mechanisms (NERC, 1988). The ESRC/NERC Joint Programme on Geographical Information Handling was a response to this.

4. CURRENT NERC GIS INFRASTRUCTURE

The Working Group also recommended that NERC should invest in current GIS technology, and in 1989 NERC made a study of the users' requirements prior to GIS purchase (Boote and Darwall, 1992). Among the key requirements identified were a spatial analysis capability, good digitization facilities, an ability to handle raster data and to integrate these with vector data, and an interface to NERC's I^2S image analysis systems. Also, because much data was already stored in existing ORACLE databases, an interface to ORACLE was required. NERC's requirements were so diverse that no one package could meet them all. NERC purchased the workstation-based LaserScan HORIZON system, and also gained access to ESRI's ARC/INFO, a package with complementary strengths. In addition, the BGS has invested in Intergraph equipment.

This range of technology used within NERC indicates that choice of system for a particular environmental science application depends to some extent on the strengths of each package for the application in question.

5. EXAMPLES OF GIS-RELATED WORK IN NERC

In order to understand the significance of the Joint Programme for NERC science, it is useful to provide a flavour of the type of work carried out by NERC in the GIS and related fields, using a range of examples from the various areas of the Council's science. These should be viewed very much as a representative sample rather than as a comprehensive survey.

In Terrestrial and Freshwater Sciences, the Institute of Terrestrial Ecology is using ARC/INFO linked with ORACLE to assist in monitoring and modelling land-use and landscape change throughout Great Britain (Lane, Howard and Cuthbertson, 1992). The team has carried out three ecological field surveys of Great Britain, in 1978, 1984 and 1990, using a sample of about five hundred 1-km squares drawn from selected environmental types (Plate I). For each sample square, data are stored in the GIS as a series of thematic overlays, including physical landscape features, agriculture and semi-natural vegetation, forestry and boundaries. The analysis is concerned particularly with the detection of change, and involves overlaying map layers from different years, before aggregating the results to provide national predictions of change. Much of the continuing publicity associated with the loss of hedgerows derives from this work.

The Institute of Hydrology is engaged in a collaborative project with International Computers Ltd to develop the Water Information System (WIS) (Finch, 1992). This is a computerized system for the storage and retrieval of water data. It provides information on water resources, water supply, sewage treatment, river management, pollution control, land drainage, fisheries and recreation. WIS records the position and extent of any object of interest, with descriptions through time of the object itself and events associated with it (Plate II). Screen graphics enable the user to roam up and down a river network, examining the related environmental data.

In the Earth Sciences, the BGS has been using the ARC/INFO GIS in the preparation of applied geological maps (London, 1992). The conventional geological map contains a wealth of information, but usually requires a geological interpretation to provide any information of practical value. Applied geological maps, on the other hand, deal with specific themes, often taken from a geological map, but integrated with related non-geological information and presented in a form that can be used by a non-specialist. These maps display such topics as the nature and extent of mineral resources, geological hazards, and suitability of land for various uses (Plate III). The GIS is used for selective retrieval of data from the geological database.

An example of a geographic data retrieval system within the field of marine sciences is the UK Digital Marine Atlas (British Oceanographic Data Centre, 1991). A daunting task facing any user of marine information is that of establishing whether or not required material already exists, and, if so, how it may be obtained. The Atlas is an electronic PC-based catalogue of all aspects of the coastline and seas around the British Isles. It contains a wide range of themes, including contour plots of physical, chemical and geological parameters, colour-coded charts of sea use, and biological and fisheries information. Information is presented as a series of summary charts which the user may browse, zooming in on areas of interest, and overlaying with other charts for comparison. All charts are accompanied by descriptive text identifying the source of the displayed data (Plate IV). The entire results of the North Sea Project, now available on CD-ROM, adopts a very similar approach.

The British Antarctic Survey has made use both of ARC/INFO and LaserScan software in an ambitious project to create the first seamless digital map of Antarctica (Thomson and Cooper, 1992). A maximum map scale of either 1:200 000 or 1:250 000 was chosen for the project, to be used where possible for all coastal areas and for the bulk of the extensive transcontinental mountain range. A considerable length of coastline had to be based on 1:1 000 000 maps and Landsat satellite imagery because no higher resolution data were available. Fig. 1 shows a small sample of the digital map.

Looking more widely, the particular strength of GIS is the power of being able to combine independent data sets which cross traditional disciplinary boundaries, and thereby solve problems that can be difficult to resolve by other means. Thus, a map of annual soil loss can be computed by combining digital maps of annual rainfall, soil erodibility, slope angle, length of slope and land cover information in a simple arithmetic overlay operation. GIS techniques have also been used extensively to model natural hazards such as volcanoes, flooding, landslides and subsidence. The outputs of such modelling are hazard maps

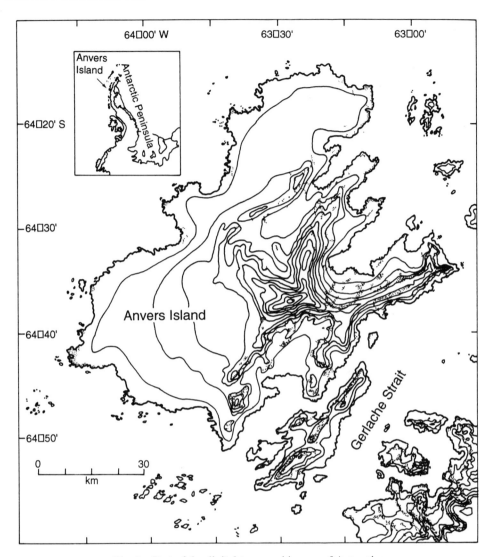

Fig. 1 Part of the digital topographic map of Antarctica

which can be used not only as a basis for land use planning but also for planning the evacuation of threatened areas.

6. THE NEED FOR RESEARCH AND DEVELOPMENT IN GIS

Inevitably, NERC has clear scientific ambitions, indeed requirements, which are well beyond the capabilities of current GIS packages. NERC's collaboration with the ESRC to fund research into innovative areas of GIS under the Joint Programme is one of the ways in which NERC has tried to address this problem. Research into the development of GIS techniques for environmental science is also carried out in specific disciplinary areas within the

NERC Institutes, and notably at the NERC Unit for Thematic Information Systems (NUTIS) (Mason, O'Conaill and Bell, 1992).

While for GIS in general R&D is needed on a range of technological issues, a number of these are of particular relevance to environmental science:

1. There is a need for the development of higher-dimensional GIS. Current GIS are essentially aimed at handling static two-dimensional data, with some extensions to three dimensions. However, much environmental science, for example geology, makes extensive use of the third vertical space dimension. Many problems also involve the dimension of time, for example the hydrodynamic modelling of the oceans. The development of higher-dimensional data structures capable of storing very large data sets, possibly global in scale, is required. As these data may often be sparse in some dimensions compared with others, methods of interpolation in up to four dimensions are also needed.

2. The modelling capabilities of current GIS need to be extended. Most current GIS are limited to simple empirical modelling. However, many environmental models are complex deterministic or stochastic models, perhaps involving the solution of sets of partial differential equations in a distributed model using finite element techniques. Many of these models are beyond the scope of current GIS.

3. Another important topic is the integration of data from a wide variety of different sensors in a GIS. This especially includes remotely-sensed data, and may involve the integration of higher-dimensional raster and vector data.

4. Research on how to assimilate spatial data into models is needed. An example problem is how to assimilate a time series of satellite altimeter images into an ocean circulation model, in order to initialize and update the model.

5. Techniques for the analysis of error propagation are required, as a consequence of integrating spatial data containing inherent errors.

6. Expert systems need to be linked to GIS to improve their capabilities in a number of ways, for example to increase their potential as decision support systems, or to improve their user interfaces.

7. Substantial improvements in user-friendliness are required, including improvements in the visualization of results.

The majority of these topics have been addressed in specific projects funded under the Joint Programme.

7. FUTURE DIRECTIONS

Looking to the future, R&D on GIS-related techniques will also continue within other NERC programmes. Of particular interest here is the Land–Ocean Interaction Study (LOIS), which is to be undertaken by the UK environmental research community over a five-year period between 1992 and 1997. LOIS will provide for the first time an integrated holistic view of how coastal ecosystems work, and how they are likely to respond to future environmental changes. LOIS will draw heavily on remote sensing and GIS techniques in performing modelling studies of the coastal zone (NERC, 1992).

Environmental problems are not just a matter for national concern; they increasingly need to be viewed at continental and global scales. A good example of a regional environmental GIS was that developed in the European Community CORINE project, which was conceived as a GIS assembling a large amount of multi-thematic environmental data. Initial topics addressed have included biotopes, acid deposition, and protection of the environment in the Mediterranean region. The role of the European Environment Agency will be of importance here.

The study of global change under the International Geosphere–Biosphere Programme (IGBP) requires that the separate parts of the earth's system be studied together in an integrated fashion in order to understand them, and to make predictions of likely future scenarios. Satellite remote sensing has made possible global coverage of many of the significant parameters, and GIS will be an important tool in this research. An example global database is that associated with the IGBP Global Change Database Project (GCDP), which is assembling various global environmental data sets, such as vegetation indices, topography, soils and ecosystems, together with a standard GIS on floppy disc, for dissemination to global change scientists worldwide.

GIS can do much for the environmental sciences. Through national and international programmes as yet unborn, as well as through continuing research, both in-house and within universities, we remain committed to the further development of GIS techniques and their application.

REFERENCES

Boote, S. and Darwall, G. (1992) GIS in environmental research. *Mapping Awareness*, **6**(1), 56–59.
British Oceanographic Data Centre (1991) *United Kingdom Digital Marine Atlas*. Proudman Oceanographic Laboratory.
Department of the Environment (1987) *Handling Geographic Information. Report of the Committee of Enquiry chaired by Lord Chorley*. HMSO, London.
Finch, J. (1992) Spatial data and GIS at the Institute of Hydrology. *Mapping Awareness*, **6**(3), 17–20.
Lane, M., Howard, D. and Cuthbertson, M. (1992) GIS, landscape pattern and ecological change. *Mapping Awareness*, **6**(4), 48–52.
London, T.V. (1992) GIS and the geological survey map. *Mapping Awareness*, **6**(2), 14–18.
Mason, D.C., O'Conaill, M. and Bell, S.B.M. (1992) Towards a 4-D GIS for environmental modelling. *Mapping Awareness*, **6**(6), 42–45.
NERC (1988) *Geographic Information in the Environmental Sciences: Report of the Working Group on Geographic Information*. NERC, Swindon.
NERC (1992) *LOIS Science Plan*. NERC, Swindon.
Thomson, J.W. and Cooper, P.A.R. (1992) A digital map of Antarctica. *Mapping Awareness*, **6**(5), 21–25.

Chapter Three

Geographical Information Systems: an ESRC perspective

H. NEWBY
Economic and Social Research Council, Swindon

The Economic and Social Research Council (ESRC) has a long-standing interest in geographical information systems (GIS), partly through its support for research in human geography but mainly through its wider responsibilities for the social sciences in general. GIS has, of course, had a profound impact on geography as an academic discipline; one might say that it has encouraged the modernization of geography and the development of a coherent set of geographical techniques for the analysis and visualization of spatially-distributed phenomena. Since 1987 the ESRC, through its £2m Regional Research Laboratory (RRL) initiative, has supported the development of eight centres of excellence within the UK. The RRLs have a remit that includes, but is not restricted to, the development of GIS applications. Nevertheless, the RRLs have provided a stimulus to the development of GIS techniques, and have extended the awareness of GIS to disciplines other than geography. The Joint Programme on Geographical Information Handling provided a specific focus for GIS research, and it is noteworthy that the RRLs have responded to the programme in a positive way, thereby demonstrating the success of the ESRC's policy.

The impact of the two programmes in which the ESRC are involved has not been solely on academic research, nor have the effects of the programmes been felt by geographers alone. Both the RRLs and the research groups involved in the Joint Programme have established links with users and vendors in both the public and private sectors. At the present time there is an explosion of interest in GIS, and it is widely accepted as a standard tool. Perhaps there is now a need for a more sober appraisal of its potential. We are all aware that GIS is now being used in a variety of applications within the social and economic sciences; for example, its use is leading to improvements in the generation and accuracy of government statistics. Its major effect has, I believe, been to modernize UK geography in particular and the social sciences in general. I have been impressed by the way in which geographers, while continually agonizing over the content, coherence and relevance of their discipline (an attitude that is perhaps peculiar to them and that is unintelligible to the rest of us), have got on with the investigation and introduction of new research methods, and have brought the new techniques into their teaching. Nevertheless, one problem that is faced in the promotion of the wider use of GIS is the 'G' in GIS. GIS is applicable to all the social

---

Geographical Information Handling – Research and Applications. Edited by P. M. Mather
© 1993 John Wiley & Sons Ltd

sciences, not just geography, and if GIS methods are to be more widely used we need to extract the technology from its home within geography departments and make it more accessible.

The introduction of GIS happened quietly and sensibly. Its use promoted interdisciplinary studies, both within the social and economic sciences (including human geography, economics, planning and sociology) and between them. The primary effect has been seen in the upgrading and enhancement of research skills and training, not just within geography but in a wide range of pure and applied areas of research. The validity of this observation is demonstrated by the range of disciplines represented at this end-of-programme conference; although it took place in the splendid setting of the Royal Geographical Society, both speakers and audience included research workers and students from disciplines ranging from the social sciences to engineering. A casual glance at a list of the first degrees of students registered for MSc courses in GIS is another clear indication of the breadth of the appeal of GIS. The Joint Programme has, of course, had a positive effect on postgraduate (MSc) training through the provision of earmarked studentships for UK students to participate in MSc courses at Edinburgh, Leicester, London and Nottingham universities.

The RRL and geographical information handling programmes are now ended. Both have been very successful, and their aims have been achieved, namely, to provide a stimulus to the development of new GIS methods and applications in the social sciences. The UK now has a leading role to play in the development of GIS in Europe; the beginnings of this movement towards greater European cooperation can be seen in the forging of links between the RRLs and the European Science Foundation. GIS research in the UK has now come of age, and must compete openly for Research Council support with other programmes and initiatives, such as the Census Research Unit at Manchester University, and the Global Environmental Network for Information Exchange (GENIE) at Loughborough and Nottingham universities. ESRC is particularly proud of its catalytic role in developing responses to the social, political and economic consequences of environmental change, through GENIE and also through the Global Environmental Change initiative. These new programmes have been developed in cooperation with other Research Councils through the Inter-Agency Committee on Global Environmental Change (IACGEC), and their success demonstrates the fact that the Research Councils actually do work together and have close relationships that are motivated by common interests rather than by political expediency or the geographical proximity of their headquarters in Swindon.

The GIS community should feel confident about the future. However, GIS specialists should begin to look outwards – outwards from the technology and the science towards the applications, and outwards from geography. There is a need to extract GIS from the geography ghetto and to make it more accessible, more user-friendly and more widely applied. After all, GIS are *tools*; while developing more and better tools we should also be asking ourselves 'what is it all for?'.

Chapter Four

Commercial impacts of the ESRC/NERC Joint Programme on Geographical Information Handling

M. J. JACKSON
LaserScan Ltd, Cambridge

The principal objectives of the Joint Programme are set out in Chapter One by Professor Paul Mather, the Programme's Evaluation Consultant. Against these objectives a budget of just over £1m was established and a three-year timetable set for implementation. As can be seen, the objectives do not specifically refer either to the commercial world or to industry. However, while there are no formal objectives relating to the flow of results and benefits out of the academic world into the commercial, the Research Councils had clearly given thought to the issue, as indicated by the inclusion of industry, and local and central government representatives on the Programme's Steering Committee. It therefore seems appropriate to evaluate the success of the Programme in terms of commercial impact against each of the four objectives identified, and to consider the issue of partnership between Government-funded research and UK industry.

Since the Steering Committee met mainly at the beginning of the Programme in 1989, a telephone survey to get up-to-date feedback from as many of the grant holders as possible was carried out. In the event nine out of the seventeen grant holders were contacted, plus the ESRC Secretary to the Committee and Professor Paul Mather who, as the Evaluation Consultant, managed to add new information on a number of the grants. Finally, a selection of commercial contacts were made, including vendors, service organizations and users.

1. OBJECTIVE ONE: TO CARRY OUT BASIC RESEARCH

To be more precise, the objective related to basic research into the area of the handling of geographic information and assumed the use of new geographical information systems (GIS) technology. The work at Edinburgh on GIS performance in multiprocessor environments by Richard Healey, Bruce Gittings and colleagues, the programme on handling four-dimensional geocoded data by Sarah Bell, David Mason and colleagues at the NERC Unit for Thematic Information Systems (NUTIS), and some of the work by Jonathan Raper at Birkbeck fall into this category. For the group as a whole, the commercial link was minimal,

Geographical Information Handling – Research and Applications. Edited by P. M. Mather
© 1993 John Wiley & Sons Ltd

though Digital Equipment Corporation's involvement in providing hardware for Edinburgh and Birkbeck's work stands out as exceptions.

In general it is difficult to track the influence of funded research of this nature on commercial development. Basic research, almost by definition, precedes commercial development, often by a considerable period of time. However, there also seems to be considerable confusion between basic research and development. Because of the lack of modern technology (referenced in this objective) there is a considerable deviation from research objectives to development to provide a platform for research. Unfortunately, in disciplines like GIS, academic programmes cannot hope to compete with the significant development teams of highly qualified staff and the sophisticated development tools that even medium-sized companies have. This leads to a situation of widely varying benefit from the programmes. Where research transforms into 'me too' development it can have considerable educational value, but the research usually suffers and commercial benefits are limited to the training. What stands out from the survey, however, was that where commercial links existed in this category of grant there appeared to be a strong advantage in terms not just of access to advanced products but also in research focus, enthusiasm, progress and the vigour with which the technology developments were passed to the outside world.

Whether it was because these latter virtues were inherent in the grant holders with respect to their fellow grant holders or whether it was the influence of industry or access to facilities is unclear. Certainly there seemed to be good research done within the programme that either came to an end with the end of the grant money or dropped to a minimal level, without any indication that the findings would progress to any commercial applications. Few commercial software organizations have the time to scour the very broad range of technical literature sources now being published and the Research Councils cannot rely on this as an adequate route for technology transfer.

In summary, the areas of basic research topics selected for grants were commercially relevant and it would appear that useful research was undertaken. However, the mechanisms to ensure that UK industry benefits from this investment, and benefits in profit terms so as to pay the taxes to repeat the cycle of investment are lacking, other than where personal initiative or serendipity wins through.

2. OBJECTIVE TWO: DEVELOP AND DEMONSTRATE GIS APPLICATIONS

The second objective was to develop and apply GIS methods to selected applications areas and to demonstrate their value. Under this category we have as examples the studies by Jeff Settle and David Mason of NUTIS with Barry Wyatt of the NERC Institute of Terrestrial Ecology, the work of John Shepherd's group at Birkbeck College, Donald Davidson of Stirling, and Ian Masser and Heather Campbell of Sheffield.

This area of the programme certainly met with some success in terms of visibility with the user community. Thus, giving one particularly good example, in conversation with Professor Masser of Sheffield University he could identify contacts or approaches from 32 local authorities, an increasing role by his team on behalf of the local authorities' management boards stimulated through the Joint Programme and contacts by several major vendors and consultancies for information on the market.

Again, the impressions gleaned by knowledge of the programmes over the last two or three years indicate significant benefit in working closely with what might be termed real-world users, whether they are in the private sector or in local or central government.

The commercial impacts have related very much more to the benefits of increasing user awareness and the stimulation of user demand rather than to the transfer of applications or methodology to the vendors. However, from the point of view of a company that has pioneered from the commercial end much of the digital mapping and GIS technology in the UK over the last 22 years, it should be stated that the creation of an educated and aware user community with realistic expectations is perhaps one of the most significant contributions in the home market that can be made on behalf of the commercial sector.

Despite the general success in this area, particularly in the areas of local and central government, an opportunity has still perhaps been lost by the programme as a whole in not selling its results in the commercial user sector. To obtain the real benefits of a programme of this nature requires significant culture changes within the academic community. While there are, or were, some natural David Bellamy's in the academic world, marketing skills and the acceptance of expenditure for transmitting results to potential users, rather than generating results for the wise and lucky to find, are largely lacking.

3. OBJECTIVES THREE AND FOUR: TRANSFER OF RESULTS FROM INFORMATION TECHNOLOGY PROGRAMMES ELSEWHERE AND GIS TRAINING

The last two objectives are taken together. Disappointingly, but honestly, little evidence was found of a systematic transfer of results from technology programmes elsewhere or a noticeable increase in numbers of suitable recruits for a development and sales organization such as LaserScan, though Edinburgh University generated five MSc dissertations from the programme. It is probably the case that from a training perspective the influence of the programme has been greater in user organizations.

4. CONCLUSIONS

The modest budget of £1m has, in the context of the Research Councils' programme, undoubtedly been beneficial, and in addition has had a positive benefit to the commercial sector. However, given the political and institutional will, a much greater benefit to UK Limited could have been achieved. The issue rests first on having a clear agenda, which from the start involves itself in ensuring commercial participation, then defining joint research – commercial priorities and finally ensuring effective and targeted marketing of the programme, an activity which must be given an adequate share of the budget. In normal commercial terms the benefit of a solid business plan ensures focus, aids monitoring and targets effort on defined benefits, which are achieved. Such an approach is not easy within a research environment, but it is possible and achievable even with the perceived negative approach that the current government has towards aiding industry.

Given these conclusions, the following recommendations suggest themselves:

1. Future research and development budgets for similar programmes should be based on plans which define objectives in terms of:
- agreed research priorities and goals;
- how the R&D will assist in meeting those goals;
- how the R&D will assist the UK economy or social/environmental welfare;
- a commercial participation plan;
- a marketing plan to promote the transfer of results to UK users and industry.

2. Similar R&D budgets should be accompanied by seed money (perhaps from DTI-funded programmes) for the implementation of results into industry or user organizations.

3. For R&D programmes to be effective for the UK economy it is necessary to equip 'critical mass' research sites with the best of UK technology.

4. Provide access to modern technology for 'blue skies' research.

5. Base selected R&D in commercial organizations.

6. Review and terminate programmes not meeting their objectives.

Chapter Five

GIS research and development: the view from Southampton

D. W. RHIND AND I. T. LOGAN
Ordnance Survey, Southampton

1. INTRODUCTION

Ordnance Survey has a vested interest in the success of geographical information systems (GIS): the more such systems are operational, the more use will be made of our data. In principle, this should also lead to better decision making and less waste of resources. The end result should be advantageous to both the taxpayer and the private sector alike. Moreover, the more efficient and functionally capable such systems are, the easier it is for Ordnance Survey to produce and maintain the topographic template of Britain. This chapter therefore sets out what Ordnance Survey's customers require and, as a consequence, the type of research interests in GIS that Ordnance Survey has and the reasons why these are important to the Survey. It also describes something of the use of GIS-type tools in Ordnance Survey. Inevitably, the Ordnance Survey view is one focused mainly on near-market R&D.

2. ORDNANCE SURVEY COMMITMENTS AND OBLIGATIONS

In the financial year 1992/93, Ordnance Survey generated about £8.5m from sales and maintenance of digital data. This is approximately 17% of all Ordnance Survey revenues. Two conclusions may be drawn from these statistics:

1. There must already be a substantial GIS enterprise in Britain.

2. GIS is very important to Ordnance Survey since government has made it very clear that the Survey is expected to continue to reduce its dependence upon the taxpayer.

The primary remit of Ordnance Survey was proposed by the Ordnance Survey Review Committee (HMSO, 1979) and accepted by government in 1984. It is essentially to maintain the 'archive' of the nation's topographic mapping up-to-date and to ensure that as much as possible of the cost of so doing is met from revenue from customers. The 'archive' is now held as the National Topographic Database. Ordnance Survey has made consider-

---

Geographical Information Handling – Research and Applications. Edited by P. M. Mather
© 1993 John Wiley & Sons Ltd

able strides in recent years to reduce its costs, increase revenues, introduce new products and operate ever more efficiently. Table 1 illustrates the changes that have occurred over a period whilst the output from the Survey has actually increased in volume and overall quality.

Table 1   Key characteristics of Ordnance Survey at the time of publication of the Ordnance Survey Review Committee Report and in April 1992

Characteristic	1979	1992
Total staff numbers	3480	2400
Marketing staff	123	185
Computer staff	73	127
Surveyors	1247	823
Drawing office staff	1002	664
Running costs in cash terms (£m at 1991/92 prices)	74	61
Overall cost recovery (%)	37	68
Number of co-publications	0	105
Number of maps in database	13 200	140 000
Sales of digital mapping (£m at 1991/92 prices)	0.015	8.4
Number of maps maintained	230 000	230 000

Ordnance Survey steers its activities on the basis of specifications made by customers of what they need now and what will be needed in years to come. The resulting Ordnance Survey plans to meet customer needs over the next five years have been set out by Rhind (1992). In essence, these are:

- A 'seamless' database of information derived from the Ordnance Survey large-scale mapping. This database needs to include all those features or themes that Ordnance Survey judges to be marketable and technically feasible to produce. It should enable the user to obtain coherent and consistent high quality data for *any* area he or she chooses.
- A reorganization and enhancement of the database to provide at least two other products. For operational efficiency reasons, these need to be integrated in, partly updated from and capable of extending the use of this seamless database. The first product is Address-Point, an up-to-date record of the postal addresses and a grid reference for every property in Britain. The great bulk of this information already exists implicitly within the Ordnance Survey map data but the various components of the address are not tied together in a usable form, nor is complete national coverage provided. Such an address database will enable Ordnance Survey to provide mapping in response to queries framed by postal address or postcode. It could also serve to meet other non-mapping purposes of users; commercial sector consultants have predicted adequate financial returns on such a product. The second product is OSCAR, a database of road centre lines, which already exists in very detailed form for a number of areas in the country but users need national cover and seem to require it in several different forms.
- Improvement of the 1:10 000 scale-type product in both urban and rural areas. Currently, this is not as up-to-date as the 1:1250 and 1:2500 mapping in many areas. The obvious way to improve its currency and economics is to 'spin it off' from the 1:1250 and 1:2500 scale data by generalization.
- The retention of as much as possible of the 1:25 000 scale mapping.
- Continuation of the 'flagship' Landranger product and production of digital versions of it, possibly to different specifications to meet different needs.

- The capability to 're-engineer the database' to meet new needs as they arise. An example of this is the need to add height information to our record of the built environment. Another is the possible need to transform the National Grid (and all mapping stored on that basis) to be compatible with position-fixing now readily obtainable from the Global Positioning System (GPS), or to make such systems provide results in National Grid coordinates (Ordnance Survey, 1992).
- Certain other 'spin-offs' from these databases and from combinations of Ordnance Survey and other data (such as Coastal Zone Mapping derived from material produced by the Hydrographic Department and Ordnance Survey, now in prototype form: see Harper and Curtis, 1993) plus new Ordnance Survey databases and satellite-derived products.

In addition to these products, the service Ordnance Survey provides in future will need:

- Improved updating capabilities. Some elements of the database ideally need to be kept more up-to-date than the Survey has hitherto achieved. Service Level Agreements, whereby Ordnance Survey guarantees that, say, x% of all new buildings, etc. will be recorded within a specified time period, may become essential, at least for many urban and peri-urban purposes.
- Price stability within reasonable limits.
- Faster delivery and better after-sales service.
- Some added value services. While we see many of these being provided by private sector Value Added Resellers of Ordnance Survey information, we can not remain simply as a purveyor of 'raw' data; we need to be closer to customers and to understand their uses of data if we are to serve them better. It is also sensible that we benefit financially from activities 'downstream' of data supply.

On a *de facto* basis, most British GIS applications operate using Ordnance Survey digital data. Many errors can arise from use of GIS without the non-expert user being aware of them. This places an obligation on Ordnance Survey to ensure adequate quality in its products and that these are in a specific form, at a particular price and delivered when customers require. The quality considerations become ever more important as more and more of other people's data becomes 'locked into position' in space through use of the Ordnance Survey 'topographic template'. As a consequence, Ordnance Survey is currently engaged in a redefinition of the quality of the Survey's products. Even if this is not research in the strictest sense, it is of importance to researchers and customers alike.

3. THE ORDNANCE SURVEY VISION OF THE FUTURE

If our focus is now largely on the next three years or thereabouts, we still require a vision of what is likely to happen thereafter. It may, for instance, take many years to bring a new product to the market-place or to enhance specifications of an existing one nationwide. Inevitably, our vision of the future is heavily based on GIS. We anticipate that Ordnance Survey will be operating in the medium and longer term in a world where:

- Much wider use of GIS-type capabilities will exist, especially by non-traditional users, though these facilities will probably be regarded as no more novel than word processing is today.

- Partially as a consequence of the growth of GIS, the use of multiple data sets in combination will also be routine. Many of these data sets will be held by different organizations.
- Computer networking for access to Ordnance Survey (and other) data will be widespread and routine. The customer will often extract data as required or it will occur on a pre-programmed basis, rather than Ordnance Survey supplying it through a human chain.
- There will be some form of a National Land and Property Information System (or Service) which operates as a gateway and bridging facility to a multiplicity of data sets and analytical tools.
- Some customers will wish to have different types of information to that presently collected (such as the heights of properties and possibly the appearance of them to facilitate three-dimensional visualization).
- Even for those customers who still want paper maps – and there will be many – printing instantaneously on demand from the database is likely to be commonplace. 'Map sheets' may well, for the professional market at least, become a meaningless concept; they will have maps covering only the area they need for any given task.
- Historical information may be routinely regarded as part of the remit of the national mapping agency and be supplied by the organization on demand through the same distribution channels as contemporary data.
- There will be both much greater international collaboration *and* competition to supply geographical information such as the topographic template.

4. WHAT ORDNANCE SURVEY NEEDS FROM RESEARCH AND DEVELOPMENT

The Ordnance Survey's needs from R&D are simply stated in general terms as follows:

- Better ways of detecting change in the landscape. Given the very high resolution of our basic mapping and the existing specification, field visits are probably unavoidable for some updating at least. The crucial factor, therefore, is to obtain even better intelligence of new built forms.
- Better ways of measuring change in the landscape. The 'low-tech' field mapping approach used hitherto involves a mixture of manual and digital methods, workstations in the field offices and much data transmission between field and headquarters. Field recording of graphical and instrumental survey directly in digital form and direct transfer from any telephone socket has the potential for significant savings in terms of effort, time and accommodation costs.
- Better validation tools, especially as the specification of some of our products is enhanced.
- Database tools to maintain integrity across multiple databases of implicitly or explicitly linked entities and to take account of within-feature variations in attribute, for example to retain a record of which points making up a coordinate string have been surveyed instrumentally and which have not.
- Tools for 'spinning off' new products from the databases, sometimes from previously unpredicted combinations of entities.
- Enhanced data distribution mechanisms to get data to customers quickly and cheaply and to retain a record of which customers have which data variants.

- Ways of storing historical data as well as contemporary versions in such a way that interrelationships between them can be established.
- Better tools for predicting how the market will develop under alternative scenarios (e.g. different pricing levels).

5. IN-HOUSE RESEARCH AND DEVELOPMENT BY ORDNANCE SURVEY

The Organization for Economic Cooperation and Development (OECD) definitions of R&D are used by Ordnance Survey as well as by all of the UK Government. In these, R&D is defined as 'creative work undertaken on a systematic basis to increase the stock of knowledge of man, culture and society, and the use of this stock of knowledge to devise new applications'. It is classified in one of four categories: basic research, strategic applied research, non-strategic applied research or experimental development. It is expected to have 'an appreciable element of novelty'.

On this definition, basic research is experimental or theoretical work undertaken primarily to acquire new knowledge of scientific principles. Strategic applied research is similar but directed towards practical but non-specific objectives. Ordnance Survey seldom tackles these types of R&D; when it does, they almost always take place in collaboration with universities or other research bodies.

About 5% of Ordnance Survey R&D can be classified as non-strategic applied research, that is, research involving the creation of fundamentally new products, processes or systems. Most Ordnance Survey R&D activities are classified as experimental development, namely, the systematic application of existing knowledge, experience and technology to the development of substantially improved or new products, processes, procedures, devices or services.

Routine system implementation, equipment testing, equipment installation and process enhancement which lacks an appreciable element of novelty is not considered as R&D. Thus most, but not all, activities in Ordnance Survey's Information and Computer Services (ICS) are formally excluded. In practice, however, the definition of some work in that part of the organization is not clear: some work on the software for Address-Point, for instance, might well be reasonably defined as R&D. Indeed, important conceptual developments have arisen from that function (and also from many other parts of the Survey). As one example, the Ordnance Survey's suggestions for a universal but unique identifier for any land parcel or property (Ordnance Survey, 1993) actually arose from within ICS.

Activities formally defined as R&D have consumed about 2.5% of Ordnance Survey's budget in recent years (thus amounting to about £2m in 1992/93). The great bulk of the R&D effort is carried out in-house. R&D activities are controlled by the R&D Steering Committee, most (but not all) are carried out within an R&D function and they are managed and executed mainly as discrete projects. A high-level science and technology committee, peopled by leading academics and industrialists, provides advice to Ordnance Survey on relevant upcoming developments.

Like all other activities in the Survey, those of the R&D staff are to support the organization's Mission and Strategic and Business Objectives. These aims are met by providing high quality advice and support for customers; providing technical inputs to the identification and development of potential new products; maintaining awareness of relevant external developments; actively seeking opportunities to improve production by adopting advanced technology and techniques; and by participating in the development of national, European and international markets for geographic data, systems and services.

During the period 1993–1996, priority so far as in-house work is concerned will be given to supporting completion and efficient maintenance of the LandLine 93, OSCAR and Address-Point databases; the Topographic Data Management Strategy; creation of high quality 1:10 000-type graphic products from basic scales data; extending GIS skills, experience and expertise; GIS standards development; development of advanced systems for database update; development, as appropriate, of new height, small scales and image products; providing users with reliable and appropriate tools for merging and using geographic data based on differing reference systems; the Domesday 2000 project; Project 93 and developments connected with the National Street Gazetteer and the National Land and Property Gazetteer.

Comparison of what Ordnance Survey needs in general terms with the in-house work will demonstrate several 'gaps'. In part, we expect these to be filled by commercial vendors. Where this is not the case, we will need to find other ways of achieving the desired ends.

6. ORDNANCE SURVEY AND OUT-OF-HOUSE RESEARCH AND DEVELOPMENT

Given all of the foregoing, Ordnance Survey has had considerable interest in ESRC/NERC-funded GIS research, especially on generalization (including that by João *et al.* and O'Conaill, Mason and Bell, this volume). Indeed, we have helped to sponsor this and other research, notably (but not exclusively) through the supply of test data. More generally, in the three years up to the end of 1992, Ordnance Survey part-funded or otherwise supported 23 university- or former polytechnic-based projects at a total cost of £286 000, excluding the supply costs of or foregone income from the data involved. These are summarized in Table 2.

One manifestation of Ordnance Survey's commitment to work with the university research community is the setting up of a summer school for research, to be held in Southampton in the summer of 1993. In the first instance, this will involve only about six researchers working on topics of interest to themselves and to Ordnance Survey for a period of up to two months, financed by Ordnance Survey. The scheme was advertised nationally and attracted a reasonable, though not overwhelming, response.

7. SOME COMPLICATIONS

Despite all of the above, one serious problem has arisen and is of growing importance. This arises from the fact that the end results of research done by or for Ordnance Survey are increasingly of commercial value. This comes about from the growing competition which the Survey faces in some markets and from the government's policy that Ordnance Survey should raise its level of cost recovery. It may therefore be commercially expedient in some cases not to publish the results of research. In this commercial environment, Ordnance Survey must clearly take steps to retain control over any intellectual property rights that arise from research funded by it: in this respect, Ordnance Survey and other government and industrial sponsors differ profoundly from Research Councils as sponsors of research and development.

The context in which government departments and agencies commission research is also evolving. The White Paper on science policy (HMSO, 1993) underlines the importance of the 'user pays and user decides' principle and exhorts the potential for privatization of some research bodies. In these circumstances, Ordnance Survey must take particular care to get best value for money – and be seen to be doing so.

Table 2 University/polytechnic research projects supported by Ordnance Survey during the period from January 1990 to December 1992

Institution	Subject	Start	Finish
University of Southampton	Optimization of geodetic levelling networks	1985	1991
NUTIS, University of Reading	Generalization	1987	1993
University of East London[1]	Classification and coding of spatial information	1987	1991
University of Nottingham	Precise local geoid	1987	1990
University of East London[1]	GPS networks	1988	1992
University of Glamorgan[1]	Automatic cartographic generalization	1988	1992
University of Newcastle	Mapping information systems for property management	1989	1992
University of Nottingham	GPS and engineering applications	1989	1992
University of Nottingham	Three-dimensional integrated geodesy and engineering applications	1989	1992
University of Hull	Parallel and object-orientated approaches in digital cartography	1989	1992
Birkbeck College	Generalization effects in GIS	1989	1991
University of Nottingham	EUREF research project	1990	1992
University of Nottingham	EUREF and engineering applications	1990	1991
University of Glamorgan[1]	Deductive database for a multiple representation spatial GIS	1990	1993
University of Nottingham	Atmospheric and multipath errors in GPS	1991	1994
Heriot-Watt University	Deductive object-orientated databases	1991	1994
University of Southampton	Parallel processing	1991	1992
University of Nottingham	Monitoring of vertical displacement of tide gauges by GPS	1991	1994
University of Nottingham	Data integration in a GIS environment	1991	1994
University of East London[1]	Domesday 2000	1991	1994
University of Edinburgh	Parallel services for corporate GIS	1992	1995
Keele University	Generic data models	1992	1995
University of Wales, Aberystwyth	AI and map data integrity	1992	1995

[1]Formerly polytechnics.

The Survey is therefore in something of a quandary: we know that innovative work can be produced effectively in the higher education sector but our requirement (in some cases) for confidentiality is at odds with the long-standing academic tradition of open publication and the need by university staff to cite journal publications from research. The practical consequence is that certain types of research are controlled more readily if done in-house or by private research organizations, though the use of non-disclosure agreements with university research entities may be helpful in some cases.

8. CONCLUSIONS

Ordnance Survey needs new products and services to continue to thrive and to meet the needs of existing and new customers. R&D findings of all types, especially in GIS, are therefore of great interest to the Survey. We recognize this and, as a result, have played a positive role in supporting research studentships and research projects to an extent appropriate to our size and status as a government department and Executive Agency. This can be expected to continue but the rationale for research being funded by Ordnance Survey – and the consequences of it – need to be understood by the research community.

REFERENCES

Harper, B. and Curtis, M. (1993) Coastal zone mapping. *Mapping Awareness*, **7**(1), 17–19.

HMSO (1979) *Report of the Ordnance Survey Review Committee*. HMSO, London.

HMSO (1993) *Realising Our Potential: A Strategy for Science, Engineering and Technology*. Cmnd 2250. HMSO, London.

Ordnance Survey (1992) *The Impact of the Global Positioning System on National Mapping Policy*. Ordnance Survey Consultative Paper no. 2, Ordnance Survey, Southampton.

Ordnance Survey (1993) *Spatial Reference Standards*. Ordnance Survey Consultative Paper no. 6, Ordnance Survey, Southampton.

Rhind, D.W. (1992) Policy on the supply and availability of Ordnance Survey information over the next five years. *Proc. Ass. Geogr. Inf. Conf*, paper no. 1.22, Birmingham, November. Also in *Mapping Awareness*, **7**(1/2), 1993.

Section I
BASIC RESEARCH

Chapter Six

Meeting expectations: a review of GIS performance issues

B. M. GITTINGS, T. M. SLOAN, R. G. HEALEY, S. DOWERS AND T. C. WAUGH
Department of Geography, The University of Edinburgh

This chapter reviews the issues relating to the performance of geographical information systems (GIS). It suggests that performance is a major issue with respect to GIS, and could present a severe bottleneck to user demands in the future. Different methods of assessing GIS performance are examined, and a systems-based critical path analysis is the chosen approach. Performance bottlenecks are identified and a variety of strategies for overcoming these are examined, including the adoption of multiprocessing and parallel processing techniques to increase processing capacity.

1. INTRODUCTION

GIS have provided an infrastructure for the examination of complex spatial problems in new and exciting ways. The technology has developed rapidly, and has quickly been taken up by a diverse user community with very different and often demanding applications. The functionality of the software has perhaps been the most notable development over the past decade, closely followed by the rapid increase in the availability of data.

As user expectations continue to rise, so the available software systems must continue to develop in order to meet these expectations. Potentially, this task is becoming more difficult. Most user perceptions of GIS are conditioned by the high technology image of computing. Consequently, expectations may exceed actual requirements or the capabilities of current technology (Calkins and Obermeyer, 1991).

It is therefore important that methodologies are developed to enable realistic assessments to be made of the performance characteristics of different GIS. Since these systems encompass data as well as hardware, software and organizational components, the potential area of investigation for performance assessment is very large. Both the complexity of the systems and the volumes of data constrain performance. In terms of data, Goodchild (1992) has identified the continuing need for research into efficient storage methods to enable analyses to be performed quickly on the very large data volumes that will become available before the end of the century. This issue is particularly critical in the context of work on global environmental modelling. In software terms, the algorithms chosen for different kinds of processing tasks, and the ways in which they are implemented, may have a major impact on performance. Related to this are the characteristics of the hardware that is used and whether or not the architecture is of a standard or more novel kind.

Given the degrees of freedom associated with the performance problem, even ignoring organizational issues, it is immediately apparent that a systematic approach is required to evaluate the relative effects of different contributing factors.

Following an elaboration of these general issues affecting performance, this chapter examines traditional approaches to performance analysis and their limitations, before proposing a new approach. This involves the adoption of a 'user-orientated' perspective, which focuses on the identification of bottlenecks that reduce performance and throughput, and an evaluation of alternative methods for circumventing these bottlenecks.

In the final part of the discussion, the lessons learned from this approach to performance analysis will be used to assess the likely performance benefits for GIS that can be derived from the use of more novel multiprocessing and parallel processing architectures.

2. WHAT IS PERFORMANCE?

Performance is an assessment of the ability to complete a particular task within a given allocation of resources. The performance of various aspects of a complete system can be assessed, including the people and organizational components in addition to the hardware and software. In terms of assessing the throughput of GIS tasks these issues are as important as the purely computational aspects. Indeed, Antenucci *et al.* (1991) suggest that performance has no specific definition in reference to computer systems, but is generally reflected by overall user satisfaction. However, it is often the computational aspects, particularly those relating to current performance rather than the capacity planning of future systems, that give rise to user frustration, yet it is potentially these aspects that can most easily be examined and the associated problems alleviated.

Wagner (1992) relates an important factor that should be considered as the baseline for any performance assessment. This is the verification of the correctness and completeness of the procedures and algorithms used. It is necessary to ensure that these do what they claim, that they deal with any special cases and that any restrictions in their use are made explicit within their operation. This is sometimes forgotten, yet it is particularly important in relation to the complex spatial processes inherent in GIS that algorithm verification has taken place. It is inappropriate to attempt to make performance assessments, especially those of a comparative nature, if this is not the case.

In computing terms, the available resources in the above definition of performance will include processor (CPU) time (the time spent by the computer actively processing data), elapsed time (the overall time for the task to complete), disk storage capacity and available memory. If we take a systems approach to GIS, we could add such tasks as digitizing time, plotter time, time required for data conversion, and so on. In this chapter we will confine ourselves to considering those factors related directly to the computational process, rather than people-related performance. Thus, we assume that the required data are loaded and available.

In a search for the most critical resource, some of the resources mentioned above can be discounted in terms of their importance; disk storage capacity could be regarded as being infinite; the cost of its provision is becoming less significant. Available memory is also of lesser importance because, at the simplistic level, the concept of virtual memory provides an almost limitless supply, and, again, 'real' memory is becoming relatively cheap. Thus, in simple terms, we can identify the critical and limiting resource as the elapsed time for the task. It is this perceived throughput that forms the user's view of system performance. This issue becomes complicated because it is also necessary to consider the perceived needs for a particular elapsed (or response) time and the actual need (Calkins and Obermeyer, 1991).

Elapsed time is influenced to a considerable extent by processor time (the power of the computer to compute) but importantly also by the ability to feed data to the processor, that is to undertake input/output (I/O). This latter influence has been shown to be particularly pronounced in the case of GIS (Gittings et al., 1991). This GIS 'data effect' is becoming increasingly important as a restriction on GIS performance. Hendley (1990) notes that many GIS applications can be characterized by the need to handle increasingly large data sets, ranging from hundreds of megabytes to possibly several gigabytes, and that the computational activity tends to be I/O dominated, with comparatively little time devoted to mathematical manipulation.

Rhind (1988), in detailing the problems that must be addressed in the handling of geographic data, also highlights the volumes of data involved and goes further in explaining the tendency for I/O domination of computational activity. Rhind states that while the types of questions asked of geographical data are conceptually simple, they may involve exhaustive searches of a large database or require multiple prior indexing. This requirement can be illustrated by examining typical GIS operations, such as polygon overlay or topology creation, where a single piece of data has to be retrieved and manipulated a number of times in the course of the operation before finally being output (Harding and Hopkins, 1993).

3. THE INFORMATION TECHNOLOGY CONTEXT FOR GIS PERFORMANCE ASSESSMENT

It is worth examining how GIS performance might compare with the performance of other systems in the broader information technology (IT) arena. This leads on to a review of the methods that can be used to assess and predict performance.

GIS shows many of the characteristics of developments in other areas of IT. Two broad trends are evident: firstly, the increasing complexity of the software that has been designed to make sophisticated tasks and analyses easier; and secondly, with the interest of the user having been awoken to the available possibilities, a thirst for tackling ever more complex problems with larger and larger data sets. GIS can be seen as the latest and most extreme example of these trends. The combination of complexity and data volume suggests that performance might become an important issue with respect to GIS.

It is interesting to compare a typical GIS workload with the computer industries' expectation of a typical 'science' workload. Digital Equipment Corporation (DEC) quotes a typical load as consisting of two multistreamed batch jobs: a quantum chemistry package capable of 'significant' I/O and a general-purpose semi-empirical molecular orbital package that generates almost no I/O (Kwag et al., 1989). Although the GIS workload is poorly understood in detail, it is characterized by significantly larger jobs, undertaking I/O an order of magnitude greater (Gittings et al., 1991). There are several possible approaches that could be adopted to assess the performance of a GIS workload.

3.1. Benchmarking

Benchmarking is often favoured when purchasing a system because it can be a very effective means of assessing performance relative to other packages (Jain, 1991). For example, the database management community has evolved a series of benchmarks that are widely used in the evaluation of software. They assess the different areas of on-line transaction processing, *ad hoc* query and update. This standardized set of benchmarks has been accepted by the database vendor community and is used to compare products effectively and fairly (DeWitt, 1985).

In addition to a comparative assessment of the ability of a system to undertake certain procedures within a specified time period, benchmarking will usually involve a largely qualitative assessment of the functionality of a system, and occasionally a validation of the correctness of the underlying operations and algorithms. To date, no such standardized approaches have been developed in the GIS market-place, although Marble and Sen (1986) have explored some of the issues, in particular those relating to the use of relational database management systems. There are a number of reasons for this. Firstly, GIS is a complex technology, being an amalgam of other related technologies together with a range of data structures and complex analytical functions that are entirely its own. It is necessary to look further than the software to assess the performance of GIS; the data and the hardware environment must also be considered. Even within the software component operations can be, and are, undertaken using a host of different, but equally valid, approaches. For example, it is difficult to compare two systems where one requires the existence of polygon topology before the data can be drawn and shaded and another does not. Clearly, for purely cartographic operations the system that does not require the topology will perform better, yet the formation of topology must be undertaken by that system at a later stage to permit, for example, spatial analysis by polygon overlay.

Such a wide range of component operations do not easily lend themselves to benchmarking. Controlling the degrees of freedom is a significant problem; not just in terms of operations, but also – importantly – in terms of data and computational environment.

The array of different data structures possible, together with the internal variability of the data themselves, makes approaches based on anything other than 'standard' data sets difficult. Generating such standardized data sets with any suggestion of typical performance is fraught with difficulties. The work of Wagner (1991) illustrates the variability of measuring the performance of a commercial GIS across even the most simplistic of synthetic data arrays.

Other problems with benchmarking relate to the fact that the computing environments for GIS can be very different. Some products are workstation-based, yet rely on a central database server, while others operate quite independently. Such rich variety makes comparisons difficult.

GIS is also a relatively immature technology, and this convinces many vendors that they should not be part of any comparative attempt to assess performance. A range of different architectures will have different performance characteristics, yet vendors have invested heavily in their particular architectures and will quite naturally be unwilling to be forced into perhaps significant changes to meet standard benchmarks before they have seen significant returns on their initial investment. To a considerable extent, these vendors are justified in maintaining their proprietary architecture if it suits the particular market segment at which their product is aimed. Thus, where benchmarking is carried out for the purposes of system comparison, the results usually remain confidential. It is not in the interests of either the consultant, client organization or vendor to release these. Marble and Sen (1986) suggest that the reasons for this relate to vendor sensitivity to the gap between announced system capabilities and actual system performance, the desire of the consultant to protect

3.2. Formal methods

Goodchild and Rizzo (1987) suggest the use of formal computer science methods for performance assessment. These methods are focused towards the design of new equipment (benchmarking requires prototypes to be built – models allow predictions to be made without this requirement and thus reduce costs). King (1990) distinguishes between using simulation models and mathematical models for this purpose.

Simulation models can be built, validated against existing systems and then modified to reflect the proposed alterations. Thus, an iterative experimental approach is adopted through running the model with different parameters. The advantage of the approach is its flexibility – arbitrary levels of detail can be included. The disadvantage is the potential complexity if a large number of parameters are used to tie the model closely to reality.

Mathematical models are based on statistical methods and rely on the assumption that it is possible to predict the performance of future tasks on the basis of a small sample of data gathered by observation of tasks of known characteristics. Although, superficially, computer systems would appear to be entirely deterministic, the random element is introduced at several levels, for example, GIS data (which are unpredictable in their form), the period of waiting for user input and data communication (which relies on a random element to overcome network collisions).

There are, however, problems with this approach. Firstly, as Goodchild and Rizzo (1987) discuss, there is debate over whether a given task can be regarded as a 'black box', without taking into account better or worse strategies for the execution of the task. This is representative of the debate between computer scientists and geographers as to whether geographical operations can be mathematically conceptualized! This approach also assumes a predictable distribution of data flow; with GIS this is inherently difficult because the data reflect the complexity of the real world. Further, significant problems are introduced when more generic approaches, which are applicable across different hardware and software platforms, are attempted.

Other problems relate to the aims of the simulation and mathematical approaches; both are better suited to workload estimation, planning future capacity and designing new generations of hardware than to the more usual problem of overcoming current bottlenecks (Morris and Roth, 1982).

3.3. Analysing the critical path

The most immediate user requirement is usually to understand and overcome bottlenecks, which affect their work in terms of response times and ability to undertake analyses. Where this is the case the methods discussed earlier are potentially too cumbersome and not sufficiently generic.

Owing to the potential number of factors that may affect the performance, it is necessary to reduce the degrees of freedom in the system. A viable approach involves the critical path analysis of a system-based perspective of the problem (Dowers *et al.*, 1990). This is based on the identification of existing bottlenecks and the prediction of future ones. Jain (1991) notes that identifying the bottlenecks should be the first step in any performance improvement project. A computer system is a collection of resources that can be seen to interface and interact in complex ways to effect the overall operation of the system. When one of these resources becomes bottlenecked, it limits the overall performance of the system (Deitel, 1984). Some bottlenecks in GIS operations occur at predictable points, regardless of the software or hardware platforms being used. Others do not. Yet the need is to identify these as the basis for subsequent attempts to improve overall performance.

This systems-based approach involves two levels of detail: firstly, an overall assessment of performance in terms of the system components (hardware and software); and then a detailed evaluation, using benchmarking techniques, to investigate the specific components targeted in the overall assessment. The techniques used involve the use of data generated from monitors on the running system. The danger of this approach is the risk of monitor artefacts (Kant, 1992), that is, the effects of the monitor interfering with the system being studied. This usually takes the form of extended runtimes, but the extent of the effect is determined by how the monitor runs; for example, it may use resources that the normal running system cannot (Morris and Roth, 1982). It is possible to reduce the effect in some circumstances (for example, the monitor data should be written to a disk that is not being used as part of the job being assessed). The residual effect is usually quantifiable and is generally insignificant when compared to the effects that need to be measured.

It has been argued in this chapter that a large number of factors are involved in GIS performance. Thus, as well as examining the capacity of the processor, the system-based methodology must be extended to include other effects relating to the hardware, networks, the operating system and the data, in addition to the application software itself. There is significant potential for interaction between these components, which can greatly complicate the picture (Gittings *et al.*, 1991; Healey *et al.*, 1991; Sloan *et al.*, 1992).

An example would be data, clearly of crucial importance to the use of GIS. Wagner (1992) suggests that synthetic data sets are required to give the necessary experimental control. While this might be true, there is a significant problem. Synthetic data sets are unrepresentative of the real world and, because of their regularity, are liable to give artefactual results determined by the algorithm used to operate on these data. Therefore, it is very difficult to generalize results from synthetic to real-world situations. Thus, for the work reported here, real data sets have been used, representing two ends of a spectrum: convoluted 'natural resource' data and relatively simple administrative boundary data. While these may be representative of their respective categorizations, there are many others and generalizations cannot be drawn easily. Significant research is still required into the effects of data on GIS performance.

Once the factors that form the critical path have been identified and quantified, this approach can be extended to form the basis of a simulation model.

4. THE FUNDAMENTAL BALANCE

It has been suggested in this chapter that a detailed investigation of GIS performance is required, paying particular attention to the interplay of processor (CPU) time and I/O. In addition to these critical factors, the effects of available memory on the computer system must also be considered.

The tendency towards I/O domination is compounded by the fact that, traditionally, GIS algorithms have been written to take account of the fact that data volume is much larger than system memory. GIS algorithms have therefore used local disks for temporary storage, so adding to the I/O overhead of a GIS application. With a high I/O overhead, disk performance plays a crucial role in determining overall application performance.

This observation confirms our thesis that, in a computational sense, the interaction of processor time and I/O is the critical performance factor. Given this definition, we can approach the problem of GIS performance by looking for both:

1. Methods for increasing the utilization of the processor by minimizing any I/O, and
2. Means of increasing the processing capacity of the computer.

There is an interdependence between these two operations. Clearly, because I/O is usually the bottleneck it must be tackled first, before any benefit from increased processing capacity can be achieved.

5. DEALING WITH I/O BOTTLENECKS

5.1. Memory-based approaches

The current trend in computing is towards much larger memory configurations, that is tens – if not hundreds – of megabytes. With the current tendency for algorithms to use temporary storage on disk, there is a need for re-implementation to properly utilize this larger memory. In addition, most platforms now support virtual memory, in which the amount of memory available to a program is larger than the physical memory available; the least used pages of memory are written to local disk until required. Because this operation is undertaken at a very low level by the operating system, rather than by the GIS algorithm, it is very much more efficient. Thus, virtual memory can ease the task of the programmer by providing a large virtual address space. However, the algorithm must be implemented carefully to ensure that the memory access is localized as far as possible; that is, the sequence of memory access is not such that pages are continually being swapped in and out from disk, since this reduces performance.

An alternative approach to re-implementing algorithms for the purpose of taking advantage of extra memory is to use data caching software. Such software retains those data blocks currently being accessed in system memory. This allows other requests for the same data blocks to be accessed via system memory and not via the disk, thus speeding up I/O read operations. Write operations incur an overhead, because, to maintain integrity, the data must be written to both cache and disk. Data caching software therefore allows GIS operations to take advantage of extra memory without the overhead of re-implementation, but with less eventual benefit, because of the I/O overhead and a processing overhead required to operate the caching activity. Potentially, both caching and algorithm re-implementation can be used together to optimize a situation where the entire data set cannot be held in memory.

5.2. Disk-based approaches

At the system configuration level, careful choice of devices and their interconnection is needed to achieve optimum performance. Any particular disk type has a limited bandwidth, that is, the maximum transfer rate (in megabytes per second) and the average access time (Deitel, 1984). In situations where disks share controllers, the controller may also be a bottleneck. Therefore, configurations where temporary disk storage is as fast as possible and closely connected to the processor will lead to better performance. In addition, particularly in multistream job loads (that is where there are several jobs running simultaneously, rather than just one large job), the storage should be split over several small fast disks, rather than one large disk, since the amount of I/O throughput will be higher. Fig. 1 shows that where a job is dependent on I/O to a single disk, an increase in the number of jobs produces a significant increase in elapsed time. This has important implications for the management of GIS in a multi-user environment.

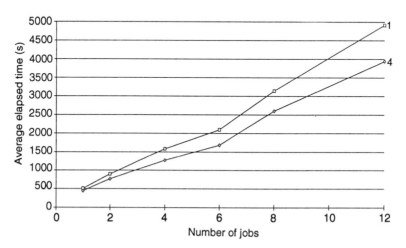

Fig. 1   The trade-off between I/O and CPU bottlenecks

Further improvements in the ability to handle I/O can be obtained by the utilization of new technologies such as RAID (Redundant Array of Inexpensive Disks) configurations or solid state disks, which can perform up to one hundred times better than magnetic disks (Cassidy and Hogan, 1991). There are a number of different RAID configurations (Alford, 1992); only some are optimized for performance; others are optimized for security of data storage (tolerance of failure of part of the array by maintaining redundant information from which the lost information can be rebuilt). Thus, care must be taken because, by introducing additional overheads, the RAID technology can, in some circumstances, actually reduce performance.

5.3.   The network effect

The workstation–server architecture is an increasingly used strategy which provides significant computational and organizational flexibility for GIS computing. The network used to connect these components often represents a performance bottleneck (Gittings et al., 1991). This is not always a function of the network itself, but can be a result of the way in which the network is used by the software package (Healey et al., 1991). Ethernet is used in the vast majority of workstation–server environments and, where there are a large number of network users, saturation does occur. In this situation alternative, fibre optic based, networking technologies may relieve this bottleneck, for example FDDI (Fibre Distributed Data Interface) (Sinkewicz et al., 1991). This is particularly the case in relation to wide area networking of GIS (Coleman and Zwart, 1992).

6. TACKLING PROCESSING BOTTLENECKS

Processing bottlenecks are the most usually perceived cause of poor performance, although, as we have established for GIS, this is often not true. Processing bottlenecks are overcome by increasing the processing capacity of the system. For this purpose, Gittings (1990) asserts the benefits of using multiple processors rather than one large processor. The primary benefit is to provide a cost effective route with enhanced computational power. There is a significant cost in providing the technology for faster generations of single-stream processors, yet there is considerable benefit in combining large numbers of commodity processors in a cooperative array. Secondary benefits relate to the extensibility of this solution; that is, future requirements need only the addition of further processors, rather than the replacement of the entire system.

The additional processors can benefit GIS through two different approaches: multiprocessing and parallel processing. However, prior to adding processors it is necessary to have established characteristic performance profiles on serial platforms, which can then be used in determining where the benefits can be gained.

6.1. Increasing throughput using multiprocessing

With multistream, multi-user job loads, multiple processors can improve throughput even for I/O bound applications. Such loads are typical on server platforms, where many users will each be running several jobs (as foreground or background tasks). This improvement occurs because the operating system can effectively schedule the work to make the best use of the available resources. Fig. 1 shows this effect. The degree of improvement is, however, dependent upon the processing requirements of the load in comparison with its I/O overhead.

When multiple processors are used in conjunction with some of the strategies outlined above for improving I/O capacity, job throughput can be increased dramatically. For example, using four processors to run four vector topology creation jobs simultaneously, utilizing different disks and data caching software, the elapsed times were reduced by a factor of between three and four compared to a single processor without using any I/O-boosting strategies.

6.2. Improving performance through parallel processing

Parallel processing involves the use of additional processors in cooperation such that each works on part of the problem, yet, together, the effect is to reduce the elapsed time for completion. Trew and Wilson (1991) demonstrate the significant variability between the array of available parallel computing architectures. These are broadly classified into the Multiple Instruction/Multiple Datastream (MIMD) architecture, where each processor acts as an independent entity, and the Single Instruction/Multiple Datastream (SIMD) architecture, where the same operation must be applied at the same time on each processing element (PE) in the processor array.

In addition, there are different approaches to subdividing the problem. Firstly, there is 'task-farm' or event parallelism, where a master processor subdivides the problem into tasks that are 'farmed out' to slave processors. The results from these slaves are assembled by another processor once every task is complete. Secondly, there is grid decomposition,

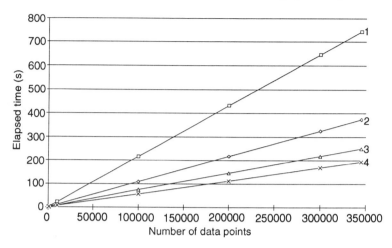

Fig. 2 The benefits of parallel processing

where the problem space is divided over an array of processors, with each processor responsible for negotiating with surrounding processors in relation to the edge effects. Lastly, there is algorithmic parallelism, where different independent application functions are placed on separate processors (Healey and Desa, 1989).

This combination requires careful redesign and reprogramming of the GIS algorithms to suit the particular architecture, but through this work it is possible to improve the processing performance of a GIS application dramatically. Fig. 2 illustrates the benefit of up to four parallel processors for a point-in-polygon operation, and this benefit can be seen to increase with the size of the data set.

It is possible to use parallelizing compilers to automate the task of parallelization, but this approach rarely maximizes the benefit from the additional processors (Gittings, 1990). Furthermore, this approach can prove difficult, as a result of data dependencies and the design of the software package. For example, Calvert (1992) found that little performance benefit was gained when attempting to parallelize a specific GIS package using automated methods.

An obvious problem with parallelization is the suggestion that customized algorithm optimization or design may be required for each of the many different parallel architectures.Trew and Wilson (1991) describe some of the developments in parallel software harnesses that offer a standardized interface between a range of different architectures and an application.

It is important to remember that even when an algorithm has been parallelized increased performance will be achieved only if no I/O bottleneck exists. It is therefore important to take advantage of the mechanisms outlined earlier in this chapter to maximize the ability to undertake I/O. Furthermore, it is now worth reviewing I/O in the context of parallel computing. Parallelization is often so effective that an I/O bottleneck immediately becomes obvious and strategies to overcome it must be explored.

Although RAID configurations provide for a data set to be spread across a number of physical disks and thus increase the I/O capacity, these are not sufficiently effective for I/O-intensive parallel applications. This is because all of the I/O must still be routed through a single disk controller, attached through one bus to the processors. Depending on the configuration, the disk may be hosted by only one processor in the array.

This sequential bottleneck can be overcome by the use of a parallel disk array, which is connected through several disk controllers, each connected to a separate processor, and thus permitting parallel I/O. Deitel (1984) clearly demonstrates the theoretical benefit for an infinite number of parallel disk servers in terms of service I/O requests. Practical implementations of this technology have been used in the specialized Teradata parallel database server (Gittings et al., 1991), but are not yet commonplace in the parallel computing market-place. A closely related technology, disk striping, has been used for some time in more traditional serial environments as a means of boosting I/O capacity. Unlike RAID, the distribution of the data is controlled by the host computer, rather than within the disk unit, and multiple disk controllers are used to maximize throughput (Davis, 1990).

7. CONCLUSIONS

The performance of GIS software is becoming increasingly important as GIS moves into a position central to the IT strategies of many organizations and large systems are created with significant data volumes. Performance assessment is, however, difficult to undertake due to the inherent complexity of the systems. This can involve the measurement of a large number of factors. It is therefore important to target these measurements precisely towards the aim of the assessment and the components of these complex systems that are being assessed and perhaps verified. The aim of the assessment may relate to either current utility or future capacity planning.

I/O is often the source of bottlenecks in GIS, exaggerated perhaps because of the architecture of some of the current generation of software packages, but there are strategies available that can be used to overcome this problem.

Processor bottlenecks are often erroneously perceived to be the major bottleneck in many GIS operations, but generally this only becomes an issue once I/O has been tackled. However, parallel processing has been shown to provide significant performance benefits for some GIS analyses. These techniques provide a flexible and cost effective means of extending processing capacity, although considerable effort may be required to design appropriate algorithms that can exploit the parallel architectures. In contrast, multiprocessing provides lesser performance benefits without the need for algorithmic re-implementation.

To achieve the best use of system resources, balanced performance is important; achieving an equilibrium between processing and I/O through, for example, use of memory. It is clear that algorithm development and optimization is necessary to match system capabilities with the required performance in both serial and parallel environments.

The systems-based critical path analysis approach has proved doubly useful through the re-application of the I/O-related findings post-parallelization.

ACKNOWLEDGEMENTS

The authors gratefully acknowledge funding provided under the ESRC/NERC Joint Programme in Geographical Information Handling. They are further grateful for the generous funding provided by Digital Equipment Corp. (under their European External Research Programme Grant UK-038) and the University of Edinburgh.

All trademarks are hereby acknowledged.

REFERENCES

Alford, R.C. (1992) Under the hood – disk arrays explained. *Byte*, **17**(10), 259–266.
Antenucci, J.C, Brown, K., Croswell. P.L. and Kevany, M.J. (1991) *Geographic Information Systems: A Guide to the Technology*. Van Nostrand Reinhold, New York.
Calkins, H.W. and Obermeyer, N.J. (1991) Taxonomy for surveying the use and value of geographical information. *Int. J. Geogr. Inf. Sys.*, **5**(3), 341–351.
Calvert, L.P. (1992) An investigation into the effectiveness of parallel processing for GIS applications. MSc dissertation, University of Edinburgh.
Cassidy, C. and Hogan, G. (1991) Improving ESE20 performance on HSC and KDM70 controllers. *VAXcluster Sys. Quorum*, **7**(1), 3–13. (Digital Equipment Corporation, Marlborough, Massachusetts.)
Coleman, D.J. and Zwart, P.R. (1992) Modelling usage and telecommunications performance in real-time spatial information networks. *Proc. 5th Int. Symp. Spatial Data Handling*, Charleston, South Carolina, 144–153.
Davis, R. (1990) Disk striping. *VAXcluster Sys. Quorum*, **6**(2), 3–16. (Digital Equipment Corporation, Marlborough, Massachusetts.)
Deitel, H.M. (1984) *An Introduction to Operating Systems*. Addison-Wesley, Reading, Massachusetts.
DeWitt, D.J. (1985) Benchmarking database systems: past efforts and future directions. *Database Engng.*, **8**(1), 2–8.
Dowers, S., Gittings, B.M., Sloan, T.M., Healey, R.G. and Waugh, T.C. (1990) Analysis of GIS performance on parallel architectures and workstation–server systems. *Proc. GIS/LIS Conf.*, Anaheim, California, November, 555–561.
Gittings, B.M. (1990) *Implementation of a Parallel Architecture within the Digital VMS Operating System*. Working Paper no. 26, Regional Research Laboratory for Scotland, University of Edinburgh, October.
Gittings, B.M., Dowers, S., Sloan, T.M., Healey, R.G. and Waugh, T.C. (1991) *Identifying the Performance Constraints on Geographical Information Systems in the VAXcluster Computing Environment*. Working Paper no. 29, Regional Research Laboratory for Scotland, University of Edinburgh, March.
Goodchild, M.F. (1992) Geographical information science. *Int. J. Geogr. Inf. Sys.*, **6**(1), 31–45.
Goodchild, M.F. and Rizzo, B.R. (1987) Performance evaluation and work-load estimation for geographic information systems. *Int. J. Geogr. Inf. Sys.*, **1**(1), 67–76.
Harding, T.J. and Hopkins, S. (1993) *Polygon Overlay Concepts*. Technical Report EPCC-CC93-10, Edinburgh Parallel Computing Centre and Department of Geography, University of Edinburgh, January.
Healey, R.G and Desa, G.B. (1989) Transputer based parallel processing for GIS analysis: problems and potentialities. *Proc. Auto-Carto 9*, Baltimore, Maryland, 90–99.
Healey, R.G., Dowers, S., Gittings, B.M., Sloan, T.M. and Waugh, T.C. (1991) Determination of the computing resource requirements for GIS processing in a workstation–server environment. *Proc. 2nd European GIS Conf. (EGIS'91)*, Brussels, 422–426.
Hendley, D.J. (1990) The scope for parallel computing in GIS. *Mapping Awareness*, **4**(5), 7–10
Jain, R. (1991) *The Art of Computer Systems Performance Analysis – Techniques for Experimental Design, Measurement, Simulation and Modelling*. Wiley, New York.
Kant, K. (1992) *Introduction to Computer Systems Performance Evaluation*. McGraw-Hill, New York.
King, P.J.B. (1990) *Computer and Communication Systems Performance Modelling*. Prentice-Hall, Englewood Cliffs, New Jersey.
Kwag, H., Von Ehren, R., Desai, R. and Walrath, D. (1989) VAX 6300 series VAXcluster performance. *VAXcluster Sys. Quorum*, **5**(1), 36–42. (Digital Equipment Corporation, Marlborough, Massachusetts.)
Marble, D. and Sen, L. (1986) The development of standardized benchmarks for spatial database systems. *Proc. 2nd Int. Symp. Spatial Data Handling*, Seattle, Washington, 488–496.
Morris, M.F. and Roth, P.F. (1982) *Computer Performance Evaluation – Tools and Techniques for Effective Analysis*. Van Nostrand Reinhold, New York.
Rhind, D. (1988) GIS – A research agenda. *Int. J. Geogr. Inf. Sys.*, **2**(1), 23–28.

Sinkewicz, U., Chang, C.-H., Palmer, L.G., Smelser, C. and Templin, F.L. (1991) ULTRIX fiber distributed data interface networking subsystem implementation. *Digital Tech. J.*, **3**(2), 85–93.

Sloan, T.M., Dowers, S., Gittings, B.M., Healey, R.G. and Waugh, T.C. (1992) Exploring GIS performance issues. *Proc. 5th Int. Symp. Spatial Data Handling*, Charleston, South Carolina, 154–165.

Trew, A. and Wilson, G. (eds) (1991) *Past, Present, Parallel*. Springer-Verlag, Berlin.

Wagner, D.F. (1991) Development and proof-of-concept of a comprehensive performance evaluation methodology for geographic information systems. PhD dissertation, Ohio State University, Columbus, Ohio.

Wagner, D.F. (1992) Synthetic test design for systematic evaluation of geographic information systems performance. *Proc. 5th Int. Symp. Spatial Data Handling*, Charleston, South Carolina, 166–177.

Chapter Seven

Simulating the effects of error in GIS

C. BRUNSDON
Department of Geography, University of Newcastle upon Tyne

AND

S. OPENSHAW
School of Geography, University of Leeds

1. INTRODUCTION

Geographical information systems (GIS) provide a powerful set of tools for capturing, manipulating and modelling map information held in digital form. GIS is, above all else, an integrating technology that allows, encourages and expects users to bring together data from many different sources through the unifying medium of geography. In many senses the power of the GIS results from this data integration function; certainly much of the value adding comes from this process. It is widely known that geographical information is subject to error and uncertainty. This applies to both the map base data, due to data capture error and generalization effects, and the associated attribute information, due to a mix of measurement error, sampling error, and the effects of categorization and aggregation. Uncertainty and error are, therefore, endemic characteristics of all spatial information, to which additional contributions are made by applying various GIS models to represent the world. Additionally, many GIS operators add their own errors to data that already contain error. This is only a problem if users ignore the effects of handling uncertain spatial information when they make use of it. The problem is that virtually all the commercial GIS in the world offer no error handling facilities and nearly all the world's supply of spatial data has been stored without any detailed error tagging.

The word 'error' is used here in the statistical sense of uncertainty as to what the precise value of a coordinate really is. In many spatial databases the values for the millions of coordinates may be precise to five, six, seven or even ten digits, or they may not be quite so precise. It may not matter much or it may be catastrophic. Once it is realized that this uncertainty characterizes all aspects of spatial information, then there is cause for concern, especially as the prevailing wisdom is to ignore it. There are various causes of uncertainty, most of which cannot be simply removed by improved data capture (see Table 1). There is also the added problem that the objects being represented in GIS are themselves difficult to define precisely, and possess what Burrough (1986) refers to as 'unseen errors' and 'natural spatial variation' (p.132). In practice, these natural sources of variation are augmented by

---

Geographical Information Handling – Research and Applications. Edited by P. M. Mather
© 1993 John Wiley & Sons Ltd

probably not inconsiderable amounts of artificially generated or unnatural uncertainties added to the data by the GIS process itself.

Table 1 Major sources of error in GIS

Data capture error
Ageing of data
Representational, generalization and simplification processes
GIS operations

One interesting fact about errors in spatial information is that often they may not matter. Under such circumstances, it is convenient to ignore them by pretending that they do not exist. The problem is that whether errors matter is dependent on both context and application. It is not an issue than can be dealt with in a global fashion. It is neither safe to assume *a priori*, on the basis of little or no relevant knowledge, that errors do not matter, nor that they do matter to the extent that an application or a spatial decision support task might be put in jeopardy. In most cases the user does not know and, worse still from a scientific point of view, has no obvious way of finding out. It follows, therefore, that the absence of error handling methods in existing GIS is a grave functional deficiency. If this situation is left unremedied, then various GIS-inspired disasters may well occur, and a new type of criminality, or GIS crime, may become notorious (Openshaw, 1993b). Certainly, as GIS diffusion gathers speed in the 1990s and the number of GIS workers increases rapidly beyond the limited supply of skilled spatial scientists, this functional deficiency could become serious. The problems are likely to be greatest in those application areas where many different small-scale, rather than large-scale, mappings are used, and when the users are more likely to be planners and resource managers than land surveyors. Now, it can be argued that the intelligent user will know not to attribute too much confidence to the accuracy of the results of spatial decision support. On the other hand, these systems are not cheap; neither are the various databases that are used. Additionally, many application areas probably have large financial implications. So it is most unlikely that end users would dare to treat GIS results as 'artistic impressions', and it is much more likely that many will believe whatever they see or wish to see in the results. Blame and litigational consequences might easily be transferred to the vendor of the systems and the databases. Table 2 outlines some of the possible consequences.

Table 2 Some consequences of ignoring errors in GIS

Selection of 'wrong' locations
Identification of fake patterns
Derivation of erroneous data
Failure to find relationships that matter
Poor decisions that cost money

This raises another important aspect. Vendors, system developers and digital map database suppliers seem to take little interest in error simulation. There are a number of possible reasons for this: (i) they require users to sign conditions that restrict or remove any legal claims upon themselves for malfunction or error; (ii) providing error simulation tools can be viewed as a negative feature of a GIS, and not an enhancement; (iii) there is no tradition in cartography and spatial decision support of error simulation; (iv) GIS developers are not aware of relevant techniques that lie outside their areas of expertise; and (v) error simulation that fails adequately to represent spatial uncertainty may be worse than none at all because it could engender a false degree of optimism about the meaningfulness of the results.

Despite these arguments, it is clear from a spatial science perspective that some means of estimating the possible effects of reasonably realistic error assumptions should be an important part of the core GIS toolkit. The results of GIS operations contain error, and some means of estimating the likely magnitude of this error should be available. Additionally, user awareness needs to be increased so that the concept can be used constructively (i.e. to avoid marginal decisions) and not destructively (i.e. 'let's scrap GIS because the data are wrong'). All data contain errors and it is important to discover how to cope with uncertain spatial information rather than to either pretend it does not exist or revert to non-GIS tools that contain even larger amounts of error (Openshaw, 1989).

The National Center for Geographic Information and Analysis (NCGIA) recognized the importance of error in spatial databases and based their first initiative on this theme. This initiative is now 'closed', but it seems that most of the problems that were identified have not been resolved. This initiative mainly focused on raising the level of user awareness of the problem; see Goodchild and Gopal (1989). Research attention was then focused on error in remotely-sensed data rather than the more general GIS case (Goodchild, Guoging and Shiren, 1992), and virtually no attention was given to the problems of error simulation and propagation.

2. ALTERNATIVE APPROACHES

2.1. Objectives

The ESRC/NERC-funded research project (1990–1992) described here set itself the objective of tackling the more general problem of error propagation in GIS. There are a number of related issues: (i) the need to provide a mechanism for estimating the possible effects of error propagation through any arbitrary series of GIS operations; (ii) the need to provide a practical means of estimating likely levels of data uncertainty in digital map data stored without any error information; (iii) provision of a set of practical tools that GIS vendors could incorporate into their systems if they so wished; and (iv) promotion of increased awareness of the problems of continuing to ignore the consequences of error in spatial databases.

2.2. Error handling via a Monte-Carlo approach

The initial approach to studying error propagation in GIS was based on Openshaw (1989), who identified a general purpose Monte-Carlo algorithm. The idea is deceptively simple and generally applicable. It involves 'wobbling' point/line segments to reflect best estimates of the underlying uncertainties in their positions in space. This process is performed upon each data source in turn, then some arbitrary sequence of GIS operations is applied. Eventually a result, or output coverage, is obtained. The 'wobbling' process would be repeated M times; for example, M might be 99. The variation in the results would be summarized as one- or two-dimensional distributions and used to provide approximate 'confidence regions'. In a polygon overlay application the results of each data wobble could be rastered, with counts kept at the pixel level of the frequency of each pixel being an acceptable site. It is fairly easy to determine 'safe' regions on the map. Openshaw et al. (1991) provide an illustration of this approach in practice.

The problems with this Monte-Carlo method include: (i) it is computer-intensive, taking orders of magnitude more time, since the GIS operations may have to be repeated many times; (ii) the 'wobbling' of line data can cause topological problems in that line segments may cross and the topology may be damaged or changed; and (iii) the error model for the wobbling is not defined and needs to be identified. In some ways these problems are now less relevant. The additional computation times are largely irrelevant as a result of trends towards faster hardware and multiple CPU workstations, and also because of the prospect of reduced numbers of Monte-Carlo re-runs due to the use of sequential significance tests. The topology problem can be overcome by switching from vector to raster for the wobbling and then back to vector, or else by remaining in the raster world. If the former strategy is used, Brunsdon and Carver (1993) and Carver and Brunsdon (1993) provide the basis for an optimal conversion. The lack of an error model is a more difficult problem. There are two possible strategies: either regard the wobbling as a form of sensitivity analysis in which the impact of relatively small data perturbations is to be investigated, without assuming they reflect levels of real error; or develop explicit error models that seek to represent plausible levels of measurement uncertainty. This aspect is investigated later. It is argued that this Monte-Carlo approach to error simulation as a form of sensitivity analysis is now a practical proposition. It is completely general purpose, unlike the method of Heuvelink, Burrough and Stein (1989), and requires development rather than research to establish it in GIS toolkits.

Finally, it should be noted that, as Openshaw et al. (1991) show, errors in a GIS environment certainly propagate and do not necessarily cancel each other out. They cannot safely be ignored. Indeed, the only 'safe' way to use GIS is to ascertain estimates of the likely effects of errors on a specific application before deciding whether they matter. Furthermore, error propagation in GIS goes beyond map data. There can easily be interactions between uncertainties in the positional information and various attribute errors. Also, the use of spatial models in decision support introduces the additional need to include model generated uncertainties (see Openshaw, 1979) as well as the digital map related ones. The Monte-Carlo approach can – in principle – cope with both.

2.3. Error handling via probabilistic epsilon errors

It soon became apparent that if the error simulation problem involved only overlay then a much simpler approach was possible. The concept of error epsilons around lines is an old idea; see for example Perkal (1966). The method creates buffers around a line with a width designed to reflect positional uncertainty in the location of the line due to measurement errors, and so on. Brunsdon et al. (1990) have generalized and extended this procedure to allow (i) the buffer width to continuously vary, and (ii) quantification of the uncertainty buffer as a piecewise quantic function. Overlay operations can then be specified that combine, using probability theory, the buffer regions around different line coverages. This has been implemented as a set of AMLs for ARC/INFO (see Carver and Brunsdon, 1992; Carver and Openshaw, 1990). These probabilistic epsilon error (PEE) commands are listed in Table 3

Table 3 Proposed PEE extensions for ARC/INFO

PAND
PNOT
POR
PXOR
PLLIP
PERASE
PIDENTITY
PINTERSECT
PUNION
PUPDATE
PPOINTDISTANCE
PNEAR

The set of PEE AMLs provides a useful toolkit for handling overlay problems. It is sophisticated in that the uncertainty region can change along a line segment. The uncertainty information can be stored as a data attribute. However, there still remains the problem of how to generate realistic estimates of positional uncertainty when none was stored with the spatial data when it was captured. The problem of error simulation is now the biggest unresolved task in GIS. By comparison, it now seems that developing error handlers is relatively straightforward, albeit computer intensive.

3. PREDICTING ERROR AND UNCERTAINTY IN VECTOR DATA

3.1. Different approaches

The focus of attention is now on developing a means of predicting the likely magnitude of measurement error in line data that was captured without any such information. It is assumed here that point data error prediction is of little interest because such errors are probably well represented by a normal distribution assumption using variance estimates that are a function of map scale. It should be fairly easy to determine these parameters via experimental methods.

Measurement error in this context is that associated with digitizing cartographic line features on paper maps. This is regarded as the most general problem that is worthy of attention. It is argued that any model that can cope with this aspect should be able to handle other variants, such as measurement error in automatic digitizers and even raster–vector conversion. It is taken for granted that any such errors will be scale-dependent. The main problem is that in general we do not know which error model to apply, as this is rarely provided or indeed known in any detail. Another problem is that measurement error cannot be assumed to be either uniform over a whole map or even uniform along a single line. It is obvious that some parts of a line are more difficult to digitize than others. Unfortunately, there is unlikely to be any simple relationship between the geometric complexity of a line and its error. For instance, it is possible that operators take more care and work more slowly with highly irregular line segments than they do with simple ones. On the other hand, cartographic generalization may be far more severe on the more complex line segments while others might be completely untouched; see João et al., this volume.

Predicting error in line data is largely a matter of building models that attempt to estimate it. The key assumption is that error is some function of the geometry of the line as this is the only class of predictor that is generally available. However, this is not a simple matter. Errors need bear little relationship to the fractal dimension of the entire line segment.

3.2. Measuring line complexity

If this cartographic line complexity approach is to be operationalized, then a very basic problem is the identification of distinct features in line-based geographical data. Taking for example the coastline of the United Kingdom, it can be seen that the north-eastern part of England has a mostly straight outline, while the western coast of Scotland exhibits much greater angular complexity. This suggests that, in terms of a human digitizer operator, the task of digitizing each of these sections of coastline will require very different physical actions. Because of this, it seems reasonable to assume that the degree of error in digitizing each of these regions will be different. It is therefore a fundamental task to identify regions of digitized lines having distinctly different characteristics before attempting to estimate the degree of error. The approach taken here is to identify these features using neural computing techniques (see, for example, Hertz, Krogh and Palmer, 1991).

To split a cartographic line into a set of distinct features, some descriptive indices must first be computed. These summarize the changes in the shape of the line as it is traversed. In order to partition the line into geometrically distinct components, different groupings or clusters of segments must be identified using these indices. Traditional clustering algorithms could be employed here, but the disadvantage with many of these is that they offer no way of determining *how many* clusters actually exist within the algorithm itself and impose structure on data that may not contain any. The former problem may be overcome by human judgement; in this case it would not be practical. In a typical error modelling application, feature partitioning would have to be applied to each digitized line segment, so that several hundred passes of the algorithm would be necessary. Clearly, some automatic form of determining the number of clusters should be employed. It is for this and other reasons that a neural net approach has been adopted.

Before carrying out the feature classification, it is necessary to define a set of line shape indices that will form the basic information to classify. Many of these have been put forward in the past (see, for example, McMaster, 1986). However, an important aspect of the indices required here is that they must provide a *local* measure of line shape. This is in contrast to *global* indices, which attempt to summarize the shape of an entire line segment. A danger with these is that in cases where the shape of the cartographic line alters drastically within the digitized segment, the global index will provide some form of aggregate or averaged description of complexity, which fails to encapsulate properly the shape of any of the features actually appearing within the line.

The indices required here will also need to be calculated for every point along the line. A simple example of such an index is the length of the line segment joining the current vertex to the last. This is illustrated in Fig. 1a and b. As the length of the linear segments making up the cartographic lines varies, so does the index. Note that as the shape of the cartographic line changes towards the left side of the diagram, there is a corresponding change in the characteristics of the local index. On a larger and more sophisticated level, the task of the neural computing algorithm is to identify such changes, and divide the cartographic line accordingly.

There are, of course, many more useful local measures of line complexity than just the length of the line segments, although these lengths may well be related to digitizing accuracy. Other statistics involve angular measures. The 'windiness' of map lines is also likely to

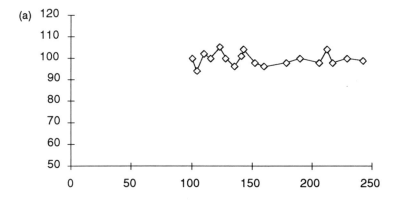

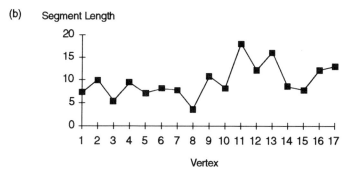

Fig. 1 (a) Line segment with locally varying properties. (b) Varying segment length along line

affect the amount of concentration and digitizing skill demanded from a human digitizer operator, and this will ultimately influence the accuracy of the digitized lines. Therefore, another important local index is the angle between successive line segments subtended at each vertex. Other candidates for local indices could be based on *changes* in the above measurements. Information about the shape of a line can also be contained in measurements of how the angle (or line length) alters from vertex to vertex, rather than in measurements of general trends in shape.

A property that all of our indices have is that of *rotational invariance*. That is, if two map lines were identical in all respects except orientation, then they would both have identical indices. This is important, since the complexity of a line should not be expected to change simply because it has undergone rotation, for example, due to an unusual choice of map transformation on the original paper document. Rotational invariance can generally be ensured if the indices are based on isotropic measures, which are also *relative* measures between vertices.

However, the shape of line features cannot be captured entirely from measurements assigned to individual vertices. For example, a line having alternately long and short segments, or a 'spike' occurring at every five or six vertices would have a very distinct and recognizable shape. Local indices based only on measurements taken at each vertex would show oscillatory properties, which would not be helpful to a clustering algorithm. There is a dilemma here: taking an entirely global viewpoint may average out distinct features within the digitized arc, but looking only at the measurements at the individual vertex level gives rise to a situation in which one cannot see the wood for the trees. There are some

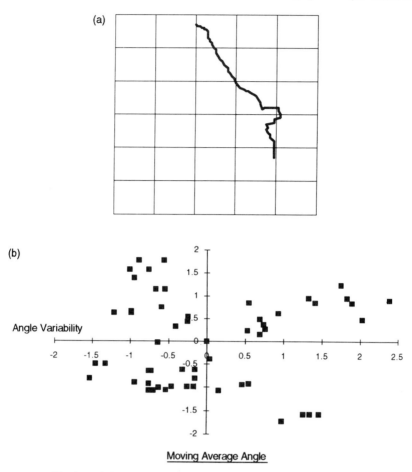

Fig. 2 (a) A cartographic line segment. (b) Complexity indices

aspects of line shape that are based on the relationship between nearby vertices, rather than each one in isolation.

One way of overcoming this is to take measurements averaged over a *moving window*, in a similar manner to techniques sometimes applied to time series. Instead of measuring the length of line segments solely at the individual vertex, the average could be taken for n segments before each vertex, and n afterwards. In this way, varying n will give a progression from the individual level measure to a global measure. Small values of n give a measure that represents the shape of *localities* on the cartographic line.

After considerable empirical experimentation, two shape measures were identified as having been found to work well in practice. These were the average angle between line segments at vertices, and the standard deviation of this quantity, both taken using moving windows with $n = 4$. The first quantity gives a measure of the general trend in angularity, while the second gives a measure of variability of angle, reflecting the higher frequency components of line shape. This is illustrated in Fig. 2a and b, showing a line segment and a scatter plot of this pair of indices. In practice, several more indices could be used, perhaps based on other measures or different window sizes, but of course there would be difficulties in showing the multivariate distributions of these graphically!

3.3. Disassembling line segments using a neural net

Using these techniques, it is possible to calculate a multidimensional descriptor of line shape for any cartographic line. However, it is more useful to divide the line up into distinct 'clusters' or features, rather than having a continuous stream of shape descriptors as the line is traversed. It is thought that this is best done using a neural computing technique. More specifically, this task is performed using an *unsupervised* neural network (Kohonen, 1988; Wasserman, 1989). In a network of this kind, the task is not of learning to provide a known response when given a set of inputs, but of providing a response 'blind' when given a set of inputs. The aim is, therefore, to find structure in the data, without any prior information other than the data itself.

Each 'neuron' in the network has a set of 'weights', which are equivalent in units to the values in the input data set. When an input is presented to the network, the neuron whose weights most closely resemble the data value is 'fired'. As the network finds structure in the data, these weight values will change. They effectively form a *caricature* of the data values, or a set of points roughly capturing the shape of the scatter plot of points for the actual shape descriptors. This is illustrated in Fig. 3, where the weights are the dark circles and the indices are the light circles. When the network has been fully 'trained', the classification task involves noting which neuron fires when the indices for each point on the line are input to it. This is effectively the classification. In Fig. 3, the classification region for each neuron is indicated by the dotted lines.

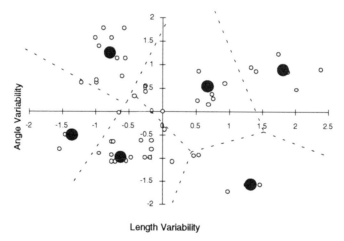

Fig. 3   Classifying complexity indices

One difficulty with this approach, which is based partly on metric distance measures in multidimensional space, is that some components of multidimensional indicators are not in consistent units. For example, indices related to line length are measured in metres, whereas those related to angle are measured in radians. Clearly, if the angular measures were entered in degrees, for example, distances on the shape-space would alter drastically, and in a non-isotropic manner. This problem is best overcome by expressing each index as a *z-score*. This involves transforming the values by subtracting the mean value, and dividing by the standard deviation. If there were a change in unit of measurement for the index here, then the standard deviation would experience the same re-scaling so that there would be no alteration in the value for the z-score.

Another major problem in this type of analysis is deciding automatically how many clusters, or features, actually exist. A theory for such algorithms, *adaptive resonance theory* (ART) has been proposed by Carpenter and Grossberg (1987a,b). In a method of this sort, when there are a certain number of neurons, a new training data point will cause one of these to fire as before. However, before the re-weighting occurs, it is checked to see whether the neuron is sufficiently similar to the data point (similarity in this case would be in terms of Euclidean distance). If it is, updating takes place, but otherwise it is assumed that *none* of the neurons in the current network was sufficiently similar. If this is the case, a new neuron is created near to this data point, and added to the network. In this way successively more neurons are added until all points in the training set may be reasonably represented. The net effectively encodes a set of exemplars that represent difficult idealized line feature types.

The criteria for deciding whether a node is sufficiently similar is often referred to as the *vigilance* of the network. The greater this is, the smaller the maximum allowable distance before a new node is created in preference to re-weighting an older one. Zero vigilance is equivalent to the standard network where no new neurons are created. The value of the vigilance can be altered during the training process. For example, a sudden increase after a certain amount of training may encourage the subdivision of some of the larger clusters that have already formed. Also, slowly reducing the value to zero may cause the system to become stable. At first, when the network is not 'familiar' with the data set, new clusters may be detected, but after some time, when much more information about the input distribution has been gained, spurious new neurons will be less likely to occur. This neural net provides an automated and highly effective means of splitting cartographic lines into arcs of similar complexity, according to a number of descriptive shape indices. Fig. 4 shows this process in operation on a line segment. Once trained, this network can be universally applied to any digital map.

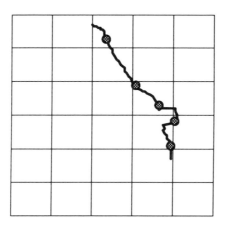

Fig. 4   Disassembled cartographic line

3.4. Quantifying digitizer error

The next stage is to attempt to predict the likely error that may be encountered in digitizing each of these features. As suggested in the previous section, line complexity, or frequency components of the line as a two-dimensional curve will have some effect on the level of error. Poulton (1974) has shown that the ability to track a moving point, and the accuracy to

Simulating the Effects of Error in GIS 57

which such a point may be tracked, will vary according to the frequency with which the line oscillates.

The main problem in this case is that it is difficult to determine the nature of the linkage between measures of line complexity for each of the features and the level of digitizing error that may be expected. In order to do this, it is first necessary to obtain information relating to digitizing error, and relate this to the measures for identified features in lines whose 'true' coordinates are known. Once a library of typical line segment shapes has been created, measurement errors can be determined by scale-specific experimentation on artificially generated samples based on accurately know arcs.

An important problem is the measurement of digitizing error. One possibility is outlined in Fig. 5. In this case, the measured error is based on the perpendicular discrepancy of the digitized line with the true line at the node points on the true line. There may be some shortcomings, particularly when there are very long segments of digitized line, but these could be overcome by introducing *pseudo-segments* by subdividing very long segments. This has the advantage of being relatively economical to calculate, while still giving a reasonable measure of digitizing accuracy at regions along the cartographic line features. This can be further enhanced by also considering the errors of the *digitized* points in addition of those of the original line.

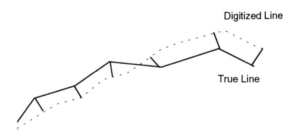

Fig. 5 Quantifying errors

A final specification must now be made, relating to the degree of error margin that should actually be allowed for when processing the map data. One possible choice might be in terms of *confidence bands* about the map line. With error information obtained on particular features within a line, it will be possible to find values that just exceed, say, 95% of all measured errors. These are effectively 95% confidence limits, determined empirically from measured error on map features of known complexity.

3.5. Modelling error: alternative one

Having established a means of quantifying line shape and also error in digitizing, and the associated error margins, the final stage is to model the association between the two. This task may also be done using supervised neural computing techniques. This approach is similar to that of regression. Typically, given a set of shape indices for a particular line feature, it will be necessary to estimate an appropriate 95% (or some other level) of error margin which may be attached to this line. It is unlikely that the relationship between the quantities will be linear, but a major problem exists in determining what functional form such an association would take. If it were known, then the problem would consist of calibrating a non-linear parametric function; since it is not, it is effectively one of nonparametric regression.

In this field, much success has recently been attributed to another form of neural network, the back propagation net (Rumelhart and McClelland, 1986). These methods are being used increasingly as an alternative to more limited, traditional, mathematical modelling methods (Openshaw, 1992, 1993a). The full description of such a network will be spared here, but its general function is to take a set of inputs and associated outputs, and to 'learn' from these, so that at some future point, when presented with a new set of inputs, it may 'guess' a set of corresponding outputs. If the inputs provided here are the line shape indices, and the outputs are the known error margins, then at some future point a network should be able to guess the appropriate error margins, when presented with a new set of values for the complexity indices.

It may also be desired to produce a network capable of producing error margins other than, say, 95%. This could be achieved by inputting the desired 'confidence level' to the network also, both in training (where the correct error margin will be incorporated into the outputs of the training data set) and subsequently in the application of the technique to new data. In this way, a network of the form in Fig. 6 may be applied to the data, after it has been disassembled into distinct features.

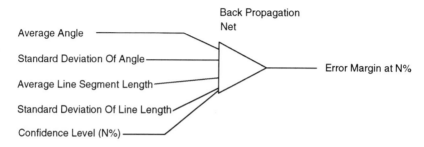

Fig. 6   Modelling error in digitized lines

Once a neural net has been trained 'off line', the resulting neural weights that represent the mapping function can easily be stored in a GIS. Neural nets are slow to train but once trained can be applied extremely rapidly to new data. A different neural net would be stored for each map scale considered relevant. Alternatively, clubs of users could combine to create training data of general interest. It is noted that the precision of this neural net error prediction is related to the size and representativeness of the training data sets employed. The principal advantage of a back propagation net approach is that it is an attempt to model the unique case. The user can adapt the error prediction technology to the circumstances most relevant to particular GIS applications.

3.6.   Modelling error: alternative two

Another approach is to associate different levels of error with different types of line feature. It seems plausible that only a relatively small number of different line feature types can actually occur in digital map databases, regardless of scale and source. These types should be identifiable by the disassembly and classification of a large sample of multi-scale digital maps. Once defined, then experimentation should yield scale-dependent measurement estimates. This is, incidently, a good way of operationalizing complex map generalization methods, such as that of Li and Openshaw, 1992. This error classification table could then be stored in a GIS as a quick means of predicting errors. This algorithm involves:

- Step 1: dividing a line into distinct features.
- Step 2: matching these features against a universal taxonomy of line shapes.
- Step 3: tagging the feature with an error band associated with the line shape class, taking into account map scale.
- Step 4: using this error information in subsequent analysis, noting that as a line is disassembled into feature types then the levels of error along the line need not be constant.

It is intended that the classification should be based on a Kohonen (1988) self-organizing map. Once training is finished, matching new features would be very rapid. The task of finding errors for each line type can be computed as previously outlined. Once the feature taxonomy has been computed, a set of error levels for each feature class will be available. These provide information about the distribution of error level for each class. As before, from this the 95% bands of error (or any other bands) can be empirically determined, by sorting the sets of error levels within each class. With some form of error tagging attached to each of the line features, the final step of carrying out the GIS operations, making use of the uncertainty margins, can take place. Whether or not there is a universal taxonomy of line types has yet to be proven; indeed, whether it is feasible to have a set of global line data error estimates held in the form of a look-up table is a matter for further, perhaps international, research. It should not be too difficult to establish its validity.

3.7. Operational aspects

The error prediction modelling is seen as a two-stage process, both of which could be automated and implemented in most GIS. Firstly, line segments would be disassembled into feature types based on their local geometric characteristics. Secondly, error estimates would be generated for line subsegment type that reflected map scale. The two alternatives differ in the linkage of the features to an associated error margin. In the first method, a back propagation neural net is used to model the linkage between the line shape indices and error. In the second, the feature is classified using knowledge based on 'typical map features' held as exemplars. The error parameters associated with these typical features are used as an estimate of the error on the specific features. This has the added benefit of allowing subjective knowledge-based error estimates to be used. This might be useful early on during the error simulation era of GIS. In addition, the levels of accuracy can be improved incrementally. These error estimates would then be stored as attributes of the data and used in either PEE or Monte-Carlo based error handlers.

The research is complete. What is needed now is development and incorporation of the technology in commercial GIS. This development task is a matter for the GIS software developers, not for academics, because the vendors will invariably want to own, copyright and protect their software investment. Hopefully, publishing this work in an open manner will help the development of practical error simulation and error handling technology in GIS.

4. CONCLUSIONS

One of the principal remaining problems concerns the empirical research needed to complete the error modelling; all that is provided here is the overall framework for doing this. Another relates to vendor reluctance to take the error in GIS issue seriously. Clearly, the admission that there is error in GIS of sufficient magnitude to require the incorporation of specific error handling methods will not come easily. Being honest about error in GIS, in spatial databases and in attribute data will not sell GIS. On the other hand, do users really think that the technology is uncertainty-free? There are also potential legal problems. It can be argued that current disclaimers by vendors about the nature of their systems will not amount to much if disasters occur because of their deliberate neglect of error in GIS. Indeed, each paper published on GIS error must make their legal position more uncertain. On the other hand, if error handlers are provided and they get it wrong, then might not the legal position be even worse?

Another difficulty is that error propagation in GIS occurs as a result of interactions between several different components: the GIS itself, the spatial database, various vendors and the user. Whose responsibility is it? Each party will in turn deny it is theirs alone, or at all. Each will blame others. So, until there is a clear agreement about need and a better understanding of litigational aspects, probably nothing much will happen. If this assessment is correct, then the onus for further scientific research to prove the case beyond all doubt returns to the academic research sector.

This chapter, with other research, has 'started the ball rolling'. The task now is to ensure that the process does not stop until practical error handling tools become widely available in GIS. If we fail in this task, then we will also be culpable and guilty of contributing to the current wilful negligence of what is likely to become the major GIS problem in the future.

REFERENCES

Brunsdon, C.F. and Carver, S. (1993) The accuracy of digital representations of 2D and 3D geographical objects: a study by simulation. In Fischer, M. and Nijkamp, P. (eds), *GIS, Spatial Modelling and Policy Evaluation*. Springer-Verlag, Heidelberg, pp. 115–130.

Brunsdon, C.F., Carver, S.J., Openshaw, S. and Charlton, M. (1990) A review of methods for handling error propagation in GIS. *Proc. 1st European GIS Conf. (EGIS'90)*, Amsterdam, 10–13 April, 106-116.

Burrough, P.A. (1986) *Principles of GIS for Land Resource Assessment*. Clarendon Press, Oxford.

Carpenter, G.A. and Grossberg, S. (1987a) A massively parallel architecture for a self-organised neural pattern recognition machine. *Comput. Vision, Graphics and Image Processing*, **37**, 54–115.

Carpenter, G.A. and Grossberg, S. (1987b) ART2: Self-organisation of stable category recognition codes for analog input patterns. *Appl. Opt.*, **26**, 4919–4930.

Carver, S.J. and Brunsdon, C.F. (1992) Practical error handling in GIS map overlay. *Proc. 3rd European GIS Conf. (EGIS'91)*, Brussels, 2–5 April.

Carver, S.J. and Brunsdon, C.F. (1993) Vector to raster conversion error and feature complexity using simulated data. *Int. J. Geogr. Inf. Sys.*, forthcoming.

Carver, S.J. and Openshaw, S. (1990) Handling error propagation in GIS: A proposal for enhancing the Arc/Info tool-kit. NERRL Research Report, CURDS, Newcastle University.

Goodchild, M.F. and Gopal, S. (eds) (1989) *The Accuracy of Spatial Databases*. Taylor and Francis, London.

Goodchild, M.F., Guoging, S.C. and Shiren, Y. (1992) Development and test of an error model for categorical data. *Int. J. Geogr. Inf. Sys.*, **6**, 87–104.

Hertz, A., Krogh, A. and Palmer, R.G. (1991) *Introduction to the Theory of Neural Computation*. Addison-Wesley, Redwood City, California

Heuvelink, G.B.M., Burrough, P.A. and Stein, A. (1989) Propagation of errors in spatial modelling with GIS. *Int. J. Geogr. Inf. Sys.*, **3**, 303–322.

Kohonen, T. (1988) *Self-Organisation and Associative Memory*, 3rd edn. Springer-Verlag, Berlin.

Li, Z. and Openshaw, S. (1992) Algorithms for automated line generalisation based on a natural principle of objective generalisation. *Int. J. Geogr. Inf. Sys.*, **6**, 373–389

McMaster, R.B. (1986) A statistical analysis of mathematical measures for linear simplification. *Amer. Cartogr.*, **13**, 103–117.

Openshaw, S. (1979) A methodology for using models for planning purposes. *Environ. Plann. A*, **11**, 879–896.

Openshaw, S. (1989) Learning to live with errors in spatial databases. In Goodchild, M.F. and Global, S. (eds), *The Accuracy of Spatial Databases*. Taylor and Francis, London, pp. 263–276.

Openshaw, S. (1992) Some suggestions concerning the development of AI tools for spatial modelling and analysis in GIS. *Ann. Regl Sci.*, **26**, 35–51

Openshaw, S. (1993a) Modelling spatial interaction using a neural net. In Fischer, M. and Nijkamp, P. (eds), *GIS, Spatial Modelling and Policy Evaluation*, Springer-Verlag, Heidelberg, pp. 147–164.

Openshaw, S. (1993b) GIS crime and GIS criminality. *Environ. Plann. A*, **26**, 451–458.

Openshaw, S., Charlton, M. and Carver, S. (1991) Error propagation: a Monte-Carlo simulation. In Masser, I. and Blakemore, M. (eds.), *Handling geographical information: Methodology and Potential Applications,* Longman, London, pp. 102–114.

Perkal, J. (1966) On the length of empirical curves. Discussion Paper no. 10, Michigan Inter-University Community of Mathematical Geographers, Ann Arbor, Michigan.

Poulton, E.C. (1974) *Tracking Skill and Manual Control*. Academic Press, New York.

Rumelhart, D.E. and McClelland, J.L. (1986) *Parallel Distributed Processing: Explorations in the Microstructure of Cognition*, vols 1 and 2. MIT Press, Cambridge, Massachusetts.

Wasserman, P.D. (1989) *Neural Computing*. Van Nostrand Reinhold, New York.

Chapter Eight

Towards a generalization machine to minimize generalization effects within a GIS

E. JOÃO
Department of Geography, London Guildhall University

G. HERBERT
Smallworld Systems Ltd, Cambridge

D. W. RHIND
Ordnance Survey, Southampton

S. OPENSHAW
School of Geography, University of Leeds

AND

J. RAPER
Department of Geography, Birkbeck College, London

This chapter concentrates on two serious problems for users of geographical information systems (GIS). The first of these is the process by which geographical data, when modified as a result of scale change or other transformation (known as generalization), produces effects (e.g. changes in length or displacement of features) that have implications for the accuracy of the data. The second problem is that existing GIS provide very little support to the user in terms both of understanding the extent of these effects and in offering means of controlling them. The authors propose the creation of a new 'generalization machine' to minimize generalization effects within a GIS and describe initial work to implement part of its capabilities.

1. INTRODUCTION

'Generalization of spatial data sets – which may be regarded as deliberately induced error – is arguably one of the most fundamental of GIS research problems' (João *et al.*, 1990). It was the acknowledgement of the importance of the problems arising from generalization that prompted the award of an ESRC/NERC research grant (Mather, 1992) on the generalization of geographic data and its effects on GIS. This chapter describes the main conclusions reached by this research project.

Generalization is an inherent characteristic of all geographical data. For instance, as the scale of a map is decreased, there is less physical space to represent the geographic features of a region: as the process continues, such features are exaggerated in size in order to be distinguishable at a smaller scale. As geographic features 'fight' for representation in the reduced map space, some features will need to be eliminated and those remaining may be further simplified, smoothed, displaced, aggregated or enhanced. In the extreme case, the map loses its geometric properties and becomes a caricature. Most of these generalization operations are carried out by trained cartographers, i.e. using manual methods. Some generalization operations, however, can also be carried out automatically, i.e. through the use of mathematical algorithms.

There are three main reasons for automatic generalization in a GIS: (i) to create a level of detail appropriate for the scale of display or analysis; (ii) to permit analysis of data at multiple levels of resolution; and (iii) to minimize data storage and input/output operations. Automated generalization can therefore be defined as the 'process which creates a derived data set with more desirable and usually less complex properties than those of the original data set' (João, Herbert and Rhind, 1991). The grounds for defining 'desirable' must relate to the specific task for which the data set is to be used and can only be defined by the user. Ideally, in a GIS, data generalization will be carried out 'on the fly' as the user requires it, unless computing resources are not sufficient to carry out the generalization or unless the same generalized data set would suffice for many purposes (João *et al.*, 1990). Both manual and automated generalization operations alter the statistical properties and the geometric characteristics of digital map data. The consequences of a generalization process are what we term generalization effects.

The generalization effects inherent in the data held in a GIS usually result either from the digitizing of previously generalized paper maps or from the subsequent use of generalization algorithms within the GIS (typically on data that has previously been generalized). From these observations, then, it can be inferred that *all* data within a GIS is generalized. Since the statistical and geometric properties of spatial data in a GIS are affected by the degree of generalization applied, so too are the results of any spatial manipulations performed on the data. Yet, though this is a fundamental area of research, very little quantitative work has been done in this field. Our research aimed to improve this situation through the quantitative measurement of the magnitude of generalization effects, and by devising ways to minimize the impact of such effects in the use of GIS. Section 2 evaluates the ways in which the generalization of geographic data affect the use of GIS. In this section, the effects that are embedded in paper maps are compared to the effects that are generated by mathematical algorithms. Section 3 outlines the ways in which generalization effects can be controlled or reduced within a GIS. The basis for a user-controlled generalization process is discussed and the abilities of a prototype intelligent system to help the user control the generalization process are presented. Finally, Section 4 contains a discussion of the results achieved by this research.

2. GENERALIZATION OF SPATIAL DATA AND ITS EFFECTS ON GIS

Most spatial data sets held in digital form are ultimately derived from paper maps or from remotely-sensed data. The generalization effects contained in them will arise from both manual and automated generalization. Generalization effects can manifest themselves in different ways, for example, by length and angularity change, by elimination of features or by displacement of features. Ultimately, a combination of these will affect GIS analysis. This is a matter of considerable significance. If the results from combining two or more data sets is a reflection of the quality of the data – rather than of the situation in the real world which the data is supposed to represent – then all subsequent interpretations and actions based upon them are liable to be flawed to some degree (João, Herbert and Rhind, 1991).

2.1. Generalization effects inherent in paper maps

Manual generalization effects embedded in paper maps are imported into a GIS when the paper maps are converted into a digital form. To define the magnitude and nature of generalization in different classes of geographical data, maps at different scales were used from a region around the town of Fronteira in Portugal and an area around Canterbury in Kent. The maps chosen were topographic and were produced by the national mapping agencies of each country. Topographic rather than thematic maps were chosen because they form the framework within which other field data are collected and/or displayed. Products from national mapping agencies were chosen because of their availability at different scales, their defined familial relationships and their widespread use. The scales studied in the British case were 1:50 000, 1:250 000 and 1:625 000. The Portuguese examples studied had a very similar range – 1:50 000, 1:200 000 and 1:500 000 in scales. In both countries the smaller scales were originally derived from the 1:50 000 scale maps or their antecedents, though they may subsequently have been updated separately.

The UK maps were obtained in digital form (i.e. the paper map conversion had been done by the Ordnance Survey) but all of the Portuguese maps were obtained as paper maps and had to be digitized. This was carried out using the semi-automatic digitizing VTRAK system produced by LaserScan. VTRAK is more accurate than manual digitizing methods and is less dependent on the individual operator. This means that the same line digitized using VTRAK at different times will be very similar, thus reducing operator variation. After all of the maps were in a digital form, different scale maps for the same region were overlaid using other LaserScan software (LITES2) and several measures of generalization effects on a feature class basis were made. The features from the smaller scale maps were compared with the same features represented in the largest scale map (1:50 000 for both countries). The 1:50 000 scale maps are themselves generalized but, for the purpose of this study, the assumption is that they are the *least* generalized and therefore are more accurate representations of the real geographical areas. The differences between individual features were measured in terms of change in line length, angularity and areal displacement. Other measurements were performed which applied to spatial domains: change in line density and in the relative position of features. These measurements are representative of the different types of quantitative descriptors of generalization developed for this project and described in detail in João, Herbert and Rhind (1990).

This study was fruitful in highlighting some of the most important facets of generalization, including some counter-intuitive results. The length of a feature generally – but not invariably – reduced as scale decreased. Occasionally, length increased with the reduction of scale when lines were displaced, and where features were exaggerated: this contradicts what certain cartographic theorists have predicted in the past. For example, Maling (1989) reported on the variability of cartometric measurements, including the 'phenomenon that the larger is the scale of the map, the greater is the length of the measured line'. The magnitude of the generalization effect also varied. Rivers tended to be more affected by generalization than roads and railways in terms of length and angularity change in data for both countries.

The lateral shift of a line when generalized was shown to be one of the most important factors in introducing spatial error due to generalization in a GIS (João et al., 1992). All feature classes in maps of both countries showed maximum displacements greater than the claimed 0.2 millimetre error limit. In absolute ground distance terms, the line displacement always increased with decreasing scale. But it was often difficult to predict which of the feature classes (e.g. road or railway) would be most affected. For both countries, the feature class that presented the highest lateral shift at one scale was not necessarily the same as that having the highest value at another. On the other hand, every different measure reflected a different degree of generalization for each feature class. For example, in the Portuguese maps the roads suffered the most displacement but it was the boundary that had the greatest length change.

Despite the fact that there was considerable inconsistency between countries and between features in the measured generalization effects (which makes it impossible to produce hard and fast rules for individual feature classes), one general pattern did emerge. This was that the more detailed a feature (measured in terms of angularity) at a larger scale map, the more likely it was to have a very high reduction of that angularity. In other words, the more 'wiggly' a line, the more vulnerable it is to generalization and the more it is 'straightened out'. A strong correlation was found between the mean angularity of each feature type at the scale 1:50 000 with the change in angularity suffered by each of those feature types ($r^2 = 0.979$). This suggests an ability to predict the amount of generalization from a given initial angularity.

2.2. Examples of the effects of generalization on GIS analysis

The findings of this study demonstrated wide variations in the magnitude of generalization effects in standard data sets. The unpredictability of the magnitude of the generalization effects will be exacerbated in a GIS analysis. For most GIS map manipulation operations, any generalization effect will be compounded when two or more features are considered simultaneously; therefore, the results will be even more unpredictable. In addition, even when individual features show apparently small generalization effects, the cumulative impact on a GIS analysis using several different generalized features in combination can be much larger. This was demonstrated by João et al. (1992). The authors carried out an analysis to determine the length of road lying within 100 metres of a river at three different scales of source map representation. Between two of the scales, the results differed by as much as 65%, caused by the simplification of the features and the changes in the relative position of the river and the road.

Most GIS analyses use a larger number of feature types. To represent this situation, a more complex overlay operation was also carried out. Its aim was to find a hypothetical site for a new countryside park within Canterbury and East Kent in the South-East of England. In general terms, the required characteristics of the park were that it be close to a river, that it did not have a major road or railway passing through it, that it was not within an existing urban area, and that it should have a certain minimum size. At the same time, it was considered important for the park to be relatively accessible by road and by railway. The specific constraints for the installation of the park were:

1. It should be within 500 metres of a river;
2. It could not be within 200 metres of a railway;
3. It could not be within 200 metres of a road;
4. It should be within 5000 metres of a railway station;
5. It should be within 1000 metres of a road;
6. It could not be within the outline of urban areas;
7. It should be at least 500 000 m^2 in area and should be a single unit: disjoint parcels summing to this area were not acceptable.

The overlay operation was repeated three times, each time using features from a different original map scale: 1:50 000, 1:250 000 and 1:625 000. All the maps were from the Ordnance Survey of Great Britain. The overlay was done using only the features common to all map scales, i.e. features that were only present in one or two of the scales were not used. Figs 1–3 show the maps that resulted from the overlay operation using the three Ordnance Survey maps. For the three scales, only two zones had an area larger than the minimum 500 000 m^2 and therefore could potentially be converted into a park on the criteria adopted. These two zones are numbered in the figures.

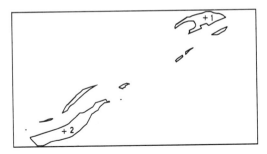

Fig. 1 Potential areas (labelled 1 and 2) for the location of a new park using the 1:50 000 scale map for the region of Canterbury and East Kent

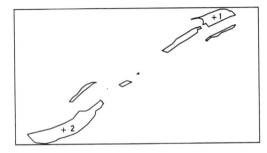

Fig. 2 Potential areas (labelled 1 and 2) for the location of a new park using the 1:250 000 scale map for the region of Canterbury and East Kent

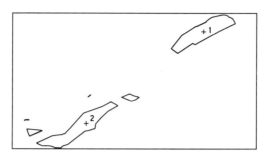

Fig. 3   Potential areas (labelled 1 and 2) for the location of a new park using the 1:625 000 scale map for the region of Canterbury and East Kent

Table 1 shows the area of the two potential zones that resulted from the overlay operation. Between the first two scales (i.e. 1:50 000 and 1:250 000) the total area of the two zones decreased by 3% and between the last two scales (i.e. 1:250 000 and 1:625 000) the area increased by 36%. In other words, depending on the scale used for the GIS analysis the land area of the parks could change by over a third.

Table 1   Area of the zones (m^2) that resulted from an overlay to determine the potential location of a new park for the region of Canterbury and East Kent

Zone	1:50 000	1:250 000	1:625 000
1	712 963	529 378	1 526 598
2	1 379 760	1 491 389	1 635 338
Total	2 092 723	2 020 767	3 161 936

It can be seen from Figs 1–3 that besides the two zones that were suitable for conversion into a park, as their area was greater than 500 000 m^2, the overlay generated a series of smaller zones which changed in area, shape and number for the three scales. The number of areas generated decreased with the reduction of scale. But the total area of these smaller zones increased by 41% between the 1:50 000 and 1:250 000 scale maps. This was followed by a 67% decrease between the 1:250 000 and the 1:625 000 scale maps.

Within the study area the relevant features used in the overlay (i.e. common across the three scales) were the following: two railways, two railway stations, one river, the outline of the city of Canterbury and thirteen roads. It was the combined generalization of all of these features that contributed to the change of the number, shape and area of the different zones for each scale used.

The crucially important test of generalization effects is their impact on the results of GIS analysis. If the generalization effects are of a magnitude that will ruin the conclusions from a 'typical' GIS analysis, then obviously they cannot be ignored. In the example described above, the possible areas and positions of a country park varied widely between the scales. At the smallest scale, part of the area labelled 1 was generalized from three much smaller areas at the middle scale. While it is possible that not all GIS analyses would be so severely affected by generalization, it is a reasonable inference that *all* users of GIS should be aware of the problems of generalization when carrying out an analysis and should seek to work with the largest scale maps available. Indeed, it is likely that generalization problems may often be much larger than the variation demonstrated in this study: use of base maps from multiple different agencies is very likely to give worse results than use of those originating from within one high quality family of maps.

2.3. Manual versus automated generalization

In addition to the generalization effects that are 'locked into' paper maps, data within a GIS can be further generalized by automated generalization algorithms. Our research has suggested that the differences caused by generalizing automatically are less pronounced than those produced by manual generalization – but they may be in addition to generalization already carried out on and 'locked into' the paper version of the map. João et al. (1992) presented evidence that generalization effects are typically greater in manually digitized paper maps than in those produced by at least one particular automated mathematical algorithm (the Douglas–Peucker algorithm – one of the most common data reduction or pseudo-generalization algorithms available in GIS). This was because the Douglas–Peucker algorithm (Douglas and Peucker, 1973) only filters the high frequency components, causing a reduction in detail of the line without the displacement that manually generalized lines typically suffer.

Although automated generalization might appear to do a better job in statistical terms, graphically the result may be inferior because automated generalization only looks at individual lines 'one at a time'; manual generalization typically considers more of the map as a whole, the cartographer moving one feature to give space to another. Automated generalization methods are poor in the maintenance of relationships between features unless these are explicitly specified through a topological statement, typically polygon adjacency. For example, the Douglas–Peucker algorithm cannot of itself cope with a situation when a contour and a river to be generalized are still in the same position relative to each other after generalization (i.e. the river crossing the contour at exactly the same point). The solution is to store the intersection point between the two features. Rhind's rule (Rhind, 1973), therefore, is to do overlay of all related features before generalization, in order to maintain fixed points in automated generalization. In general terms, where care is taken in maintaining the geographic relationship between features, GIS analysis is liable to be more reliable, if less graphically pleasing, when using lines that resulted from automated generalization. However, if any sort of pictorial analysis is needed, then maps derived from manual generalization would often be better because they would be clearer visually.

It is manifestly important that automated generalization algorithms do a 'good job', both statistically and graphically. This has been reflected in the recent development of new algorithms that try to mimic the results produced by manual generalization methods. For example, Li and Openshaw (1992) proposed a new set of algorithms for line simplification that is better than existing counterparts at producing results equivalent to those produced by manual methods. The authors called this approach 'the natural principle of objective generalisation'. Other researchers (Müller and Zeshen, 1992) have also proposed new 'natural' algorithms for the generalization of area-patches. The development of these new types of algorithms reflects a dissatisfaction with existing pseudo-generalization algorithms within the GIS community. Visvalingam and Whyatt (1990), for example, pointed out other problems arising from the widely used Douglas–Peucker algorithm, such as the creation of lines that are too 'spiky'.

Parallel to the development of new algorithms, it is fundamental to devise mechanisms that can provide greater user control over effects resulting from automated map manipulation processes. Existing GIS systems do not offer sufficient support to users wishing to use generalized data or to generalize existing data further. In particular, users often have to take important decisions relating to generalization that affect the accuracy and 'quality' of the data, without any guidance on the possible consequences of these decisions for the data or access to any technology for identifying the resulting spatial data uncertainties. As Openshaw, Charlton and Carver (1991) have shown, errors contained in digital map data propagate by GIS operations, and therefore any errors due to generalization are going to spread to newly created maps.

A GIS user faces considerable difficulty because different algorithms cause different effects on the data, and the magnitude of these effects varies for different features. McMaster (1987) analysed nine generalization algorithms in terms of the changes produced when simplifying lines. For the same percentage of coordinates eliminated, the different algorithms caused different changes on the data. The author concluded that the Douglas–Peucker algorithm caused less overall distortion (e.g. less areal displacement) than did the other line simplification algorithms. In addition – and as confirmed earlier in this chapter – the degree to which generalization algorithms affect different geographical features represented in a map can vary for different feature types, i.e. the application of an algorithm at the same tolerance to two different types of feature can produce different results for each feature when measured in terms such as the percentage of length lost (Herbert, João and Rhind, 1992).

The different effects on different digital features when the Douglas–Peucker algorithm is applied at varying tolerances is demonstrated in Fig. 4. The graph shows the percentage of length of the rivers and contours lost after application of successively larger tolerances at intervals of 10 m. The rivers and contours were digitized from the 1:200 000 scale Portuguese map of the Fronteira area. It can be seen that, for the same tolerance level, the percentage

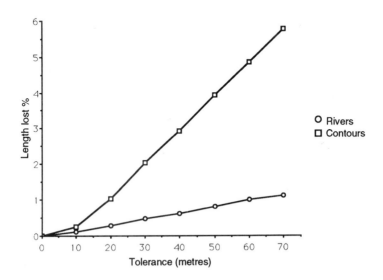

Fig. 4   Percentage of length lost using the Douglas–Peucker algorithm, for the contours and rivers of the 1:200 000 scale Portuguese map for the region of Fronteira

of length lost is very different in the case of rivers and contours. Although there are obviously differences to be expected between individual rivers or contours, it is interesting that when they are grouped together for comparison at feature level there is such a large difference between the two feature types. Moreover, this difference is apparent after relatively few observations.

Herbert, João and Rhind (1992) observed that the loss of length shown for the group of contours in Fig. 4 occurred despite the fact that the contours lost a smaller proportion of their points than the rivers; this suggested that length change was a better indicator of the amount of detail being lost, while points loss served simply as a measure of the reduction in data storage. The results obtained, using such a simple measurement as length change, suggest that automated generalization techniques need to take differences between features into account, just as much as the manual techniques employed by cartographers do.

The generalization problem is widely acknowledged in academic studies. Other researchers, such as Goodchild (1980), Blakemore (1983) and Müller (1991), have assessed qualitatively the impact of generalization effects on data quality. Goodchild (1980) described some of the types of data errors that can occur in both raster and vector systems due to the effects of generalization, such as in area and point estimation, and in the creation of a number of spurious polygons. Blakemore (1983) considered that generalization effects create potentially serious geocoding and retrieval errors. Müller (1991) pointed out that generalization can alter data quality by decreasing location and attribute accuracy, and by affecting completeness and consistency within the data. Yet, despite the acknowledgement of the generalization problem within academic circles, *no mechanisms* of any kind have been developed or proposed within commercial GIS to minimize the problems arising from generalization or even to warn users of its likely effects.

3. CONTROLLING AND REDUCING GENERALIZATION EFFECTS

Existing GIS systems do not offer adequate support to users wishing to generalize data. For example, when generalizing using the Douglas–Peucker algorithm, the user needs to specify a *tolerance* on which the simplification is based (and which will determine how many points of the line will be eliminated). Depending on the geometric characteristics of the lines and their geographic distribution, the choice of this tolerance can be difficult. Too small a tolerance might not simplify the lines as required, and too large a tolerance might make the lines spiky and, eventually, overlapping. The user can only determine the best tolerance value by a trial-and-error procedure involving re-plotting maps, which can be time-consuming and frustrating, especially with very large coverages (João, 1991). In addition, different types of features will be affected differently by different tolerance values. Thus, while the spread of GIS may increase the use of spatial data, it can also result in many relatively inexperienced users trying to produce adequate results without fully understanding the implications of generalization.

Ideally, we would seek to automate generalization as much as possible for the following reasons:

- Not all users will be skilled in cartographic generalization procedures – the number of GIS workers (and hence map-makers) without cartographic training is growing rapidly.
- Not all of them will appreciate the effects of generalization carried out within 'black box' GIS or already enshrined in data sets.

- In some fields, at least, it is impossible (or at least difficult) to carry out generalization visually (e.g. with remote sensing data, where 'real world' spatial objects have to be inferred from samples from a continuous distribution).
- In a GIS context, it is often as important to take account of the statistical consequences of generalization as it is of the graphic ones (and these may not co-vary).

Due to its complex, diverse and non-deterministic nature, the generalization process is difficult to automate, in particular if one is attempting to mimic a manual procedure. It is this complexity that has motivated much of the research effort being directed towards the use of concepts and techniques from the field of artificial intelligence (AI). The difficulties involved in making substantial progress in this area include:

- Generalization is itself a broad description for multiple, interacting processes. Add to this the many different purposes, map scales and patterns of data that can be involved, and automation of any substantial process becomes a major task involving an appropriately high commitment of resources.
- Much of generalization is not about a single option to reach an objective or about a well-defined series of steps. Automated systems need to be constructed around the idea that more than one acceptable solution may be available for a particular problem.
- Those who see knowledge-based systems as a potential solution come up against the difficulty of formulating rules for generalization that can be said to be generally applicable. In cases where rules do exist, they often relate to the practices of a particular organization rather than being universal rules for all countries and organizations. The other aspect of this problem is that some practitioners find it very difficult to explain their actions as a set of easily explicable rules.

Because of the complexity of the generalization process, there is an absence of effective automated programs (Herbert and João, 1991). Despite the substantial amount of research on the further automation of the generalization process in recent years, much effort has been directed towards particular tasks, for example, national atlas production (Richardson, 1989) or text placement (Freeman and Ahn, 1984; Cook and Jones, 1990).

However, given the power of contemporary leading-edge GIS technology and our expectations of near-future improvements to equipment, we believe that there is scope for considerable advances to be made in the automation of the process. In essence, there are two alternative ways forward. The first is to develop and perfect a completely general-purpose fully automated operator-invariant generalizer (e.g. Li and Openshaw, 1992). The second is to develop a knowledge-based (expert systems) approach that will allow users to cope better with the problems until good and completely automatic generalization is a practical and reliable proposition. It is the latter approach that is emphasized here.

Our approach is to develop programs incrementally, permitting the inclusion of further details on the effects of generalization as knowledge increases and enabling the user to set constraints on the degree of generalization. Central to our philosophy is the initial design of a generalization machine which copes with all the known generalization processes, in combination wherever appropriate, but with modules that are progressively implemented as resources and knowledge permit.

3.1. User-controlled generalization by setting constraints on the extent of generalization

The extent to which generalization can be carried out automatically and the necessary involvement of skilled manual intervention is (and has long been; see Rhind, 1973) a matter of debate. Several authors, such as Weibel (1991), Mackaness and Beard (1990), and Müller (1991), have argued for the inclusion of human intelligence in the automated cartographic generalization process. Human input has been suggested in those cases where the automation of generalization fails; that is, as generalization is an extremely complex procedure, the user has to intervene to help the system perform the generalization. However, user intervention on automated generalization based on controlling the *extent* of generalization effects in the first place has not previously been suggested.

The principle behind the expert system software development was to increase user control over the generalization process while at the same time letting the GIS do the basic work involved. In order to put the user in control of the process, he or she needs to specify his/her objectives and then let the system choose the generalization algorithms, and respective tolerances, appropriate to the job. The area of work chosen – generalization using *line* generalization algorithms – forms only a small part of the generalization process but the aim was to develop principles that could be extended to the automation of other aspects of generalization.

The prototype that was developed concentrated on selection of appropriate tolerances for generalizing lines based on preferences expressed by the user. The user was allowed to set preferences and constraints for the generalization process for different features. For example, the user could ask the system to select automatically a generalization tolerance that allowed for only a maximum of 0.5% length change for the selected feature class. Fig. 5 shows the control panel dialogue through which the user could set parameters on the extent of generalization desired.

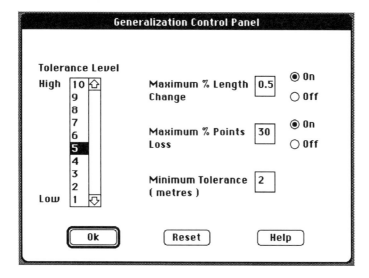

Fig. 5   Control panel for setting preferences and constraints for the generalization process

If the user wanted to set no specific constraints on the process, then the scale for tolerance levels (on the left of the control panel) could be used to give a broad indication of the desired degree of generalization. The bottom of this scale represents the minimum tolerance for which a default value is calculated on the basis of removing redundant or very fine detail at the output scale. The user can change this value, if desired, by entering a new figure. The top of the scale approximates to the level of tolerance necessary to remove 90% of the points in a line, and is therefore only to be used if 'heavy' point reduction is the principal aim of generalization. Any selections between these minimum and maximum levels will be calculated according to the total range of possible values. This means that the scale is a flexible one, rather than one mapped on to a permanent set of values.

The constraints on point or length reduction are optional but, if used, they override the tolerance level setting if generalization on that basis threatens to exceed these constraints. By using these two constraints, the user can be given a means of setting more specific targets for the system. These constraints therefore set upper generalization limits that could be used by the program. Setting a low figure for permissible length loss (e.g. less than 1%) guarantees preservation of a significant degree of detail in the output data set, while the limit for points loss can be used to control loss of points more precisely than by using the tolerance level setting.

After generalization has been carried out, the user will normally have available a graphical representation of the simplified data set. However, he or she also needs other reporting facilities to verify whether his or her requirements have been met. The system reports back on changes resulting from generalization by presenting how far it has been able to meet the user's specifications by giving specific figures for length change, points loss, etc. that the features suffered. In addition to this, there may be occasions when a user wishes to know for a given set of measurements (examples being length or vector displacement) what changes have taken place to the data so that there is a record of the extent to which a particular file has been altered. If some effects of generalization are unavoidable, there is potential for compensation if the extent of these effects is made known to those who are using the data, as a basis is provided for assumptions about the accuracy of the generalized data set.

3.2. Elements of an intelligent system for generalization

The approach taken here maintains the user as the main agent in specifying objectives and setting constraints. The focus of automation then becomes one of finding the best ways in which the user can do this and then of letting the system decide the best way of achieving the user's requirements. This latter aspect means that there must be at least some rudimentary decision making mechanism built into systems for generalization to enable choices to be made between alternative actions. Even more important is a form of *learning mechanism* so that the knowledge a system has of different situations or strategies is progressively increased. This means that such a system need only be provided with the structure of the knowledge it requires, rather than having to be pre-programmed with every rule for every possible situation. The prototype employed a basic method of storing results from the generalization of different feature types and then using these stored results as the basis for making future decisions on tolerance levels. This type of mechanism may represent the only means of overcoming the potentially vast number of rules that could be compiled to build an 'intelligent' generalization system capable of coping with both a variety of tasks and differing data sets.

An ideal for an automated system dealing with generalization is data-independence: the system needs to have the capability of dealing with many different types of data. The difficulty here is that much of the knowledge on how to generalize is intimately linked to the specific type of data concerned. However, through the development of a knowledge base that can be augmented at runtime, we have attempted to tackle the problem of coping with a wide variety of data types. Such an approach is viable provided that the system initially has adequate knowledge about the data or can gain it through initial training by a user. In the case of line generalization, the criteria chosen for analysis are common to all feature types (only the parameters differ), which makes it easier for the program to carry out analysis without the need for user intervention. The extent to which this is true for other parts of the generalization process needs to be tested.

A general solution to minimize generalization effects (at least of a statistical nature) would be for users to use a single topographic base wherever possible, retain data as detailed as possible and generalize automatically 'on the fly'. The extent of manual generalization effects would be minimized by the use of larger scale maps than apparently necessary for any one application. The control of the effects resulting from automated generalization processes would be achieved by the development of a generalization machine with capabilities similar to, but extended from, the prototype built for this research.

3.3. Application-specific intelligence using search processes

An alternative to the above design and rule-based systems would be to adopt a much more computationally intensive approach. Once the user has provided his or her preferences (Fig. 5), the generalization challenge becomes one of selecting a generalization method and then seeking to estimate associated parameters that produce the desired results, ideally via an intelligent search process. The search should not assume that there is a single optimal solution. In mathematical terms, the solution is unlikely to be a convex objective function, nor is any objective function that incorporates the user's preferences likely to be linear. Hence there is a preference for non-standard mathematical techniques such as genetic algorithms (Goldberg, 1989) or simulated annealing. Another complication arises when the task involves evaluating many different generalization procedures, each with an estimation of associated parameters. In both cases, the generalization process is specific to the user's data or application. This sort of procedure is computationally intensive, but recent trends in hardware performance suggest that this may not be a limitation in the future.

Ultimately, it might be possible to develop a hybrid machine design that combines a knowledge-based approach (in areas where there is applicable knowledge or where the task is trivial) with learning the search procedure for other situations. But all the alternative generalization procedures would have to be evaluated in an ordered sequence informed by the expert system. Additionally, as the results of approximately optimal generalizations are known, so the production rules can be updated, allowing a degree of machine learning.

A final design principle concerns how best to handle generalization failure. How should the machine report the consequences of a failure to meet the specified generalization objectives? The expert system may fail to meet the stated objectives. At this point the generalization machine could start a search for a solution. A final end state also needs some consideration: there may well be multiple different, equally good, generalization solutions to the user's preferences. Accordingly, it would be better to obtain from the user not merely a constraints set but also an objective function, so as to minimize data volume and meet stated preferences.

4. CONCLUSIONS

The primary objective of this research project was to understand, anticipate and predict the effects that generalization has upon the results of using analytical techniques within a GIS. We have demonstrated that generalization of data materially affects the results we get out of a GIS. Depending on the original scale of the source maps, such simple measures as length or area will have different values and the geographical position of individual features will differ. Any generalization effect is also compounded when you look at more than one feature simultaneously. So, even if features individually show generalization effects that might seem irrelevant, when different features are used in a GIS analysis the final generalization effect is a combined one. This 'combined action' can cause very large effects in an way that is not predictable deterministically. It is, therefore, fundamental that all users of GIS should be aware of the problems of generalization when carrying out an analysis.

The implications of this study are that we can not 'reverse engineer' generalization with confidence, though one might be able to deduce limits to the magnitude of generalization effects. Some features are more affected in certain particular ways than are others. Generalization effects depend on what property the user is concerned with and not just on the feature type. This also shows the importance of user input in automatic generalization in controlling the effects of generalization.

The generalization effects contained in most existing data sets derived from existing maps will consist of the combined effects from both manual and automated processes. Differences caused by generalizing automatically are less pronounced than the ones resulting from manual generalization, but may exacerbate the previous generalization carried out on the paper version of the map. While manual generalization effects might be reduced by the use of larger scale maps, it is also important to devise mechanisms that can provide greater user control over effects caused by automated processes. This would be the case especially with the advent of scale-free GIS (see Guptill, 1989). Such automated generalization of spatial data should involve at least a *control facility* in which the user can specify the maximum permissible effect (e.g. the maximum change in line length) and a *reporting facility* by which the system can report back to the user on the extent of changes that have been produced in the data. The automation of generalization within a GIS should therefore be approached as an *optimization process*. Generalization facilities available within a GIS should allow the selection of generalization algorithms that will *minimize* selected generalization effects and therefore *optimize* the process from the point of view of the user.

The prototype software developed has demonstrated the practicability of this principle, albeit in a limited domain. The knowledge base for the system is mainly statistically based and is centred on the distinctions between the way in which line simplification algorithms affect different features at various levels of tolerance. The Douglas–Peucker algorithm was used although system development has not been algorithm-dependent. The objective has been to give the user a greater control of generalization effects and also to make him or her aware of what has happened to the data. The approach taken has emphasized the simplification of the user's task while at the same time increasing his or her understanding of the generalization occurring through the addition of a quality report. The role of the user is still very important to the process: the objectives to be met are specified by the user but the system is then able to do much of the work involved in translating these objectives into practice. AI techniques have been used in a supportive role to the user and also in a way that allows the system to increase gradually its knowledge base and therefore the informed basis of its operations.

Our conclusion, based on the work carried out in this project, is that such an operational, full-scale, generalization system is now technically and intellectually feasible: resources are the only serious constraint on creating it. Since generalization effects are central to the use of GIS and since we have demonstrated that there are considerable advantages in automated over manual generalization techniques, we see the creation of a more complete generalization machine as a high priority on the global GIS research agenda.

ACKNOWLEDGEMENTS

Thanks are due to the Ordnance Survey, who provided the data for part of this research, and to the Instituto Geográfico e Cadastral for permitting use of their maps and providing information on how they are compiled. The UK Economic and Social Research Council and the Natural Environment Research Council funded this research under grant GST/02/490. We would like to acknowledge the assistance of LaserScan Ltd, Cambridge, for providing scanning facilities, LITES2 and VTRAK software, and for offering helpful advice during the project.

REFERENCES

Blakemore, M. (1983) Generalization and error in spatial data bases. *Proc. Auto-Carto 6*, 313–322.
Cook, A. and Jones, C. (1990) A Prolog interface to a cartographic database for name placement. *Proc. 4th Int. Symp. Spatial Data Handling*, Zürich, 701–710.
Douglas, D. and Peucker, T. (1973) Algorithms for the reduction of the number of points required to represent a digitized line or its caricature. *Can. Cartogr.*, **10**, 112–122.
Freeman, H. and Ahn, J. (1984) AUTONAP – An expert system for automatic name placement. *Proc. Int. Symp. Spatial Data Handling*, vol. 2, Zürich, 544–569.
Goldberg, D. (1989) *Genetic Algorithms in Search Optimisation and Machine Learning*. Addison-Wesley, Reading, Massachusetts.
Goodchild, M.F. (1980) The effects of generalisation in geographical data encoding. In Freeman, H. and Pieroni, G. (eds), *Map Data Processing*. Academic Press, New York, pp. 191–205.
Guptill, S. (1989) Speculations on seamless, scaleless cartographic databases. *Proc. Auto-Carto 9*, Baltimore, 2–7 April 1989, 436–443.
Herbert, G. and João, E. (1991) Automating map design and generalisation: A review of systems and prospects for future progress in the 1990s. SERRL Working Report no. 27, Birkbeck College, London.
Herbert, G., João, E. and Rhind, D.W. (1992) Use of an artificial intelligence approach to increase user control of automatic line generalisation. *Proc. 3rd European GIS Conf. (EGIS'92)*, Munich, 23–26 March, 554–563.
João, E. (1991) The role of the user in generalisation within GIS. In Mark, D. and Frank, A. (eds), *Cognitive and Linguistic Aspects of Geographic Space*. Kluwer, Dordrecht, pp. 493–506.
João, E., Herbert, G. and Rhind, D.W. (1990) The measurement and control of generalisation effects. SERRL Working Report no. 19, Birkbeck College, London.
João, E., Herbert, G. and Rhind, D.W. (1991) A strategy for the measurement and control of generalisation effects. *Proc. EUROCARTO IX*, Warsaw, 11–13 June, 32–47.
João, E., Rhind, D.W., Openshaw, S. and Kelk, B. (1990) Generalisation and GIS databases. *Proc. 1st European GIS Conf. (EGIS'90)*, Amsterdam, 10–13 April, 504–516.
João, E., Herbert, G., Openshaw, S. and Rhind, D.W. (1992) Magnitude and significance of generalisation and its effects. *Proc. 3rd European GIS Conf. (EGIS'92)*, Munich, 23–26 March, 711–721.
Li, Z. and Openshaw, S. (1992) Algorithms for automated line generalisation based on a natural principle of objective generalisation. *Int. J. Geogr. Inf. Sys.*, **6**(5), 373–389.
Mackaness, W. and Beard, K. (1990) Development of an interface for user interaction in rule based map generalisation. *Proc. GIS/LIS'90 Conf.*, Anaheim, California, November, 107–116.
Maling, D. (1989) *Measurements from Maps. Principles and Methods of Cartometry*. Pergamon Press, Oxford.

Mather, P. (1992) The U.K. ESRC/NERC Joint Programme on geographical data handling: Progress report. *Proc. 3rd European GIS Conf. (EGIS'92)*, Munich, 23–26 March, 474–482.

McMaster, R. (1987) The geometric properties of numerical generalisation. *Geogr. Analysis*, **19**, 330–346.

Müller, J.C. (1991) Generalization of spatial databases. In Maguire, D.J., Goodchild, M.F. and Rhind, D.W. (eds.), *Geographical Information Systems: Principles and Applications*. Longman, Harlow, pp. 457–475.

Müller, J.C. and Zeshen, W. (1992) Area-patch generalisation: a competitive approach. *Cartogr. J.*, **29**, 137–144.

Openshaw, S., Charlton, M. and Carver, S. (1991) Error propagation: a Monte Carlo simulation. In Masser, I. and Blakemore, M. (eds.), *Handling Geographical Information: Methodology and Potential Applications*. Longman, Harlow, pp. 78–101.

Rhind, D.W. (1973) Generalisation and realism within automated cartographic systems. *Can. Cartogr.*, **10**, 51–62.

Richardson, D. (1989) Rule based generalisation for base map production. *Proc. Natl. Conf. GIS – Challenge for the 1990s*, Ottawa, 718–739.

Visvalingam, M. and Whyatt, J. (1990) The Douglas–Peucker algorithm for line simplification: re-evaluation through visualization. *Comput. Graphics Forum*, **9**, 213–228.

Weibel, R. (1991) Amplified intelligence and rule-based systems. In Buttenfield, B. and McMaster, R. (eds.), *Map Generalization: Making Rules for Knowledge Representation*. Longman, Harlow, pp. 172–186.

Chapter Nine

Datum transformations and data integration in a GIS environment

A. H. DODSON
Institute of Engineering Surveying and Space Geodesy, University of Nottingham

AND

R. H. HAINES-YOUNG
Department of Geography, University of Nottingham

National mapping by the Ordnance Survey of Great Britain is based on the local OSGB36 datum, which for historic reasons is both inconsistent and inhomogeneous. By contrast, position data from the Global Positioning System (GPS) is based on the World Geodetic System (WGS84), which is a consistent global geocentric datum. Coordinates from one datum cannot be related to the other without the use of mathematical transformations, which, for OSGB36 in particular, may vary across relatively small areas. This research examines the issues of datum inconsistency, which are of increasing importance in the UK given the wider availability of GPS data. The project was undertaken in association with the Ordnance Survey, whose transformation algorithms were tested by field survey. The results suggest that the Ordnance Survey transformations can be used to position GPS points on large-scale maps based on the OSGB36 datum to within the range of error associated with such maps, although the algorithms did not remove all of the systematic variation associated with OSGB36. The methodology developed here will be useful to organizations such as utilities wishing to locate plant using GPS and display these position data accurately on a digital 1:1250 map base.

1. BACKGROUND

The availability of position information supplied by the GPS of navigation satellites represents an important opportunity for the GIS community. GPS gives access to positional information in near and real time, and provides the basis of accurate and rapid survey. Position data from GPS will therefore be used increasingly in association with existing spatial databases, both for update and for a range of new applications. There is, therefore, a need to develop methods for the integration of GPS data with our existing map information in order to ensure consistency of our digital databases.

Geographical Information Handling – Research and Applications. Edited by P. M. Mather
© 1993 John Wiley & Sons Ltd

Unfortunately, the integration of GPS with the spatial databases that are now widespread in the GIS community is not a trivial matter. In many countries the potential exists for serious inconsistencies between data derived from these different sources. Inconsistencies may arise because of differences in the definition of the geodetic datums themselves, and as a result of problems of internal consistency within many of the older, local datums that are in widespread use for national mapping. In this chapter we consider the particular problems faced in Great Britain.

Geodetic coordinates, on which most national mapping is based, are related to a coordinate datum for computational purposes. Such a datum is conventionally an ellipsoid of rotation,

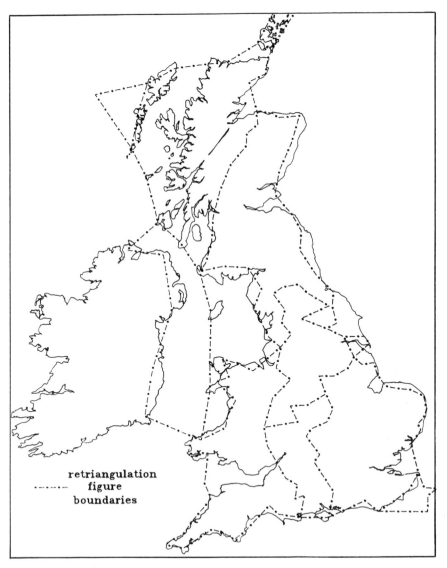

Fig. 1 OSGB36 computation figure boundaries (based upon the Ordnance Survey map with the permission of the Controller of Her Majesty's Stationery Office, © Crown copyright).

thereby approximating to the shape of the earth's surface, which includes the size and shape of the ellipse, the orientation of the axes and the position of an origin. If the centre of the ellipse is given an origin at the earth's mass-centre, then the datum is said to be geocentric. The GPS datum, WGS84, is just such a datum. It uses the WGS84 ellipsoid, defined by global satellite observations (Defense Mapping Agency, 1987). By contrast, most national mapping is based on datums that have a local origin. In Great Britain, for example, both digital and analogue maps are based on the OSGB36 datum, which was established between 1935 and 1951. This datum uses an Airy ellipsoid, with an origin defined by the mean fit to the eleven stations of the earlier Principal Triangulation (Bordley and Christie, 1989; Ordnance Survey, 1967).

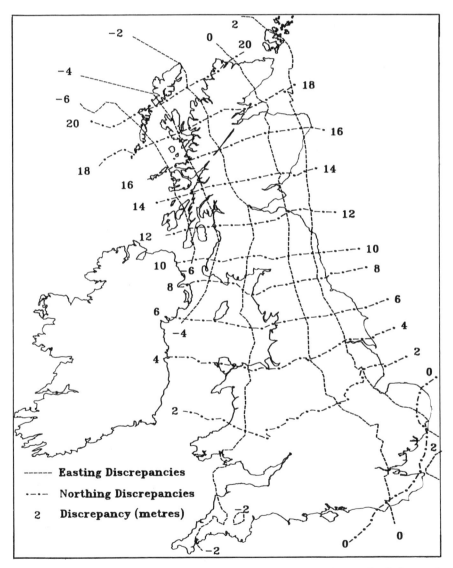

Fig. 2 Systematic differences between OSGB36 and OSSN80 (based upon the Ordnance Survey map with the permission of the Controller of Her Majesty's Stationery Office, © Crown copyright).

When faced with the problem of bringing together information collected on different map datums, the user must have information about the scale and rotation differences between the two. Unfortunately, the translation of data from one datum to another is not always achievable by the application of a simple transformation algorithm when one datum is not internally consistent. This is the case in Great Britain (Bordley and Christie, 1989).

The computation (or 'readjustment') of the observations for the retriangulation on which OSGB36 was based was not a rigorous one because of the difficulty of making all the necessary calculations in the pre-computer age. The computation was carried out in blocks or 'figures' (see Fig. 1) with the result that the solution is not homogeneous. Additionally, because of the difficulty of making distance measurements in the days before electronic distance measurement (EDM) techniques, few baselines were observed in making the readjustment, so there were also systematic scale errors present in the solution.

The Ordnance Survey has long recognized the problems associated with OSGB36, and has produced a rigorous readjustment of the primary triangulation, including EDM and satellite (Doppler) observations to reduce systematic errors. This readjustment is known as Ordnance Survey Scientific Network (OSSN) 80 (Ashkenazi *et al.*, undated). The overall trend and size of the systematic difference between OSGB36 and OSSN80 is shown in Fig. 2, but local variations from the trend still exist. All second order triangulation stations have been re-computed by the Ordnance Survey in OSSN80, by local adjustment to the rigorously computed primary network. Although OSSN80 provides a consistent datum for scientific purposes and can easily be related to WGS84, national mapping is still based on OSGB36. With the advent of GPS, the problem of how users can best integrate data held on these two very different map datums arises.

2. AIMS

This project was, therefore, designed to consider these issues of datum inconsistency, which are of increasing importance in the UK, given the availability of positional information from GPS. The specific research goals were:

- To develop and test a methodology for transforming and integrating GPS data with map data held on OSGB36.
- To develop a suitable interface within a GIS environment to allow non-expert users to work with Ordnance Survey digital map data on the OSGB36 datum, and to integrate these data with other information held on scientific datums, such as WGS84.
- To explore the implications of the use of GPS in the development and maintenance of digital map databases in the UK.

3. METHODOLOGY

3.1. Background

There are three methods of using GPS to obtain position information (Dodson and Basker, 1992). The first involves the use of the 'pseudo-range' or code observable and gives an instantaneous solution for the coordinates of the GPS receiver. This method can give accuracies of between 10 m and 100 m, which are satisfactory for navigation purposes. The accuracy of this navigation solution can be improved to around 5 m, by using the second general method, on-line differential GPS positioning (Blanchard, 1991; Dodson, Haines-Young and Roberts, 1992).

Datum Transformation and Data Integration 83

The third and most accurate method of using GPS involves the measurement of the phase of the carrier signal. This observable can only be used to determine the relative position between two or more GPS receivers. At present, it involves post-processing of the data and so does not allow a real-time solution. In the near future telemetry links and improved processing algorithms will allow near real-time solutions. The accuracies that can be obtained depend on many factors, but 1 cm in 10 km is achievable.

Only this last method of using GPS to achieve engineering survey accuracy was considered in this project, because it is with this type of application that problems of datum inconsistency are most likely to be observed. Given the magnitude of the positional differences that may arise (Table 1) between positional data derived from GPS and OSGB36 map data, the analysis that follows focuses on the 1:1250 map scales.

Table 1 Possible position errors associated with different GPS techniques

Possible error		GPS observations	Pseudo-range: single point	GPS differential (<100 km)	Phase GPS differential (<100 km)
GPS		True	200 m	5.0 m	0.1 m
		At 1/50 000	4 mm	0.1 mm	<0.01 mm
		At 1/10 000	20 mm	0.5 mm	<0.01 mm
		At 1/1250	160 mm	4.0 mm	<0.1 mm
Datum		True	500 m	5.0 m	5.0 m
		At 1/50 000	10 mm	0.1 mm	0.1 mm
		At 1/10 000	50 mm	0.5 mm	0.5 mm
		At 1/1250	400 mm	4.0 mm	4.0 mm

3.2. Using GPS for positioning in an OSGB36 map base

The use of GPS to obtain relative position provides only differences in coordinates between stations. If a base station is at a point of known OSGB36 coordinates, then the position of the second point will essentially also be obtained in OSGB36, but the value obtained may well differ from its correct OSGB36 value. Two factors control the magnitude of the error involved. Firstly, the difference vector observed will be in WGS84, and therefore any systematic scale and orientation differences between datums will be important. Secondly, particularly if the base station is at some distance from the observation point, local differences in OSGB36 will manifest themselves.

In order to overcome these problems of data integration, it would clearly be better to base all mapping on a consistent datum such as OSSN80. With such a consistent datum, vector differences will be consistent with WGS84. The only problem comes in fitting the coordinates obtained into a map base. For this purpose a series of mathematical transformations between OSSN80 (and hence WGS84) and OSGB36 must be available.

A likely use of GPS positioning, which illustrates some of the transformation and integration issues, is that by utilities companies, who might use the technology for fixing the location of their plant and for finding it again. Although GPS can be used without relating position data to other existing (and mapped) features, it is likely that there will be a requirement to integrate such information with existing (digital) mapping systems. How should this integration be achieved?

Firstly, the accuracy requirements will determine which method of GPS positioning is appropriate; in most cases, this is likely to be differential (rather than single point) positioning.

There will therefore be a need for a reference (base) station (or stations) to be established, which could be conveniently located, for example, as a permanent unmanned station at a utility company's head or regional office. The coordinates of this point would then initially be determined precisely, using GPS (static) observations from local Ordnance Survey triangulation stations using known OSSN80 coordinates.

Detail positioning could then proceed (by measuring relative position vectors) at any time, using any number of GPS receivers, over a wide area surrounding the reference station. Certainly, a GPS accuracy of less than 10 cm could be achieved (using the phase observation method) over ranges of at least 50 km. It is therefore important, if the positions thus obtained (in the OSSN80 system) are to be integrated with existing mapping, that the available OSSN80 to OSGB36 transformations have corresponding accuracy over similar ranges.

3.3. Operational methods

The testing of the set of transformations between OSGB36 and OSSN80, and its development into a general method for integrating GPS data collected on WGS84 with OSGB36 data, formed the focus for this study. If a set of transformations between OSSN80 and OSGB36 is available, the procedure proposed for integrating GPS data into an OSGB36 map is to establish the base station at a known OSSN80 point (or less satisfactorily at a known OSGB36 point transformed to OSSN80), to observe the vector differences in WGS84, which for most purposes can be added directly to the OSSN80 coordinates, to obtain the new positions in OSSN80. Local transformations can then be used to obtain OSGB36 coordinates, which should then be consistent with the existing map.

4. RESULTS

4.1. Ordnance Survey transformations

This project was undertaken in collaboration with the Ordnance Survey of Great Britain, whose main input was the development of a new transformation between OSSN80 and OSGB36. The solution derived was to calculate separate transformation functions for each triangle defined by the second order triangulation of Great Britain (Dyer, 1990). The vertices of each triangle were defined by control points in known OSGB36 and OSSN80 coordinates and a Delauney triangulation covering the entire national grid was constructed, giving a unique solution made up of the 'most equilateral' triangles possible. Functions were computed so as to produce continuity across triangle boundaries, with the result that a continuous, differentiable rubber-sheet transformation between OSSN80 and OSGB36 is available for each of the triangles covering the national grid.

The transformation was implemented by calculating shifts in terms of eastings and northings for a set of point making up a regular 1 km × 1 km grid for the whole of Great Britain, and using these data to interpolate the actual shift for points located within the framework thus defined. A 1 km × 1 km resolution for the grid was found to be the most effective in terms of minimizing data storage volumes without jeopardizing transformation accuracy.

The grid-based format allows the parameters to be stored in a raster format that is easily referenced by the kilometre square containing the point to be transformed. If (x,y) are the

Datum Transformation and Data Integration

OSGB36 coordinates of a point of detail to be transformed, and if (a_i,b_i) are the calculated deviations between OSGB36 and OSSN80 at the corners of the kilometre grid square in which it falls, then the OSSN80 coordinates (X,Y) are given by

$$X = x + v[ua_3 + (1 - u)a_2] + (1 - v)[ua_4 + (1 - u)a_1]$$
$$Y = y + v[ub_3 + (1 - u)b_2] + (1 - v)[ub_4 + (1 - u)b_1] \quad (1)$$

where

$$u = (x - x_0)/1000$$
$$v = (y - y_0)/1000 \quad (2)$$

A grid-based solution also allowed a computable inverse, so that it is possible to transform from OSSN80 to OSGB36 with the same accuracy as from OSGB36 to OSSN80.

4.2. Field tests of GPS positioning

In order to examine the problems associated with attempting to integrate data held in different datums, a series of field trials were undertaken in the Nottingham area. The sites chosen for study were open in character, to ensure good satellite visibility, with well-defined ground features that could be located easily on 1:1250 digital Ordnance Survey maps. In each survey a number of identifiable points were positioned using the kinematic GPS method of observation, which consists of establishing an initial base station in the survey area with approximately 30 minutes of observation, and then measuring at other points for approximately 2 minutes each.

The groups of surveys differed as follows:

- Survey 1: Single base station with a short baseline (1 km) between base station and survey points.
- Survey 2: Two base stations differing in distance to survey site (<1 km and 9 km), with one baseline crossing a transformation block boundary.
- Survey 3: Two base stations at different distances to survey site (<8 km and 40 km) with one baseline crossing both a transformation block and an OSGB36 computation block boundary.

The results of the three survey campaigns are shown in Tables 2, 3 and 4. For the purposes of evaluating the Ordnance Survey transformation, a comparison with untransformed data is given.

Survey 1 represents the 'best possible case' scenario, with a short survey baseline and comparison of survey data with map coordinates for a recently redeveloped urban area. Table 2 shows that when using the Ordnance Survey transformation to convert GPS to OSGB36 the difference between the surveyed position and the corresponding map location was about 0.39 m, which is within the limit of map accuracy published by Ordnance Survey (see results set b). Comparison with the data obtained using the untransformed solution, however, shows that the two techniques differ little in their performance (compare results sets b and c). This is to be expected, since there will be little variation in the OSGB36 datum over such a short distance.

Table 2 Results of Survey 1

	Vector difference	Easting difference	Northing difference
(a) Transformed GPS (OSGB36) to GPS (OSSN80)			
Average	2.04	0.15	2.04
Standard deviation	0.01	0.03	0.01
Maximum	2.07	0.19	2.06
Minimum	2.03	0.11	2.03
(b) Transformed GPS (OSGB36) to map (OSGB36)			
Average	0.39	−0.15	0.15
Standard deviation	0.13	0.28	0.24
Maximum	0.63	0.33	0.52
Minimum	0.24	−0.62	−0.28
(c) Untransformed GPS (OSGB36) to map (OSGB36)			
Average	0.41	−0.17	0.18
Standard deviation	0.15	0.30	0.24
Maximum	0.68	0.31	0.56
Minimum	0.24	−0.68	−0.24

Survey 2 (Table 3) shows the effects of increasing baseline length. Using the short baseline to calculate positions of the survey points, the two techniques showed little variation in their performance, with the same mean difference in position between surveyed and map coordinates of 0.38 m for both the untransformed and transformed solutions (results sets b and e). Those calculations based on the longer baseline, however, showed mean differences of 0.58 m and 0.45 m, respectively (see results sets a and d). Thus, the growing discrepancy in the OSGB36 datum is shown to become significant, even at a range of less than 10 km.

Table 3 Results of Survey 2

	Vector difference	Easting difference	Northing difference
(a) Untransformed GPS Sandiacre (OSGB36) to map (OSGB36) (9 km baseline)			
Average	0.58	0.41	−0.29
Standard deviation	0.29	0.35	−0.03
Minimum	0.27	0.02	−0.69
Maximum	0.88	0.84	−0.33
(b) Untransformed GPS University (OSGB36) to map (OSGB36) (1 km baseline)			
Average	0.38	0.00	−0.12
Standard deviation	0.14	0.36	0.22
Minimum	0.11	−0.40	−0.52
Maximum	0.54	0.42	0.11
(c) Untransformed GPS Sandiacre (OSGB36) to untransformed GPS University (OSGB36)			
Average	0.45	0.41	−0.17
Standard deviation	0.02	0.01	0.02
Minimum	0.42	0.39	−0.20
Maximum	0.47	0.42	−0.15
(d) Transformed GPS Sandiacre (OSGB36) to map (OSGB36)			
Average	0.45	0.25	−0.22
Standard deviation	0.25	0.35	0.22
Minimum	0.17	−0.14	−0.62
Maximum	0.73	0.68	−0.04

Table 3 (*continued*)

	Vector difference	Easting difference	Northing difference
(e) Transformed GPS University (OSGB36) to map (OSGB36)			
Average	0.38	0.01	−0.11
Standard deviation	0.14	0.36	0.22
Minimum	0.11	−0.39	−0.52
Maximum	0.53	0.43	0.12
(f) Transformed GPS Sandiacre (OSGB36) to transformed GPS University (OSGB36)			
Average	0.27	0.25	−0.11
Standard deviation	0.02	0.01	0.02
Minimum	0.24	0.22	−0.13
Maximum	0.29	0.26	−0.08

The trend observed in the second survey campaign was confirmed in the third and final set of survey results (Table 4). The results suggest that with the even longer baseline the mean difference in performance between the transformed and untransformed solutions becomes more marked. Once again, the shorter baseline calculation showed only a small difference in the methods (results sets b and e). However, for the long baseline data set there was an average difference of 0.69 m without transformation compared to 0.27 m when the data were transformed (see results sets a and d).

Table 4 Results of Survey 3

	Vector difference	Easting difference	Northing difference
(a) Untransformed GPS Retford (OSGB36) to map (OSGB36) (40 km baseline)			
Average	0.690	0.397	0.561
Standard deviation	0.121	0.058	0.127
Minimum	0.559	0.266	0.394
Maximum	0.904	0.471	0.772
(b) Untransformed GPS University (OSGB36) to map (OSGB36) (8 km baseline)			
Average	0.486	−0.248	−0.414
Standard deviation	0.128	0.053	0.133
Minimum	0.263	−0.374	−0.584
Maximum	0.629	−0.174	−0.197
(c) Untransformed GPS University (OSGB36) to untransformed GPS Retford (OSGB36)			
Average	1.168	−0.645	−0.973
Standard deviation	0.009	0.019	0.009
Minimum	1.157	−0.688	−0.984
Maximum	1.186	−0.630	−0.960
(d) Transformed GPS Retford (OSGB36) to map (OSGB36)			
Average	0.266	−0.233	0.021
Standard deviation	0.042	0.058	0.127
Minimum	0.216	−0.364	−0.146
Maximum	0.364	−0.159	0.232
(e) Transformed GPS University (OSGB36) to map (OSGB36)			
Average	0.428	−0.112	−0.411
Standard deviation	0.136	0.053	0.133
Minimum	0.198	−0.238	−0.581
Maximum	0.589	−0.038	−0.194

(*continues*)

Table 4 (*continued*)

	Vector difference	Easting difference	Northing difference
(f) Transformed GPS University (OSGB36) to transformed GPS Retford (OSGB36)			
Average	0.448	0.120	−0.431
Standard deviation	0.012	0.018	0.009
Minimum	0.430	0.078	−0.442
Maximum	0.459	0.136	−0.417

Moreover, a significant feature of the results is that in both the second and third surveys, points were observed from two different base stations. The resulting mean differences in the two GPS positions for the same point were 0.45 m (untransformed) and 0.27 m (transformed) for Survey 2, and 1.17 m (untransformed) and 0.45 m (transformed) for Survey 3. These results show the improvement obtained when using the transformation without the added complication of the map error affecting the comparison. Since it is quite possible that, in the operational scenario, points may be positioned from different regional base stations, this comparison is of particular importance.

4.3. Development of the GIS interface

In order to process the GPS survey data, a set of macros was developed to handle the position information and display it against a digital 1:1250 Ordnance Survey map base within the LaserScan LITES2 and HORIZON GIS packages. The macros interfaced with the shift data for a regular 1 km × 1 km grid, on which the Ordnance Survey transformation was based, which was stored in an INGRES database. Once displayed on the Ordnance Survey map base, survey data were documented with information on their origin and positional accuracy. The macros, which allowed the integration of the GPS survey into the existing OSGB36 digital map base, were menu driven, and of a design suitable for the non-expert user.

A particularly important feature of the system was the ability to label line and point features according to the origin of the survey information. The need to attach statements of quality to the data items held in a digital database has been widely discussed (Burrough, 1986), and particular emphasis has been placed on the analysis of how errors may be propagated in subsequent analysis (Goodchild and Sucharita, 1989). A review of recent literature shows that in contrast to the types of error associated with spatial data processing, relatively little attention has been directed to the factors affecting the quality of the position information derived from different survey methods, despite the potential impact that such errors might have.

A primary requirement of a GIS is that there is consistency in the geodetic coordinate datum used for the referencing of spatial data held on the system. Correct relationships cannot be displayed or measured unless all attributes are related to the same reference datum. In this context, the implication of the results presented here is that despite the high positional accuracies that can be achieved using GPS, when these data are integrated with a digital map based on the OSGB36 datum, the positional uncertainty is about the same as that cited for the original map as a result of the inconsistencies within the OSGB36 datum (i.e. about 0.4 m). The very precise position fixing of GPS is thus lost to the user once the data are displayed in the OSGB36 datum. The situation cannot be improved by transforming the original OSGB36 map to either the OSSN80 or WGS84, unless the map detail were also resurveyed using a method such as GPS that would give more precise position fixing than traditional techniques, since the same inconsistencies in the original map datum apply.

Thus, using map detail to locate features on the ground relative to each other will be as accurate whether the features were surveyed using GPS or traditional techniques. If the original GPS survey data are used to relocate features using the same position fixing techniques as used in their generation, then users will have to face the implication that not only must two sets of coordinates be held on the system, but the location of the feature on the ground may be at variance with that implied on the map by up to 1 m at the 1:1250 scale. In a practical situation, features such as pipelines and other underground plant surveyed using GPS may appear on the wrong side of boundary lines when displayed on an OSGB36 map base. The management, integration and use of position information collected on different map datums will therefore be a complex undertaking.

5. DISCUSSION OF RESULTS

The results suggest that using the transformation algorithm developed by the Ordnance Survey, GPS data can be integrated into an existing digital OSGB36 map database more accurately than if no transformation is applied. Points surveyed by GPS were located within about 0.2–0.5 m of their map location, which is within the limits of map accuracy for large-scale plans published by Ordnance Survey.

Detailed inspection of the differences obtained demonstrates that the transformation need not be applied over very short (approximately 1 km) distances, due to the small variation in the transformations over such distances. However, over distances greater than this, the application of the transformation becomes increasingly important.

Further consideration of the results indicates that, despite the success of the transformation in reducing the average of the difference between survey and map position, there remains a systematic trend in the differences observed. In Survey 3, for example, the mean vector difference between positions for the eight survey points obtained from each of two base stations was 0.45 m, which is constituted from an absolute difference of 0.43 m (standard deviation = 0.009 m) for northings and 0.12 m (standard deviation = 0.018 m) for eastings. The small standard deviations compared to the magnitude of the mean differences emphasizes the systematic nature of the remaining error. These differences cannot be attributed to the carrier phase GPS measurements, which are generally accepted to be accurate to within an order of magnitude better. The implication of the systematic trends is that the transformation has not completely resolved the datum inconsistencies within OSGB36.

6. CONCLUSIONS

The project was successful in quantifying and testing a methodology for allowing the integration of GPS positional information into an OSGB36 map base. The methodology developed is of a kind that can be used by utilities and other agencies wishing to exploit GPS technology for position fixing in the context of a GIS application at the 1:1250 or 1:2500 map scales. The major limitation of the study is that testing has only been possible in a limited geographical area and that the pattern of remaining systematic variation in the transformation algorithm at national scales is not yet understood. More work is required in this area.

The significance of new methods of position fixing such as GPS is appreciated by the Ordnance Survey, which in a recent policy statement stated:

> the market currently seems to require from Ordnance Survey ... the capability to re-engineer the database to meet new needs as they arise, [an example of this] is the need to transform the national grid (and all mapping stored on that basis) to be compatible with position fixing now readily obtainable from the Global Positioning System.
>
> (Rhind, 1992)

This is not simply a matter for the Ordnance Survey, however, since all GIS users who have located the position of plant and other features on existing digital maps will also need to transfer their data. An understanding of the positional implications of such a re-engineering of their digital databases will be essential. This project has made a contribution to this end.

ACKNOWLEDGEMENTS

The authors acknowledge the receipt of a grant under the NERC/ESRC Joint Programme on Geographical Information Handling (ESRC Ref No. A-505-25-5017), and thank Ordnance Survey for the valuable input of their expertise and data. We also wish to acknowledge the contribution to this work made by Andrew Roberts. The views presented here are solely those of the authors.

REFERENCES

Ashkenazi, V., Crane, S.A., Preiss, W.J. and Williams, J.W. (undated) The 1980 readjustment of the triangulation of the United Kingdom and the Republic of Ireland OS(SN)80. Ordnance Survey Professional Paper, New Series, no. 31.

Blanchard, W.F. (1991) Differential GPS. *Proc. Tutorial on Satellite Navigation*. Royal Institute of Navigation and Institute of Engineering Surveying and Space Geodesy, University of Nottingham, 15–17 April.

Bordley R.F. and Christie, R.R. (1989) Ordnance Survey data transformation procedures. *Proc. Semin. Coordinate Systems and Reference Datums*. Institute of Engineering Surveying and Space Geodesy, University of Nottingham, 6 December.

Burrough, P.A. (1986) *Principles of Geographical Information Systems for Land Resource Assessment*. Clarendon Press, Oxford.

Defense Mapping Agency (1987) *Supplement to Department of Defense World Geodetic System 1984 Technical Report*. DMA TR 8350 2-A/B, vols. 1 and 2.

Dodson, A.H. and Basker, G. (1992) GPS: GIS problems solved? *Proc. AGI'92 Conf.*, Birmingham, November. Association for Geographical Information, London, 1.12.1–1.12.7.

Dodson, A.H, Haines-Young, R.H. and Roberts, A. (1992) Positioning in a GIS: Datum definition and dangers. *Proc. Mapping Awareness '92 Conf.*, London, February. Blenheim Online, London, pp. 335–346.

Dyer, D (1990) Transformations for digital map detail. Ordnance Survey Interim report.

Goodchild, M. and Sucharita, G. (eds) (1989) *The Accuracy of Spatial Databases*. Taylor and Francis, London.

Ordnance Survey (1967) *History of the Retriangulation of Great Britain, 1935–1962*, vols. I and II. HMSO, London.

Rhind, D. (1992) Policy on the supply and availability of Ordnance Survey information over the next five years. *Proc. AGI'92 Conf.*, Birmingham, November. Association for Geographical Information, London, 1.22.1–1.22.8.

Chapter Ten

The object-based paradigm for a geographical database system: modelling, design and implementation issues

M. F. WORBOYS
Department of Computer Science, Keele University

K. T. MASON
Department of Geography, Keele University

AND

B. R. P. DAWSON
Centre for Land Use and Environmental Studies, University of Aberdeen

This chapter discusses work done during the period of the grant held at Keele University by members of the departments of computer science and geography as part of the ESRC/NERC Collaborative Programme on Geographical Information Handling. We consider some of the issues related to the object-based paradigm as it applies to geographical information systems (GIS). In particular, we discuss an underlying model for spatial objects in two dimensions and how it impacts upon GIS design and implementation considerations. The paper also briefly describes some of the conceptual 'spin-offs' from the basic work, which has led to generic models of temporal GIS.

1. INTRODUCTION

In the raster–vector debate in GIS, there are no winners or losers. As Couclelis (1992) vividly expressed it: 'People manipulate objects (but cultivate fields)'. Both views of the geographic world are natural and necessary. Indeed, the raster–vector debate is part of a much wider issue in the history of ideas, which may be expressed using a variety of metaphors, including field versus object, wave versus particle and plenum versus atom. In some fields of endeavour, a unification of the two notions has long been present (c.f. 'wavicle' in quantum theory). However, maybe because of technical difficulties, a unified system is not yet available for GIS users, although it is possible to imagine how a unified generic model might be constructed.

The work described in this chapter is placed firmly in the object camp. The world is modelled as a collection of independently-existing self-contained entities, each having an identity independent of any of its properties, which may themselves fluctuate over time and space.

The object-based approach immediately faces us with several questions, among the most important of which are:

- What are the basic collections of object that populate the geographic spaces in which we are interested?
- How can we construct an information system that efficiently and effectively holds and manipulates these object collections?

To answer the first question, we must identify object classes, not only in terms of the properties of the constituent objects (for example, coordinates of a point object) but also by the operations that these object classes admit (for example, calculation of the area of a polygonal object). Collections of objects must also be classified and structured (for example, into an inheritance hierarchy). What will result will be a generic object model for geographic space. For the second question, the generic object model is transformed into a representation that is computationally tractable. This may involve approximation and discretization. The representation must be natural and capable of resulting in an efficient system. An implementation strategy may then be selected.

Traditionally, object-based models are implemented in hybrid systems with separate database and topology engines. Currently, work is in progress to implement the object-based model as a fully object-oriented unified GIS. In the work described here, an alternative approach has been taken, implementing the object-based approach using a unified enhanced relational system.

Research at Keele University, which formed part of the ESRC/NERC Collaborative Programme on Geographical Information Handling, was a collaboration between the departments of geography and computer science. The principal objective of the collaboration was to carry out basic research into enhancements to current relational technologies which would make the handling of geographic information more effective. In particular, the research team investigated issues arising from taking the unified approach to geographic data handling, for which all data (spatial and non-spatial) is held in and managed by the same relational database management system (DBMS). The conceptual framework upon which the work was based was a formal object-oriented model of spatial information constructed by one of the investigators and fully discussed in Worboys (1992a). An experimental implementation of this formal model using an enhanced relational DBMS was performed. The main research findings relate to the construction of the formal model, representation of the model using appropriate data structures and implementation and performance of the resulting system.

This chapter is divided into six sections. Section 2 describes unified and hybrid approaches to geographical information management. Section 3 sets the work in the context of international research efforts and publications in this area. In Section 4, a generic conceptual object model for spatial information is outlined. Section 5 describes the representation of the spatial object classes using appropriate data structures. It also discusses the structural questions that arise as a consequence of holding the data in a relational database, and deals specifically with implementation and performance. The final section discusses extensions to the work and draws general conclusions.

2. UNIFIED AND HYBRID APPROACHES TO GEOGRAPHICAL INFORMATION MANAGEMENT

The fundamental significance of the database to GIS and the attendant influences of database theory and practice have focused considerable attention on DBMS, the importance of which has been progressively realized. The necessity of a system for managing data derives from a requirement to confront such problems as backup, recovery, data integrity, concurrent update and data security, as well as making provision for data sharing and integration between applications and for managing distributed databases. These are issues that are common to all database systems; GIS are no exception. Consequently, the considerable effort that has already gone into developing standard database tools to address such problems in conventional database environments can be put to good advantage by adopting the same methodologies for GIS purposes.

Standard DBMS provide many solutions to the issues identified above. They employ a range of structures and mechanisms to store, manipulate and retrieve data from a database. However, although examples can be found of the use of various logical DBMS models (network, relational and hierarchical), in GIS, as elsewhere, it is the relational model (Codd, 1970) that has come to dominate the field. In such systems all the data are simply represented in tables of rows and columns. The relational approach has a number of advantages over other systems, including flexibility in data retrieval, modifiability, ease of use and a sound theoretical foundation. The use of proprietary systems also encourages a useful degree of standardization and comparability in data handling.

The majority of GIS implementing a relational processing model typically adopt a dual database or 'hybrid' model whereby attribute data and their spatial references are stored and managed in independent structures (Bracken and Webster, 1989). In such systems, the common approach is to maintain the attribute data in a conventional DBMS but to organize and manipulate the geographical data using conventional file handling techniques and bespoke software. This is the approach adopted by, for example, ARC/INFO. This 'geo-relational' type of approach provides a degree of flexibility and a potential for data integration above that which is available from systems storing all aspects of their data in specialized file structures (for instance, IDRISI). However, the hybrid approach is not in itself entirely advantageous. In particular, in failing to maintain an entire map base within a single DBMS, coordinate data are not subject to the same rigorous management as might be applied to the attribute data. Because of this, the handling of topographic data lacks the quality of control that the data demands and is ill served in such fundamental DBMS facilities as data security, integrity control, multiple user access and concurrency management. Furthermore, data are essentially 'locked-up' in the system and are not easily shared with other applications – a significant shortcoming in view of the hard-won nature of such information and the common need to integrate GIS into an overall corporate information system. Finally, and perhaps most importantly, if benefits are to be derived from distributed database technology, then GIS must be developed that deal with the particular complexities of distributed database management without being hindered by an approach that stores some data outside the DBMS–and thus inaccessible to any distributed functions provided therein (Batty, 1992).

For these and other reasons, the project described here sought to explore the development of a single data model, in which attribute and topographic geometry are stored under the same relational architecture. In this way, an application-independent description of the objects in the database is obtained, and all data definition and manipulation is controlled by a DBMS, such functions being removed from the application. Unfortunately, a number of problems are known to inhibit the development of such systems, largely as a result of the special characteristics of spatial data. For instance, the common support for only fixed-length fields in relational systems poses a major problem for the storage of segments of unequal length, in particular the considerable duplication of primary key attributes in each relation, thereby wasting large amounts of space when storing variable length line geometry. One possible solution lies in the use of the 'long' or 'bulk field' data type offered by a number of DBMS, in which arrays, lists and data types may be accommodated in a single record (Charlwood, Moon and Tulip, 1987; Waugh and Healey, 1987). This means, however, that first Normal Form is violated as each column value is not atomic. Also, few functions are supported for the long data type compared to other supported data types, and difficulties are often experienced in loading and manipulating data in a long field.

Another problem in adopting existing relational database technology derives from the fact that the potentially enormous number of retrievals for a given query is problematical for systems optimized for much smaller numbers. This is a common difficulty arising from the large volumes of data associated with geographical information, and is exacerbated by normalization of the database and the resultant complex join operations during retrieval – a process which is itself hard on system performance. In seeking solutions to some of these problems, the array fetch mechanism has been suggested as one possible method of improving access time while avoiding the use of non-normalized tables (Sinha and Waugh, 1988). Using this technique, rows containing coordinate data are retrieved as batches rather than as a series of separate items.

The interfaces to existing relational systems, including database query languages, do not adequately provide for the retrieval of spatially-oriented data, lacking as they do the functionality to handle a query based upon such fundamental spatial operations as adjacency, overlap and containment. A summary of the range of operators required for spatial query languages is given by Guptill (1986). This lack of standard spatial operators in languages such as SQL is perceived as a major hindrance to the use of proprietary DBMS for the manipulation of geographical data. A view worth exploring is that the use of an existing language is preferable to designing an entirely new language for GIS, and, indeed, it is a common view that existing languages should be extended. There are, in fact, a number of systems in existence where the query language has been extended to incorporate appropriate spatial operators.

Despite the difficulties outlined above, the need for GIS to have at least the facilities of a standard DBMS is a powerful one and the industry's standardization upon relational database technology is similarly persuasive. There are at least two major projects, both in the USA, that are exploring the suitability of extensible relational databases as unified management systems for geographic information. The Starburst project, from the IBM Research Laboratory at Almaden, California, has been examining extensions to standard relational database systems (Lohman *et al.*, 1991). The Postgres project (Stonebraker and Dozier, 1991) is applying extensions to the INGRES relational database system for managing data associated with global change research. A further project is discussed in Abel (1989). Work on further extended standards for SQL is also taking place. Features such as user-defined functions will add further possibilities for an extended relational database as a unified geographical data management engine.

3. OBJECT MODELLING OF GEOGRAPHICAL INFORMATION

There is a growing body of literature on the application of the object-oriented methodology to geographical information. This can be divided into two areas: object-oriented data modelling and object-oriented database management. The latter is still in the early experimental stages: no proprietary GIS can claim to be fully object-oriented at this stage. However, object modelling has already opened up many possibilities for geographic information

(Egenhofer and Frank, 1987, 1989; Worboys, Hearnshaw and Maguire, 1990a,b).

Previous work has been done on appropriate spatial object classes and operators. For example, there has been much discussion on the set of spatial relationships. Freeman (1975) declares a set of thirteen operators: left of, right of, beside, above, below, in front of, behind, near, far, touching, between, inside and outside. Feuchtwanger (1989) lists the following operators: adjacency, proximity, subdivision, overlap, nearest neighbour and subregion. Egenhofer and Frank (1987) list neighbour, inclusion, distance and direction. Peuquet (1984, 1988) lists distance, direction and the Boolean set operators of intersection, union and complement. Worboys, Hearnshaw and Maguire (1990b) set out a general approach to the use of such objects and operators in a spatial query language. Two further references that are particularly relevant to this work are those by Manola and Orenstein (1986), which takes a functional data modelling approach, and Güting (1988), which constructs a many-sorted algebra, including spatial sorts.

4. A GENERIC OBJECT MODEL FOR PLANAR GEOGRAPHICAL INFORMATION

4.1. Motivation

An object model for a geographic database has as its constituents spatially-referenced objects. A spatially-referenced object has attributes, at least one of which is a spatial object (see Fig. 1). For example, an object of type *road* might have an attribute of type *arc* representing its centre line, other spatial attributes representing the limits of its metalling, intersections, etc., as well as non-spatial attributes, such as name and classification. It follows that the modelling of geographic objects with spatial references requires as a foundation a model of purely spatial objects. Such a model is described in detail in Worboys (1992a). Here, we outline the major features of the model.

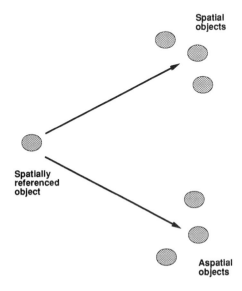

Fig. 1   Spatially referenced object

4.2. Definition of spatial object classes

We assume, for simplicity, that all the spatial objects to be considered are embedded in the coordinatized Euclidean plane, R^2, with the usual topology. Fig. 2 shows an inheritance hierarchy of spatial object classes, extending the analysis of Güting (1988). The class *spatial* is defined to be any point set embedded in R^2. This class is clearly too all-embracing for most practical applications. It includes fractals and topologically pathological point sets. It may be partitioned into disjoint subclasses, *point* and *ext*. Class *point* has instance variables, *x*-coordinate and *y*-coordinate, and models single points embedded in the plane. Class *ext* is intended to model any extended objects embedded in two dimensions, and consists of sets of points. Class *ext* has as disjoint subclasses *1-ext* and *2-ext* of one- and two-dimensional extents, respectively.

The most general class of one-dimensional objects considered here is *arc*. The intuitive notion of an arc may be grasped by imagining a finite piece of string thrown onto a table. It may cross over itself a finite number of times, and may or may not have its ends touching.

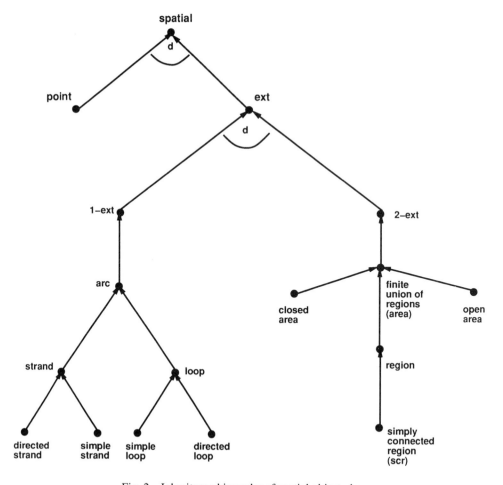

Fig. 2   Inheritance hierarchy of spatial object classes

Formally, it is a one-dimensional object, which may be segmented into a finite collection of homeomorphic images of a finite straight-line segment. The class *strand* is a subclass of *arc* where each member does not have joined ends. The class *loop* is a subclass of *arc* where each member has its ends touching, and therefore is closed up. Simple arcs are those that do not cross themselves. An arc may have a direction or an orientation.

The most general two-dimensional object class considered here is the class *area*, members of which are disjoint finite unions of regions. A *region* is a homeomorphic image of a disk embedded in the plane with a finite number of holes, cuts (arcs subtracted from the interior) and punctures (points subtracted from the interior), and which may include all, some or none of its boundary. Examples of further specialized classes are *closed area*, *open area*, and *simply connected region*.

4.3. The definition of spatial operations

Table 1 shows some of the primitive operations that may be defined upon the spatial object classes. The operators are divided into four groups: set-orientated, topological, metric and Euclidean, each one dependent upon a richer substructure of the cartesian plane than those that precede it.

Table 1   Classification of spatial relationships

Group	Operator	Operand	Operand	Resultant
Set-oriented	equals	spatial	spatial	Boolean
	member	point	ext	Boolean
	subset	ext	ext	Boolean
	intersection	ext	ext	ext
	union	ext	ext	ext
	difference	ext	ext	ext
	cardinality	set (spatial)		cardinal
Topological	interior	area		open area
	closure	area		closed area
	boundary	area		1-ext
	components	area		set (region)
	extremes	strand		set (point)
	begin	directed strand		point
	end	directed strand		point
	inside	point	simple loop	Boolean
	clockwise	oriented loop		Boolean
Metric	distance	point	point	$\mathbb{R}$
	length	arc		$\mathbb{R}$
	perimeter	region		$\mathbb{R}$
Euclidean	bearing	point	point	$(0-2\pi)$
	area	area		$\mathbb{R}$

Spatial objects may be treated as pure sets (collections of objects). The set-orientated operations that result are *equals*, *member*, *subset*, *intersection*, *union*, *difference* and *cardinality*. Topological properties of the embedding of the spatial objects in the plane are needed for the second group of operators. Among the topological operators are *interior*, *closure*, *boundary*, *extremes*, *end*, *inside* and *clockwise*. Metric (related to distance) and Euclidean (related to angle) properties of space allow the definition of operators in the third and fourth subclasses.

5. REPRESENTATION AND IMPLEMENTATION OF THE OBJECT MODEL USING THE EXTENDED RELATIONAL MODEL

5.1. Introduction

In order that the object model can be realized computationally, a process of discretization must take place. This process converts continuous objects which may have no computational representation to computationally tractable types. This is done by processes of abstraction, generalization and approximation.

For the discrete space, assume an underlying plane coordinatized by the computational real numbers, into which are embedded the spatial objects as:

- Points.
- Piecewise linear arcs formed by concatenating a finite number of bounded straight line segments.
- Polygonal areas, with or without punctures and piecewise linear cuts, whose boundaries are piecewise linear arcs.

Assume also an approximation function from the general object space to the discretized object space.

All discretized objects may be represented by either a single vertex (in the case of a point) or a sequence of vertices, which define an extent in the case of an arc or its boundary in the case of an area. The key idea is to hold not only raw geometric information, but also higher level object information. Two fundamental data structures are employed, one to hold data on the geometric primitives (vertex data), and the other to hold high-level object information.

5.2. Implementation and system configuration

Implementation of the discretized generic object model was performed on an extended relational DBMS. The standard proprietary relational system ORACLE was chosen for the underlying DBMS. Extensibility was provided by the Pro*C interface to ORACLE.

It has been necessary, in the process of implementation, to:

- Examine the data structures to be employed.
- Establish the relationship between the ORACLE tables and the data being operated upon.
- Apportion the processing between ORACLE and the Pro*C parts of the system.
- Determine how the generality of geometric processing can be reduced to the minimum number of functions in the implementation.
- Demonstrate the operation of the system using real data in the form of maps of administrative and post-code sector boundaries.

The configuration of the system is shown schematically in Fig. 3. Example data was drawn from the hybrid system ARC/INFO by means of a dual process which took vertex

The Object-based Paradigm for a Geographical Database

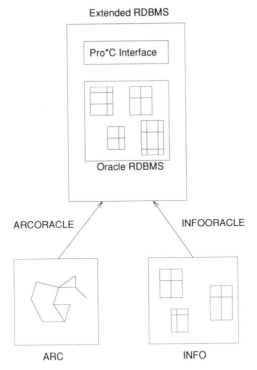

Fig. 3 Configuration of the extended relational database

data and topological data from the ARC and the INFO components and transformed them into a set of tables in ORACLE. Since the user interface currently provided by ORACLE is insufficient for handling spatial data, it was enhanced by means of a set of procedures, written in the C language, and linked into the system using the Pro*C interface to ORACLE.

5.2.1. Implementation of spatial objects

Point data are represented by single vertices. Strands are represented by a set of vertices organized as a chain. Areas are represented by collections of loops that define their outer boundaries and any internal holes. The object structures, besides storing the location of the corresponding vertex chain's head, contain data that are meaningful in the context of an object; its unique identifier, its types, set membership information and a flag indicating the state of the topology for the object.

5.2.2. Implementation of spatial operations

All operations eventually reduce to the processing of vertex coordinates, so it is these that must be extracted from the database. Operations are performed in two stages:

1. Retrieval of raw data using SQL queries to the ORACLE database.

2. Further processing carried out by custom-built C language routines.

We discuss the second stage briefly. The algorithm used to process a spatial object is selected on the basis of the type of the object being processed and the spatial relationship between the two objects being operated upon. The low-level geometric routines are common to all of the methods. The differences between methods occur in the way in which the low-level routines are called. It was found that only three functions, *point-in-polygon*, *point-on-line* and *line-on-line*, were necessary to implement the majority of functions. What is critical for each of these three, therefore, is that they are robust, that the data operated upon are properly prepared, that the routines are able to produce a variety of results dependent upon the relative positions of the objects being operated upon and, lastly, that they operate as efficiently as possible given the other criteria.

5.3. System performance

In order to benchmark the system, three test scripts were devised, one each for region union, intersection and difference. Each script loaded different region data from the database and all involved some union operations.

5.3.1. ORACLE access

The time taken to retrieve the data from ORACLE was at least 44% of the runtime of the script, rising to 90% for the union test script. Differences between the complexity of the regions and other factors make comparisons difficult.

5.3.2. Topology building

The second major subsystem is concerned with the construction of the topology of the objects, and determining their juxtaposition. These operations typically take up about 55% of the processing time of the program, and include the three key geometric functions mentioned above. We suggest that system optimization should concentrate on these functions. However, considering the large number of calls made to them (99.8% of all function calls are to these routines), the calling overhead may limit the possibilities for optimization.

5.3.3. Geometric processing

Once the data have been prepared, the object methods execute rapidly (taking approximately 0.001% of the total runtime of the program), suggesting efficiency. The creation and extension of object lists are also operations that run so quickly that even a sample of 1000 queries could not yield reliable timing results.

6. CONCLUSIONS AND FURTHER DEVELOPMENTS

6.1. Summary and evaluation

A major objective of this work was to investigate the implications of using an extensible relational DBMS for spatial data handling. The implications of maximizing the use of the extensible system for geometric data processing were sought. The strategy was to analyse the generic structure of two-dimensional geographic data as a set of abstract data types with associated operations. The high-level model of spatial objects was constructed according to this object modelling approach. Data structures were devised that would support the object classes and operations.

It is certainly the case that the unified approach to spatial data handling is conceptually more satisfying, since it is not required to make a distinction at the top level between spatial and non-spatial components of the system. A unified approach brings with it the considerable benefit of sheltering under the same umbrella all the data, with consequent advantages for integrity, security and reliability. The disadvantages relate mainly to system performance. With improved hardware, algorithm design, and data management and retrieval facilities, we feel that this issue will become less critical (although it is still unclear whether the advantages will outweigh the disadvantages). Also, with improved query interfaces (such as new SQL standards), it will be more natural to express spatial relationships in the relational model. Data will be accessible in an application-independent way by many users. As spatial databases grow in size and usage, this facility will become increasingly important.

6.2. Extensions to the generic model

During the period of the Joint Programme, in parallel to the work described here, one of the authors investigated some extensions to the model described above. This is very briefly discussed and references are provided.

6.2.1. *Expressing more topology explicitly*

A limitation of current vector-based data models is their restricted power to represent complex patterns of spatial phenomena. For example, varieties of connectivity are not explicitly distinguished, neither are all types of holes explicitly recognized. Further work on a data model which explicitly captures such complex topological configurations is described in Worboys and Bofakos (1992).

6.2.2. *A generic model for spatiotemporal information*

A second type of limitation of current models for geographic data modelling is that geographic space is better viewed as being populated with objects having not just a spatial reference but also a spatiotemporal reference. Work is progressing on the development of a model which allows spatial and temporal references (Worboys, 1992b,c).

REFERENCES

Abel, D.J. (1989) SIRO-DBMS: A database tool-kit for geographical information systems. *Int. J. Geogr. Inf. Sys.*, **3**(2), 103–116.
Batty, P. (1992) Exploiting relational database technology in a GIS. *Comput. Geosci.*, **18**(4), 453–462.
Bracken, I. and Webster, C. (1989) Towards a typology of geographical information systems. *Int. J. Geogr. Inf. Sys.*, **3**(2), 137–152.
Charlwood, G., Moon, G. and Tulip, J. (1987) Developing a DBMS for geographic information: A review. *Proc. Auto-Carto 8*, Baltimore, Maryland, 302–315.
Codd, E.F. (1970) A relational model of data for large shared data banks. *Commun. ACM*, **13**(6), 377–387.

Couclelis, H. (1992) Beyond the raster–vector debate in GIS. In Frank, A. U., Campari, I. and Formentini, U. (eds.), *Theories of Spatio-Temporal Reasoning in Geographic Space.* Springer–Verlag, New York, 65–77. Lecture Notes in Computer Science, vol. 369.

Egenhofer, M.J. and Frank, A. (1987) Object-oriented databases: database requirements for GIS. *Proc. Int. GIS Symp.: The Research Agenda*, vol. II. US Government Printing Office, Washington DC, 189–211.

Egenhofer, M.J. and Frank, A. (1989) Object oriented modelling in GIS: Inheritance and propagation. *Proc. Auto-Carto 9*, Baltimore, Maryland. American Society of Photogrammetry and Remote Sensing, Falls Church, Virginia, 588–598.

Feuchtwanger, M. (1989) Geographical logical database model requirements. *Proc. Auto-Carto 9*, Baltimore, Maryland. American Society of Photogrammetry and Remote Sensing, Falls Church, Virginia, 599–608.

Freeman, J. (1975) The modelling of spatial relations. *Comput. Graphics and Image Processing*, **4**, 156–171.

Guptill, S.C. (1986) A new design for the US Geological Survey's national digital cartographic database. In Blakemore, M.J. (ed.), *Proc. Auto-Carto*, vol. 2. Royal Institution of Chartered Surveyors, London, 10–18.

Güting, R.H. (1988) Geo-relational algebra: A model and query language for geometric database systems. In *Extending Database Technology*. Springer-Verlag, Berlin, 506–527. Springer-Verlag Lecture Notes in Computer Science, vol. 303.

Lohman, G., Lindsay, B., Pirahesh, H. and Schiefer, K.B. (1991) Extensions to Starburst: Objects, types, functions and rules. *Communs ACM*, **34** (10), 94–109.

Manola, F. and Orenstein, J.A. (1986) Towards a general spatial data model for an object oriented DBMS. *Proc. 12th Int. Conf. Very Large Databases*, Kyoto, Japan, 328–335.

Peuquet, D.J. (1984) A conceptual framework and comparison of spatial data models. *Cartographica*, **21**(4), 66–113.

Peuquet, D.J. (1988) Toward the definition and use of complex spatial relationships. *Proc. 3rd Int. Symp. Spatial Data Handling*, Sydney, Australia, 211–223.

Sinha, A.K. and Waugh, T.C. (1988) Aspects of the implementation of the GEOVIEW design. *Int. J. Geogr. Inf. Sys.*, **2**(2), 91–99.

Stonebraker, M. and Dozier, J. (1991) Large capacity object servers to support global change research. Sequoia 2000 Technical Report, no. 1, University of California at Berkeley, Berkeley, California.

Waugh, T.C. and Healey, R.G. (1987) The GEOVIEW design. A relational database approach to geographical data handling. *Int. J. Geogr. Inf. Sys.*, **1**, 101–118.

Worboys, M.F. (1992a) A generic model for planar geographic objects. *Int. J. Geogr. Inf. Sys.*, **6**(5), 353–372.

Worboys, M.F. (1992b) A model for spatio-temporal information. *Proc. 5th Int. Symp. Spatial Data Handling*, Charleston, South Carolina, USA, 602–611.

Worboys, M.F. (1992c) Object-oriented models of spatio-temporal information. *Proc. GIS/LIS'92 Conf.*, San Jose, California, 824–835.

Worboys, M.F. and Bofakos, P. (1992) A canonical model for a class of areal spatial objects. Technical Report, TR92-15, University of Keele.

Worboys, M.F, Hearnshaw, H.M. and Maguire, D.J. (1990a) Object-oriented data and query modelling for geographical information systems. *Proc. 4th Int. Symp. Spatial Data Handling*, Zürich, 679–688.

Worboys, M.F, Hearnshaw, H.M. and Maguire, D.J. (1990b) Object-oriented modelling for spatial databases. *Int. J. Geogr. Inf. Sys.*, **4**(4), 369–383.

Chapter Eleven

Spatiotemporal GIS techniques for environmental modelling

M. A. O'CONAILL, D. C. MASON AND S. B. M. BELL
Natural Environment Research Council Unit for Thematic Information Systems (NUTIS), University of Reading

A wide range of environmental process simulations would benefit from the development of geographical information systems (GIS) capable of coping with the additional dimensions of vertical space and time, and having extended spatial modelling facilities. This chapter summarizes a project in the first of these areas, namely the development of techniques for handling spatiotemporal data in up to four dimensions. The key topics of data structures, visualization and interpolation have been studied. The techniques have been implemented within a computational testbed, allowing parallel processing to cope with large data sizes.

1. INTRODUCTION

One area of spatial data processing in which GIS have a rather restricted use is environmental process simulation. This is partly due to limitations in GIS technology. For example, consider a distributed, physically-based process model predicting the state of the atmosphere in three-dimensional space and time. It is unlikely that existing GIS could be of use other than peripherally in this modelling process. However, there seems to be no overriding reason why GIS should not be developed to assist in this type of modelling, in a way that involves close integration of GIS and model. This project is concerned with the development of spatiotemporal GIS techniques to support such environmental modelling.

The limitations of current GIS for environmental modelling become apparent when one considers the types of models involved and the data associated with them. A wide range of environmental models are discussed in depth by Goodchild, Parks and Steyaert (1993) and Farmer and Rycroft (1991). The models considered cover all areas of natural environmental science, including earth science, atmospheric science, oceanography, and terrestrial and freshwater science. A particular example in oceanographic modelling is the problem of combining time-dependent ship and satellite data in order to improve numerical models of the ocean. This problem is currently being addressed in the World Ocean Circulation Experiment (WOCE). WOCE is aimed at understanding the present state of the ocean circulation through observations and the development of models (Smythe-Wright, 1991).

These models will provide a basis for predicting climate change. The traditional method of obtaining a spatial description of the physical oceanography of a region is to conduct a hydrographic survey, in which a ship carries out vertical profile measurements at a number of discrete stations, typically separated by about 20 km. Among the variables measured are ocean temperature and salinity, using conductivity–temperature–depth (CTD) casts. Three-dimensional fields of measured variables can be computed from these. The advent of satellite sensors such as the Advanced Very High Resolution Radiometer (AVHRR) makes possible the synoptic mapping of variables at a single level (the surface). There is a need to investigate the relationship between remotely-sensed surface signatures and subsurface ocean features in order to be able to combine these data sets effectively, and thus make possible the production of improved fields, which in turn will allow a better understanding of the behaviour of mesoscale (~20 km) ocean features such as eddies and fronts. The eventual goal is to be able to assimilate the derived fields into numerical models of the ocean. Plate V shows a three-dimensional projection of ocean temperature data from an AVHRR image and a contemporaneous CTD cast. The undefined areas in the AVHRR image are due to the presence of clouds, which give rise to invalid surface temperatures; these locations have been masked out.

The techniques developed in this project have been evaluated by applying them to some existing environmental research programmes, including the one mentioned above. These involve the use of types of models and data that are representative of those used in a much wider set of environmental science modelling projects. They include one example from each of the fields of earth science, meteorology and oceanography. The problems were provided, respectively, by the British Geological Survey, Reading University Department of Meteorology and the Institute of Oceanographic Sciences (Rennell Centre). The earth science problem involves an assessment of the potential of GIS for acting as an interactive aid in the development of a hydrodynamic model of a pollution plume moving beneath the earth's surface. The plume may be due to the pollution of groundwater by salt or radioactive material. The GIS would be used to compare the model output with the actual behaviour of the plume, in order to determine model parameters. The meteorological problem involves the simultaneous visualization of global atmospheric model outputs. The aim is to take data for two horizontal distributions available at discrete time intervals and display their temporal development. The two distributions may be the same variable at different heights in the atmosphere, or different variables. The continuous time evolution of two variables, possibly acting as tracers, may reveal properties of the model that are not evident in two-dimensional snapshots, and may thus increase understanding of atmospheric flow.

Analysis of the characteristics of the data gathered in these projects indicates that, in contrast to the two-dimensional static data sets handled by most current GIS, it is necessary to link models to data sets of up to four dimensions, the additional dimensions generally being those of vertical space and time. These data sets are generated by a wide variety of different sensors, and include remotely-sensed data from satellites and aircraft that provide a two-dimensional synoptic view, allowing the possibility of extrapolation from site-specific measurements to the whole area under consideration. There is a need to integrate data from different sensors, which may be of different dimensionalities, resolutions and accuracies (for example, AVHRR and CTD ship data for sea temperature). The data sets involved may be very large, because a data set which is relatively small in two dimensions may become very large with the addition of a further two dimensions. For example, the output of some four-dimensional meteorological model runs can be measured in gigabytes of data. Data may also be sparse in some dimensions compared with others, as with geological borehole data and CTD casts, where sampling is frequent in the vertical and sparse in the horizontal. This highlights the fact that it is often difficult to gather observations for all points in a higher-dimensional space. While environmental models most commonly involve the use of gridded data, some may also require the identification of objects within the data, and the creation of attributed vector data for these. An example of a four-dimensional object might be a phytoplankton bloom, tracking of which in space and time is required.

As regards the characteristics of the environmental models, many are complex deterministic or stochastic process models, which may operate in three-dimensional space and time. Deterministic process models describe processes in terms of known physical and chemical laws. Modelled output variables are obtained by solving differential equations describing the process. The models may be distributed, finite element or finite difference models, in which the space in which the process operates is subdivided into a number of regions such that modelled variables are relatively homogeneous within each region, but differ between adjacent regions. The output from one region forms the input to adjacent regions, and vice versa. Stochastic process models model the average outcome of a large number of events, rather than the outcome of just a single event. Models may also be composed of linked submodels (as when a hydrodynamic model describing water movement is used to drive an ecological model describing water quality).

2. DEVELOPMENT OF SPATIOTEMPORAL GIS TECHNIQUES

Only a brief consideration of these data and model types is necessary to see why current GIS have a restricted use in the environmental sciences. Current GIS generally deal with digital maps, which contain data of two spatial dimensions that are static in time. Also, they generally allow only simple empirical modelling, and many of the models discussed above are far beyond their scope. What is required is a GIS 'shell' capable of supporting higher-dimensional data sets and having extended facilities for mathematical modelling.

The development of such a shell requires research and development on a range of specific topics within these two broad areas. This project has concentrated on work in the first of these areas, namely the development of GIS techniques for handling four-dimensional data, and also a computational testbed for investigating these techniques. The research should allow confident decisions to be taken on the design and implementation of a full four-dimensional GIS. Because of the wide potential scope of the project, work has concentrated on the key topics of data structures, visualization and interpolation. Further details are given in Mason, O'Conaill and Bell (1993), Mason, O'Conaill and McKendrick (1993), and O'Conaill, Bell and Mason (1992). Within these topics there are a number of issues. Research is needed into volume- and time-orientated data structures (see Jones, 1989; Langran, 1989), for both gridded and object data. In the area of data visualization, facilities are required for studying the output of models as well as the raw data, and for examining differences between these. A basic problem here is the difficulty of visualizing four-dimensional data on a two-dimensional display. Important additional facilities not present in most current GIS would be visualization of the third dimension and the dynamic nature of the data. As higher-dimensional data are often sparse in one or more dimensions, there is a need for the GIS to incorporate methods of interpolation in up to four dimensions to allow the user to infer values between sample points, in order to visualize the natural variation within the data, and, importantly, to initialize models.

The computational testbed is a distributed system consisting of a variable number of Sun Sparc workstations and an optional Meiko Box transputer array with eight T800 transputers. The vast volumes of data often associated with four-dimensional GIS problems make some form of parallel processing attractive if computations are to be performed in a reasonable time at reasonable cost. The aim is a workstation environment allowing interactive, possibly computer-intensive, modelling of environmental processes, with the modelling and graphical analysis being performed on the same system. The demonstrator is implemented as a number of independent communicating processes where each process provides a distinct functional subset of the system. Provided the interface between processes remains constant, any process may be easily modified or replaced. In addition, processes may be assigned to any computer platform, thereby making use of any unused computer resources on the network. Disk access is via the host Sun. Processors may communicate data over an Ethernet high-speed computer network using Meiko CSTools message-passing primitives (Meiko Ltd, 1990). Other utilities used include Sun Open Windows graphical user interface toolkit and UNIRAS graphics libraries. Code is written in the C programming language.

2.1. Data models

It is necessary to support more than one data model for GIS-based environmental modelling. Many environmental models involve the use of gridded data, for which volume- and time-orientated data structures based on the voxel and its four-dimensional analogue are appropriate. It is also necessary (though less important) to handle higher-dimensional object data, and to be able to interrelate object and volume data.

2.1.1. Volume data

A decision was made to treat the spatial and time axes in the same way in the volume data model. The volume data are assumed to be embedded in four-dimensional Euclidean space. Nothing appears to be gained by hard-wiring the human psychological view of a separate space and time into the data model. For example, a four-dimensional data set can be displayed as a time series of three-dimensional (x,y,z) volumes, or alternatively as a series of (x,y) and time plots at fixed values of z. In some cases a user may find it easier to extract information from the latter type of display than the more commonly used former type, and the data structure allows the flexibility to make this comparison. This does not mean, however, that different ways of treating space and time will not sometimes have to co-exist, for example the use of different interpolation algorithms for space and time.

For a number of reasons, a hierarchical data structure has been chosen as the method of storing volume data (Samet, 1990a, b). Hierarchies often allow data compression of volume data, for example of undefined volumes within sparsely-filled four-dimensional data sets, or of model output, which is often generalized to a degree (though it should be noted that for busy images which fill the space faithful storage of the data set in a hierarchy may increase storage requirements). Hierarchies also provide a simple way of producing scale-variable volume data. Often four-dimensional data sets are very large and global views are attainable only at low resolution, but there is usually a requirement to view some part of the data set at high resolution (Gahegan, 1989, Callen *et al.*, 1986). Hierarchies also allow fast data access and the identical treatment of all dimensions.

The particular hierarchical data structure employed for storing volume data is the bintree (Tamminen, 1984). This has a branching ratio of two, and can be used in spaces of arbitrary dimension (Samet and Tamminen, 1988). In two dimensions, if one considers a $2^n \times 2^n$ array (where n is a positive integer) comprised of unit square pixels which are either black or white, the bintree is obtained by recursively dividing the array into two halves alternately horizontally in y then vertically in x until maximal blocks of either black or white are obtained. The process is represented by a tree in which the root node represents the entire array, each non-terminal (grey) node has two sons, and the terminal (leaf) nodes correspond to uniform blocks. The bintree representation of the binary array of Fig. 1a is shown in Fig. 1b. In practice, the colours associated with the array may be multi-valued rather than binary. Grey nodes hold the average colour of their descendants in the hierarchy. Thus, each level in the bintree represents a view of the data set at a different level of generalization, with the full leaf set representing the data set at the highest resolution, and the root node representing it at the lowest resolution. The four-dimensional bintree is the four-dimensional analogue of the two-dimensional case, with the four-dimensional space covering a region of $2^n \times 2^n \times 2^n \times 2^n$ pixels. The bintree representation adopted is a linear bintree form analogous to the linear quadtree of Gargantini (1982).

A memory management system similar to that described in Callen *et al.* (1986) has been developed for bintrees. This allows bintrees constructed from large images to be handled without the necessity for the complete bintree to be core-resident in the host Sun workstation. The bintree nodes are stored on disk in segments, where each segment is a user-defined number of disk pages. Only those segments currently being operated on are stored in memory, and segments are swapped in as required and swapped out on a least-recently-used basis. This approach is feasible because much of the processing required is local.

Two algorithms are available for building a four-dimensional bintree from raster data. Which of these is used for a particular four-dimensional image depends on whether or not

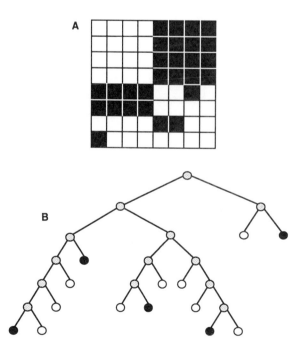

Fig. 1   Bintree representation of a two-dimensional binary array. (a) Binary array; (b) bintree

raster data exists for all points in the four-dimensional space. If the four-dimensional raster image is incomplete (if it is formed from a number of two-dimensional AVHRR images and one-dimensional CTD casts, for example), a 'naive' construction algorithm along the lines of that described by Shaffer and Samet (1987) for linear quadtrees is used. In the less likely event that a full four-dimensional raster data set exists, as may be the case with model output, an algorithm analogous to that discussed by Holroyd and Mason (1990) for building a linear quadtree may be used instead. A discussion of the compressions achieved for the environmental data sets is given in Mason, O'Conaill and Bell (1993).

2.1.2. Object data

Objects occur in the environmental science data sets, and are of great interest to the user. Examples of objects are pollution plumes in geophysics, and fronts and eddies in oceanography and meteorology. Objects can vary in dimension up to four dimensions. Environmental scientists need to be able to define, store, manipulate and display objects, attach attributes to them, and relate them to the volume database.

There is a small but growing body of literature addressing the development of object models of spatiotemporal information (see, for example, Worboys, 1992). In this study effort limitations have dictated that only a simple object model is employed, but in any event four-dimensional objects are as yet used in only simple ways. Objects are limited to being simply-connected. The boundary of each object is stored, but no internal or external topological relationships with other objects are captured. An existing spatial database management system (DBMS) based on the network model (Mason, Oddy *et al.*, 1993) has been adapted to handle objects. A four-dimensional object boundary is stored as a sequence of boundary codes of two-dimensional (x,y) cross-sections, the boundaries being treated as a set of related attributes. A boundary is stored for each of the two-dimensional (x,y) cross-sections within each three-dimensional (x,y,z) cross-section at each time value. In principle, this DBMS has been designed to store super-objects made up of sets of objects, so that in the future complex four-dimensional objects, such as an ocean eddy which, after a certain time, splits into two or more sub-eddies (some of which may subsequently recombine), may be handled.

There are also facilities for cross-relating object and volume databases, for instance to determine which four-dimensional volume nodes make up a particular object, or conversely which objects intersect a particular volume node.

2.2. Visualization

Querying the volume database, for example for the purpose of display, is carried out in parallel using the transputer array. A simple variable resolution window search facility is provided. This allows the display of two- and three-dimensional cross-sections at a user-specified level in the hierarchy in either static or animated form (the latter may also be 'canned'). The user must specify the coordinates of the four-dimensional window from which nodes are to be selected, and the resolution and colour variable range required. The windowing algorithm returns all nodes within the window at the specified level. Prior to the windowing process, a loader process divides the disk-based bintree into equal sized portions and transmits each portion to a transputer in the array (Mason, 1990). This ensures that to first approximation the transputers are equally loaded. Each transputer then runs an instance of the windowing algorithm on its allotted portion of the four-dimensional bintree, and returns a set of nodes that satisfy the user's request. These are sent to the workstation-housed user interface for display, using the UNIRAS graphics package running on the host.

The result of a true four-dimensional windowing into the database is a time series of three-dimensional spatial volumes. These can be displayed using a split screen facility with zoom capability or as an animation sequence. The user can also query the database using any subset of the four axes by fixing one or more variable values. For example, Plate VI shows surface resistivity data from a pollution plume displayed as an (x,y,time) plot at a fixed z value ($z = 0$). Plate VII illustrates the generalization capability of the bintree using a two-dimensional cross-section of atmospheric temperature in the meteorological data set.

Regarding visualization of object data, a display of the volume data within an object interior may be made by extracting from the volume database on the transputer array all those nodes at a specified level that are tagged as intersecting the particular object.

2.3. Interpolation in sparsely sampled spaces

Interpolation is an important facility in the analysis of spatially-referenced data, because it allows a user to obtain a greater understanding of the data by providing a synoptic view from a set of scattered data points. This may be used for improved visualization of the data, or as an interpolated field for the initialization of a mathematical model.

A feature of data in higher-dimensional spaces is that they are often sparse in some dimensions compared with others. Geological applications involving the use of borehole and well data provide well-known examples, as sampling is typically dense along each borehole's vertical profile, but the boreholes are distributed relatively sparsely in the horizontal plane. Another example is provided by the oceanography data set. In order to prevent the user drawing inaccurate inferences from sparse data, it is an advantage if the interpolation procedure is capable of providing error estimates on the interpolated values. An important method of interpolation falling into this category is kriging. This is a form of weighted local averaging based on the theory of regionalized variables (Matheron, 1971; Olea, 1975). It is an optimal interpolator in the sense that each interpolated value is unbiased and its estimated variance is a minimum.

Estimation by kriging can be performed for points or over blocks. While punctual kriging produces the best estimates at points, these can, in some cases, be subject to local discontinuities that may obscure longer range spatial variation. Block kriging avoids this by computing averaged estimates over areas or volumes (Burgess and Webster, 1980b). One of the characteristics of block kriging is that as the size of a block increases its estimation variance decreases.

Block kriging usually involves a subdivision of space into a regular grid structure, with the size of a block being the same at all points. In sparsely sampled spaces, this typically leads to a variable estimation error being associated with each block, with blocks near to observed values having smaller estimation errors than blocks farther away. An alternative strategy in such situations is to allow the sizes of blocks to vary in such a way that a block's interpolated value is just significantly different from that of its neighbour, given the block errors. This would give the advantages of increased spatial resolution over the constant size block method in many regions, and conversely reduced spatial resolution in regions where the constant size block resolution was unjustified. It would also lead to block estimation errors that were more similar over the space. There would tend to be smaller blocks and higher spatial resolution near to observed values, while farther from observed values blocks would tend to be larger.

This type of variable decomposition of space can be achieved using a hierarchical data structure such as the bintree. A block kriging algorithm has been developed in which the leaves of the tree are treated as the blocks, for which interpolated values and their errors are calculated (Mason, O'Conaill and McKendrick, 1993). The technique is illustrated using the oceanographic data set, with the kriged variable being ocean temperature.

Kriging first requires the spatial dependence of the property to be encapsulated in a semi-variogram (Burgess and Webster, 1980a). The semi-variogram is then modelled mathematically. The interpolated value at the centre of a block is a weighted average of the observed values in the neighbourhood of the block. A data subset selection algorithm is used to select the 16 or so observed values nearest to the block centre. The weights are calculated using a knowledge of the semi-variances of the residuals between the observed values themselves, and the averaged semi-variances between the observed values and the points in the block. It turned out that, although simple kriging could be used in the plane of the sea surface, it was necessary to use universal kriging in the vertical direction because of the gradual decrease of temperature with increasing depth (Webster and Burgess, 1980). Universal kriging takes account of local trends in data when minimizing the error associated with estimation. The outputs of the kriging process are the estimates of the weights, the drift parameters (for universal kriging), and the estimation variance of the block's interpolated value.

The block kriging algorithm is basically a split-and-merge algorithm along the lines of that developed by Chen and Pavlidis (1979) for segmentation of textured images in computer vision. Initially, the bintree consists of a single leaf covering the whole space, with an undefined colour and error. The first stage involves splitting leaves in the bintree recursively until they are, if possible, no more than B pixels along any of their sides, and preferably less than this (B is taken to be the length of block sides in the constant volume kriging method). Provided a sibling leaf generated during splitting is sufficiently close to enough observed values, its colour record is changed from undefined to its interpolated value and error. The basic criterion used to split leaves less than or equal to B along any side (but larger than a single pixel) is that the two sons of the leaf should have interpolated values that are significantly different statistically, given their respective errors. The main function of the second stage is to merge adjacent leaves wherever possible. Two sibling leaves are merged to form their father leaf if they either both have undefined colours, or both have defined colours that are not significantly different.

Plate VIII shows the interpolated ocean temperature values obtained by applying split-and-merge interpolation to the three-dimensional oceanographic data set shown in Plate V. The variable resolution introduced by the method can be clearly seen, with high resolution being present near temperature gradients in the plane of the sea surface.

3. CONCLUSION

A number of techniques for handling the spatiotemporal data sets associated with a selected set of environmental models have been studied, using a computational testbed allowing parallel processing. The volume data structure employed gives the advantages of fast data access, a scale-of-two generalization capability, and simplified database query software because all dimensions may be treated identically. Used in conjunction with a split-and-merge algorithm, the data structure has also allowed the development of a method of interpolation that may provide increased resolution compared with constant size block kriging.

The development of a GIS shell for environmental modelling requires not only the development of spatiotemporal GIS techniques, but also research and development aimed at extending the spatial modelling capabilities of GIS. Priority topics in this area include research on linking models to heterogeneous data sets (for example, determining how to assimilate a time sequence of satellite altimeter images into an ocean global circulation model); research on error sources and error propagation for the complex nonlinear models usually involved in environmental science; the development of modelling tools to make models easier to construct and use; the linking of GIS with statistical packages to ease the determination of model parameters; and the use of expert systems in conjunction with GIS and models in order to increase the decision support capabilities of GIS.

ACKNOWLEDGEMENTS

Thanks are due to the environmental science collaborators, in particular Peter Challenor and Trevor Guymer of the Institute of Oceanographic Sciences (Rennell Centre), Brian Hoskins of Reading University Department of Meteorology, and Richard Ogilvy of the British Geological Survey. Thanks are also due to Toby Baldwin for programming assistance. This project was partly funded under NERC Contract no. F60/G6/12.

REFERENCES

Burgess, T.M. and Webster R. (1980a) Optimal interpolation and isarithmic mapping of soil properties. I. The semivariogram and punctual kriging. *J. Soil Sci.*, **31**, 315–331.

Burgess, T.M. and Webster R. (1980b) Optimal interpolation and isarithmic mapping of soil properties. II. Block kriging. *J. Soil Sci.*, **31**, 333–341.

Callen, M., James, I., Mason, D.C. and Quarmby, N. (1986) A test-bed for experiments on hierarchical data models in integrated geographic information systems. *Proc. Workshop Spatial Data Processing using Tesseral Methods*, NERC Unit for Thematic Information Systems, University of Reading, 193–212.

Chen, P.C. and Pavlidis, T. (1979) Segmentation by texture using a co-occurrence matrix and split-and-merge algorithm. *Comput. Graphics and Image Processing*, **10**, 172–182.

Farmer, D.G. and Rycroft, M.J. (1991) *Computer Modelling in the Environmental Sciences*. Oxford University Press, Oxford.

Gahegan, M.N. (1989) An efficient use of quadtrees in a geographical information system. *Int. J. Geogr. Inf. Sys.*, **3**(3), 201–214.

Gargantini, I. (1982) An efficient way to represent quadtrees. *Communs. ACM*, **25**(12), 905–910.

Goodchild, M.F., Parks, B.O. and Steyaert, L.T. (1993) *Geographic Information Systems and Environmental Modeling*. Oxford University Press, Oxford.

Holroyd, F.C. and Mason, D.C. (1990) Efficient linear quadtree construction algorithm. *Image Vision Comput.*, **8**(3), 218–224.

Jones, C. B. (1989) Data structures for three-dimensional spatial information systems. *Int. J. Geogr. Inf. Sys.*, **3**(1), 15–31.

Langran, G. (1989) A review of temporal database research and its use in GIS applications. *Int. J. Geogr. Inf. Sys.*, **3**(3), 215–232.

Mason, D.C. (1990) Linear quadtree algorithms for a transputer array. *IEE Proc.*, **137E**(1), 114–128.

Mason, D.C., O'Conaill, M.A. and Bell, S.B.M. (1993) Handling four-dimensional geo-referenced data in environmental GIS. *Int. J. Geogr. Inf. Sys.*, submitted.

Mason, D.C., O'Conaill, M.A. and McKendrick, I. (1993) Block kriging using a hierarchical data structure. *Int. J. Geogr. Inf. Sys.*, submitted.

Mason, D.C., Oddy, C., Rye, A.J., Bell, S.B.M., Illingworth, M., Preedy, K., Angelikaki, C. and Pearson, E. (1993) Spatial database manager for a multi-source image understanding system. *Image and Vision Comput.*, **11**(1), 25–34.

Matheron, G. (1971) The theory of regionalized variables and its application. Les Cahiers du Centre de Morphologie Mathematique de Fontainebleau, no. 5.

Meiko Ltd (1990) *Programmer's Introduction to Meiko CSTools*. Meiko Ltd, Bristol, UK.

O'Conaill, M., Bell, S.B.M. and Mason D.C. (1992) Developing a prototype 4-D geographical information system on a transputer array. *ITC J.*, **1992**(1), 47–54.

Olea, R.A. (1975) Optimum mapping techniques using regionalised variable theory. Series on Spatial Analysis, no. 2, Kansas Geological Survey, Lawrence, Kansas, USA.

Samet, H. (1990a) *The Design and Analysis of Spatial Data Structures*. Addison-Wesley, New York.

Samet, H. (1990b) *Applications of Spatial Data Structures*. Addison-Wesley, New York.

Samet, H. and Tamminen, M. (1938) Efficient component labelling of images of arbitrary dimension represented by linear bintrees. *IEEE Trans. Pattern Analysis and Machine Intelligence*, **10**(4), 579–586.

Shaffer, C.A. and Samet, H. (1987) Optimal quadtree construction algorithms. *Comput. Vision, Graphics and Image Processing*, **37**(3), 402–419.

Smythe-Wright, D. (1991) *Ocean Circulation and Climate*. UK Natural Environment Research Council, Swindon.

Tamminen, M. (1984) Comment on quad- and octrees. *Communs. ACM*, **27**(3), 248–249.

Webster, R. and Burgess T.M. (1980) Optimal interpolation and isarithmic mapping of soil properties. III. Changing drift and universal kriging. *J. Soil Sci.*, **31**, 505–524.

Worboys, M.F. (1992) A model for spatio-temporal information. *Proc. 5th Int. Symp. Spatial Data Handling*, Charleston, South Carolina, 3–7 August, 602–611.

Chapter Twelve

Development of a generic spatial language interface for GIS

J. RAPER AND M. BUNDOCK
Department of Geography, Birkbeck College, London

This chapter examines the user interface 'problem' within geographical information systems (GIS) and reviews research and development in this field. Working from the database towards the user, this research has defined the components of a generic user interface (known as UGIX), the method of communication with the database and the screen metaphors appropriate to this form of interaction. This chapter proposes that GIS interfaces should be user-customizable and should be built on top of a spatially-extended form of structured query language (SQL). This architecture will make it possible to implement UGIX with existing spatial databases when they are able to support the emerging SQL3 standard.

1. DEFINITION OF THE PROBLEM AND OBJECTIVES

The GIS user interface 'problem', stated simply, is that people who wish to gain access to spatially-related information through a GIS are often frustrated by aspects of the interaction procedure. This sense of frustration leads to under-use of GIS or outright unwillingness to use it, in spite of the scope it offers for the storage and analysis of spatial information. As such, the whole user interaction process has come under close scrutiny, since it is clear that improvements must be made in this area before GIS will be exploited to its full extent. Solving this problem has become a major priority in GIS; however, it is a multifaceted problem, involving the form of communication with the underlying database forming the archive of information in the GIS, the 'look and feel' of the user interface, and the motivation and knowledge of the user.

This research project aimed to recognize the user interface problem as a whole through an end-to-end design, starting with the underlying software architecture, and passing through the interface to the GIS users. The direction of movement from database and tools to users in this context is crucial: in this work the research problem has been seen as a transformation in the accessibility and understanding of data in the massive conventional databases that have been populated with spatial data in the last decade. Other researchers have suggested that such databases (Smith, 1993) and their query languages (Egenhofer,

1992) are not expressive enough to store and manipulate the spatial phenomena of interest. Their research agenda has been to move from users and their conceptions of space to define the database: while this perspective is equally valid, it deals with a wider class of users and technology that does not yet exist in the database field. As such, research proceeding from users' conceptions to the database must be concerned with the longer term. This research focuses on a set of problems that exist now by virtue of the widespread implementation of relational database technology and the complex GIS currently in use.

Since the GIS user interface problem is so complex, research has focused on certain elements of the problem that were seen to be tractable and of key importance. The following section aims to define the context in which the specific research was carried out in this project.

1.1. The problem definition

The worldwide success of GIS in the last 10 years has created a huge interest in the tools and insights that it provides, generating a demand for GIS use in many application sectors and at many levels in organizations (Maguire, Goodchild and Rhind, 1991). In the early phase of this development, GIS products were competing with each other almost exclusively on functionality terms, i.e. in terms of the algorithms and procedures they were able to carry out. Then, as it became clear that the early growth of GIS could not be sustained without 'better' user interfaces, interest turned to this aspect (Raper, 1991). New GIS user interfaces based on window–icon–menu–pointer (WIMP) interfaces for personal computers and workstations changed user interaction from exclusively command language-based to largely mouse- and window-based. However, despite these developments, users still complain that they find GIS 'difficult to use', and GIS is still seen as a skill to be acquired by intensive and time-consuming training. In order to discover how to approach this problem it is necessary to look in detail at all the processes involved in interaction, starting with the knowledge that the users themselves bring to the interaction.

In the sense that GIS provide tools to manipulate rigorous models of spatial phenomena, they have proved attractive to people holding spatially related information (Raper and Maguire, 1992). The pressures to adopt and use GIS are many and various, and include press coverage, sales promotion, legislation, professional practice and information technology updating. For some or all of these reasons, interest in GIS has risen steadily during the late 1980s and early 1990s, leading to widespread implementation of GIS in organizations. However, in many organizations, despite the potential for GIS to become an across-organization 'horizontal' methodology, it has become a skills-orientated 'vertical' technology, since only a few users have been able to build up the skills needed.

In this context, the user interface to GIS has become of pivotal importance in the motivation of individuals to commit to GIS as a means of handling spatially-orientated data and making spatially-orientated decisions. Although there are relatively few accounts of 'failed' GIS implementations (an exception is Openshaw *et al.*, 1990), there is some evidence that user motivation and the interface environment are closely interconnected. It is a premise of this research that huge productivity increases can be achieved in the use of GIS if user interface environments can be improved within a context where GIS are seen as an integral part of the information strategy.

This assessment clearly defines two categories of user: firstly, the 'spatially aware professional' (SAP) (Raper and Bundock, 1991), who is a user of spatial information as part of his or her business or job function; secondly, the wider general public. While it is possible to establish that the SAP is motivated to use a GIS either through self-interest or job design, the public form an infinitely larger heterogeneous group with no necessary interest in GIS at all. In this research only the former group has been addressed since its concerns are immediate and well defined. In particular, the design of a GIS user interface for SAP use benefits from a knowledge of their job objectives, the way professional language describes space and the data likely to be held. Hence, this work has aimed to define a generic GIS user interface environment likely to tackle the problems of the SAP user.

1.2. A critique of existing GIS user interfaces

At the time of writing in early 1993 most commercial GIS user interfaces have been ported to a window environment on a PC (Macintosh or Windows) or workstation (Motif), making them much more visually attractive than the old command-driven text interfaces. Several vendors have developed new interfaces such as GENIUS from Genamap, SPANMAP from SPANS and ARCVIEW from ARC/INFO, the latter two of which have been marketed as executive information systems (EIS), specifically aimed at 'non-technical' users of GIS. However, although these products represent significant advances, they still suffer from a number of problems, detailed below.

1.2.1. Visualization of the database

User interfaces in GIS have long lacked good visualization tools for the spatial database. Ideally, the user should be able to see the entities and their interrelationships graphically expressed, to enable queries to be formulated more easily. This problem has been tackled in the Macintosh and Windows user interfaces by the use of dialogue boxes presenting files listed in alphabetical order, and new GIS user interfaces such as ARCVIEW have emulated this technique. However, the alphabetical listing does not convey the spatial interrelationships, not does it enable the user to distinguish between the different attribute and geometric characteristics of databases. The list approach is also appropriate to flat files and not to relational databases with complex primary/foreign key relationships. Such list-orientated views of the spatial database are also inevitably orientated towards system concepts such as maps of geometry and attributes in tables rather than addressing the actual spatial phenomena in the real world with which the users are concerned.

Several visualization tools have been developed to present the relationships among relational tables (e.g. Data Works from Harlequin), which visualize the tables in a 'data space' to avoid the limitations of the list format. Such extensions to the presentation of information need to be adapted to the spatial environment. The poor quality of help systems for many GIS has also become a major drawback for many user interfaces (Raper and Green, 1992). Frequently the help is simply a formal statement of the command syntax and arguments, and not an explanation of its actual usage.

1.2.2. Control over GIS tools

One of the most pervasive problems in existing GIS interfaces is the blurring of the distinction between goals, tasks and system functions in the language and process of interaction with the system. This means that the user cannot easily appreciate the structure of the user environment. In this work the following definitions were used:

- Goal: a user target for spatial data manipulation expressed in terms of application-specific outcomes, e.g. finding which stands of trees in a forest will come to maturity in each of the next 5 years.
- Task: a spatial data manipulation procedure expressed in terms of system implementable steps, e.g. searching for certain spatially referenced items in a database and displaying the results in a map.
- Function: a low-level system operation to manipulate spatial data, e.g. plotting a symbol at specified x,y coordinates on an output device.

In this scheme tasks and functions refer to system operations, while goals apply to conceptual operations, which are conceived of without reference to a computing environment. In most cases the translation of the desired activity into a computing environment occurs between the task and the function: at present, training is required to teach this skill.

Using this terminology, most GIS offer a command language composed of functions that are spatial tools and algorithms of various kinds. These commands are often modifiable with arguments and the complete expressions used are complex and often obscure. As part of a general movement to improve this situation a number of commercial systems have begun to offer graphical user interfaces (GUIs) built using WIMP techniques. However, these developments have illustrated the difficulties inherent in assigning icons or menu items to functions, i.e. while the range of options available is now stated, the system structure is still no easier to understand. In particular, it is difficult to convert a goal into a task made up of the appropriate functions. Added to these implicit difficulties are the problems of overfilling the screen with icons or creating very long menus, the use of inappropriate screen metaphors and the lack of activity indicators to indicate the status of an operation to the user.

The process by which the user links the concept and implementation can lead to confusion and to users making errors in they way they specify operations. A well-designed user interface permits the customizer or expert user to link the appropriate user language for space to the system architecture in a way that is transparent to the end users. In other words, the interface should allow the user to manipulate objects that are meaningful in terms of the application, for example subdivide a parcel rather than split a polygon.

Above all, making a spatial query in a GIS is far too complex, with users requiring training to learn how the system describes data that they themselves may have been responsible for collecting or defining. Given the relational database management system (DBMS) basis for most commercial (vector) GIS, the use of SQL has been common in system designs. Many vendors have developed their own extensions to SQL to handle spatial concepts; however, few of these developments have recognized the evolution of SQL as an international standard and supported the SQL2 and SQL3 developments (Bundock and Raper, 1992).

1.2.3. Customizability

Commercial GIS software packages are normally designed to be fairly general purpose in nature – they are not designed for a specific well-defined application within a particular organization. Consequently, they need to be adapted to fit the specific application and user requirements of the organization within which they are implemented. This adaptation of the as-supplied system is termed *customization*. The term *customizability* is used to describe the ease and extent to which a system may be customized. The objectives of the customization process (see Bundock and Raper, 1991) are to provide a system for the user that:

Development of a Spatial Language Interface for GIS   117

- supports both the data model and functionality demanded by the application requirements;
- presents to the end user an interface specific to the user's application, language and experience;
- is uncluttered by non-required functions, icons and menus;
- is easy to learn and easy to use.

At present, the base products delivered by GIS vendors are little more than a box of low-level spatial tools. These general-purpose tools do not directly satisfy the user's functional requirements, which are determined by organizational and application-specific objectives. Furthermore, the tools will often have little meaning or applicability to the end user, who must be educated in the language, interface and conceptual model supported by the product. The customizability of existing GIS products is poor, especially in the areas of database design and implementation, task definition and user interface design and development. The result is that effective customization of a GIS product to satisfy corporate GIS requirements involves enormous expense and effort. This problem is so acute that it is not unusual to see organizations struggling to use a system that is uncustomized and uncustomizable with the available resources. Effectively, the application requirements are largely discarded so that the functions that the GIS supports actually become the application implemented.

The customization process incorporates all the normal stages of the familiar systems development life cycle, including analysis, design, construction, implementation, operation and planning. The analysis stage incorporates both data analysis (resulting in the development of data models) and function analysis, which involves both process and event analysis. The design phase incorporates logical and physical database design, task design and user interface design. Construction refers to the actual development of the physical database, tasks and user interfaces, while implementation is concerned with delivering the working system to the user environment.

Most (if not all) existing GIS software exhibits poor customizability. At present, the customization process can best be described as:

- time consuming;
- technically difficult, requiring technical expertise in multiple languages and subsystems;
- expensive;
- yielding a poor result.

These problems can be attributed in part to the poor quality of (or sometimes absence of) suitable tools within the GIS to support customization and also to the lack of an integrated analysis, design and software engineering methodology that is appropriate to spatial applications. The use of methodologies based on commercial applications is viable, but without a clear understanding of spatial/GIS concepts they are likely to yield poor results.

1.2.4. Non-transferability of skills

Each GIS product on the market today incorporates its own distinctive environment, being substantially different from virtually all other available products. Each system tends to have its own unique command language, icon set, menu organization and form layouts. The methods of interaction with the system vary considerably, even for such simple actions as selecting an object, obtaining help information or indicating confirmation of an action.

Each vendor tends to use their own set of jargon, often in a manner that is inconsistent with other GIS vendors.

Even worse, the underlying system architectures show through and must be understood by the user before effective system usage is possible. In the absence of any other strong conceptual model for the system (as might be presented in a fully customized environment), the underlying architecture (files, layers, coverages, points, lines and polygons) becomes the mental model understood by the user. The application problem, for instance, forest resource management, becomes mapped to the problem of manipulating the components of the GIS architecture (i.e. the coverages, polygons etc.). Consequently, the skill set acquired by a user is specific to the jargon and architecture of a particular product. Since each GIS uses different jargon and different architectures, the user's knowledge of one system is not readily transferable to another. This is clearly not desirable in an era of rapid software and system development.

1.3. Previous research

Significant research has been completed in GIS user interface design, largely through community actions initiated by the US National Center for Geographic Information and Analysis (NCGIA), the European Community ESPRIT II programme and the NATO Advanced Study Institutes (ASI), all of which held study meetings and conferences.

The NCGIA established two Research Initiatives which generated new research in this field, namely Initiative 2 'Languages of Spatial Relations' and Initiative 13 'User Interfaces for GIS'. The Initiative 2 research programme culminated in a meeting in July 1990 under the NATO ASI programme, entitled 'Cognitive and Linguistic Aspects of Geographic Space'. Mark and Frank (1990) present an edited collection of the papers contributed to this meeting under six major headings:

1. Geographic space – examinations of the fundamental concepts of geographic space employed in GIS.
2. Cultural influences on the conceptualization of geographic space assessments of the cultural input to the concepts of space employed in GIS.
3. Wayfinding and spatial cognition – research on aspects of spatial cognition used to establish concepts of space.
4. Cartographic perspectives – research into mapping as language, e.g. feature definition and user understanding of maps.
5. Formal treatment of space in mathematics – statements of the construction and limits of the representation systems underlying GIS.
6. User interfaces and human–computer interaction (HCI) – research into HCI in a GIS context, in particular the development of metaphors for use in GIS user interfaces and debates about the expressiveness of the SQL database query language.

The NCGIA Initiative 13 Specialist Meeting on 'User Interfaces for GIS' was summarized by Kuhn *et al.* (1992), who identified three key groups of issues for research:

1. Structuring GIS tasks. Considerable research has been carried out in the HCI field into task analysis, for example by Rasmussen into human behaviour, which he classified into

skills-based, rule-based and knowledge-based, or under the goals, operators, methods and selection (GOMS) model, which focuses on *post hoc* task analysis. There now seems to be general agreement that most GIS are not organized according to any principles of task analysis, but according to the syntax of operations on the concepts of space which the GIS implements.

2. How people think about space. This research partially revisited work done under Initiative 2, but in addition examined the meaning of cartographic concepts such as the 'layer' and data models for GIS, and the level of understanding of these concepts in relation to GIS by the user. This latter point can be seen as crucial since the congruence between the internal structure used by a GIS and the external presentation is the essential transformation of knowledge undertaken when using a GIS.

3. Interface styles and paradigms. Work on the communication channels available to GIS user interfaces has been expanded considerably in the multimedia revolution, and now extends to the new field of virtual reality.

By contrast, the work undertaken in the ESPRIT II project concentrated on the performance of the mechanisms available to query GIS and the issue of whether SQL was the right language to use for spatial queries (Ausiello, 1991). Research on the use of tools such as data cubes as alternative query formats was discussed (Frank, 1991), although these ideas have not been implemented.

Work has also been undertaken on the human factors associated with GIS interaction (Medyckyj-Scott and Hearnshaw, 1993). The Usable Spatial Information Systems (USIS) project at Loughborough University has attempted to assess the 'usability' of GIS using questionnaire surveys based on the Digital System Usability Scale (DSUS) and observation of users. Preliminary conclusions (Medyckyj-Scott, Davies and Byrne, 1993) indicate that although users state that they have an overall positive impression of GIS, they frequently then go on to cite difficulties with jargon, file handling, lack of feedback and inconsistencies between modules. These findings confirm the need to examine the GIS 'experience'.

1.4. The objectives of the UGIX research project

Since the GIS market is still relatively immature, little user pressure is being exerted on vendors to define and agree standards, especially in user interfaces. The result is that each GIS product tends to have its own data model, set of jargon, macro or customization language and, perhaps most importantly, user interface, such that knowledge gained in using one product is rarely of use when migrating to a different product. Whereas other parts of the computer industry have developed and adopted standards for data management, communications, operating systems, languages and user interface presentation and behaviour, the GIS industry is still plagued by vendors offering proprietary hardware solutions, user interface environments and interaction languages.

The first attempt made at Birkbeck College to develop an improved GIS user interface was the HyperArc project to create a graphical front end to the ARC/INFO GIS (Raper, Linsey and Connolly, 1990). Implemented using HyperCard on the Apple Macintosh and communicating over a network to a VAX or UNIX host system running ARC/INFO, HyperArc enabled a user to carry out some generic GIS tasks without specific knowledge of the ARC/INFO command language. HyperArc was designed as a tool for those who needed to use ARC/INFO (perhaps because it is the 'corporate' GIS available in their organization) but who do not have the time or training to become familiar with the command language at the functional level.

The subset of the ARC/INFO commands that could be accessed through HyperArc were selected to allow the user to carry out the following tasks:

- Browse maps and attributes
- Query the attribute database
- Perform map composition
- Perform basic analysis functions such as 'buffer', 'integrate', etc.

In practice, the most commonly-used commands are usually the browse and query commands, although the user's access to analytical functions is not limited as long as this can be done in an inherently 'safe' manner. 'Safe' is defined as meaning that the user cannot delete existing data.

HyperArc uses ARC/INFO system-specific terminology such as 'coverage' (meaning map sheet in digital form), or 'clean' (meaning create topologically structured geometry from 'spaghetti' vector data), but explains and illustrates these terms for the user with diagrams and descriptions. HyperArc was tested by an ARC/INFO user (Bruce Wright of the US Geological Survey) who helped to define the required functionality. Display and query functions are the most frequently required, along with integration tools. HyperArc is also integrated with the GISTutor hypertext tutorial developed at Birkbeck College, enabling the user to obtain background information on the concepts underlying any of the HyperArc options. The creation of HyperArc was an extremely valuable exercise in identifying the important design factors for a GIS user interface of this kind. In particular, it showed the difficulties inherent in using system terminology, and in coupling an interface to the underlying GIS. Hence, HyperArc pointed the way towards generic interfaces (Linsey and Raper, in press).

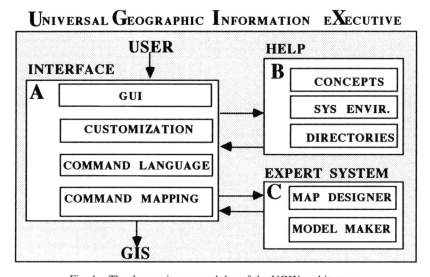

Fig. 1   The three primary modules of the UGIX architecture

1.5. Components of UGIX

The idealized design of UGIX as originally envisaged by Rhind, Raper and Green (1989) contained three main modules. The first of these contains the GUI, dialogues, and command language and command mapping. The second comprises a 'help' and information system for GIS, and the third module is an expert system shell or high-level system access module (Fig. 1).

2. THE UGIX SYSTEM: REQUIREMENTS AND DESIGN

Given the practical user interface experience with HyperArc and the detailed analysis of the user interface problem given above, a set of requirements was rapidly established for UGIX. These can be divided into the assessment of the SAP user's needs, the components of the language of interaction and the means of communication with the underlying spatial database and GIS.

2.1. Identification of users

GIS technology is being applied to a wide spectrum of problems, in research, in small or self-contained projects, and as an integral part of corporate applications. The applications for which the GIS is used may vary from a single, well-focused application such as route planning to a generic information system aimed at supporting a wide variety of applications from a common interface, for example corporate applications. GIS products and software aiming to support a single, well-defined, limited scope application may have their user interfaces optimized for that hardware environment, software environment, application and targeted user group. User interfaces for such systems need not be totally flexible and customizable since the applications are tightly focused and limited in scope. At the other end of the range, a corporate GIS may be expected to support a large range of applications, be flexible enough to be readily applied to new applications, access multiple heterogeneous corporate databases, utilize existing corporate hardware environments (e.g. existing PCs, workstations and communications), and be customized to support both tightly focused applications and general-purpose analytical applications.

A single system installed within an organization may be used by a wide range of users, including GIS professionals, drafting personnel, applications-specific personnel (for instance, surveyors, electrical engineers), clerical personnel and even customers of the organization. The technical abilities of GIS users will vary widely, from technically competent professionals able to solve complex problems, to individuals of below average ability employed to do repetitive, uninteresting tasks.

The confidence with which individuals use a workstation varies widely. Teenagers and young adults with experience of PCs, portable games machines and the like generally have a high level of confidence with GIS workstations and will enthusiastically explore a system to see what it can and cannot do. Other users may be extremely timid of new technology, scared that they may in some way 'do something wrong'. Hand–eye coordination required to drive a mouse or tracker ball varies considerably from one individual to another, although repetitive use will improve an operator's expertise radically. The personal interest that an individual has in using a GIS may also vary – some users may view their job simply as a way of making enough money to pay the bills, with little or no interest in what they do during work hours. At the other extreme are users whose jobs are also their hobby or primary interest.

The frequency with which an individual may use a system will range from once in a lifetime to all day, every day. People arriving at the customer enquiry desk of a large utility organization or local authority may find themselves confronted with a GIS workstation at which they might be able to enquire about planned developments or property valuations. People employed by that organization may use the GIS every day to perform either a small number of tasks with high frequency (e.g. recording asset maintenance information) or a wide range of tasks, each of which may take a significant amount of time, for example planning a new development.

Each of the aspects of GIS usage described above has a significant impact on user interface design. Systems designed to address tightly focused applications in a stand-alone environment may have an essentially static, uncustomizable user interface. GIS products aimed at addressing a wide range of corporate applications must support extreme flexibility in user interface customization. SAPs with a high degree of confidence in using GIS technology and computer hardware are able to tolerate visually complex user interfaces with a large number of available commands, while those of lower ability or people with infrequent system usage will require very simple user interfaces with a limited number of options and a high level of prompting and explanatory comment.

We may classify users into a number of groups, where within each group the user interface requirements are more consistent. The design of UGIX attempts to address the needs of the following categories of users:

- The inexperienced, infrequent and/or task-limited user
- The SAP and expert user
- The system customizer and system administrator
- The system developer

Each category of user has a different, although sometimes overlapping, set of requirements. For example, inexperienced users require a task-orientated interface to guide them through a particular application-specific task. Ease of use and an intuitive user interface are primary requirements. Such users would generally have a limited set of jobs to perform. These would utilize a limited set of lower level functions and operate on a restricted section of the database. The customized environment must restrict access to only the data and functions required to perform the user's tasks. The tasks may include various functions, including data interrogation, capture, maintenance, report generation and map production.

The analyst, engineer or expert user requires a much less restricting interface. Typical system usage might be one-off analysis (e.g. in a utility environment, mains fault/burst analysis such as the number of bursts per 100 m per year). Read-only access to much of the database is required, plus the ability to create test data sets to perform complex modelling and 'what-if' analysis. Virtually unrestricted access to the functionality is required, to allow experimentation and analysis via a number of alternative techniques. The user interface must again be easy to use, but direct access to the functions is required, sometimes in a command-line manner. Furthermore, the ability to link a number of functions together to perform a complex query repetitively, and then make that available at the GUI level, is required. The functions accessible to the expert include all those available to the novice, plus analytical functions and user interface design functions to support modification of the user's own environment.

The customizer and database administrator (DBA) must be able to synthesize the end user environments for the other users. The creation of a task-orientated GUI for the infrequent user must be achievable in a short time frame. The integration of computer aided software engineering (CASE) tools, object-oriented analysis/design tools and a graphical customization environment within the GIS would assist the customizer and DBA in performing their functions. If the CASE tool was already primed with knowledge of the spatial data types and spatial operators supported by the GIS, significant simplification of the functional analysis, data analysis, data modelling and database design stages of customization could be achieved.

The style of interface offered must satisfy the user's requirements for ease of use, in particular, access to functionality. The use of GUIs is becoming more common, not only for end user applications, but also within customization and development environments. Consequently, the design of UGIX proposes a GUI for task-orientated applications, common GIS functions (e.g. spatial and aspatial interrogation and analysis), access to the help system, customization functions (such as task specification or user interface design) and DBA functions (e.g. access control specifications, database schema definitions and backup). In addition, a separate layer within UGIX would offer a command line facility to perform the same functions. It is intended that a command line interface would be based on SQL-SX, a spatially extended version of SQL. SQL-SX is defined for the interactive command line/fourth generation language environment and as a set of extensions for embedded SQL within third generation language environments.

2.2. Definition of common functional components

In order to specify better the functional requirements of a user interface we need to classify the functions in some way. Using a top-down functional decomposition approach, we can classify GIS functional components in terms of:

- System definition
- System administration
- Database interrogation and display
- Data input and maintenance
- Analytical functionality
- Output functions for softcopy and hardcopy production.

Using functional decomposition, each of the high-level functions may be decomposed into a small number of sub-functions. Recursive decomposition may be applied until a function is easy to comprehend or implementable.

The system definition function supports all those processes required to define the target GIS environment. It includes functions to define the application objects in terms of both their data structure and behaviour, defining how objects are to be displayed, standard report formats, hardcopy output specifications and the user interface to be supported for the various types of system user. The functionality required to support the system definition function can be provided in a CASE environment, which itself is capable of creating an environment supporting the other functions through a GUI communicating with the underlying database. This latter environment forms the basic UGIX architecture.

2.3. The UGIX architecture

A basic principle of the UGIX design was that improving access to existing GIS could be achieved by converting the function-orientation of the native system interface (primitive and implementation-specific operations) to a task-orientated interface (sequences of low level spatial operations) usable by a SAP. However, implementation of such an interface in a generic way requires a new form of software architecture that is independent of specific implementations, does not enforce a particular data model, and adheres to the standards in the user community that are most crucial to the success of GIS in heterogeneous computing environments.

To achieve this aim, a layered model was suggested that protects the user interface from the actual implementation mechanisms provided by each GIS vendor. Each layer within the model will perform a particular job and have a well-defined interface to the layers both above and below. Some of the layers within UGIX will be able to communicate directly with the underlying GIS at a matching level. The overall architecture for UGIX is similar to the Seeheim model (Green, 1985). The requirement for a high bandwidth communications channel from the GIS application is supported to allow efficient graphics display and manipulation. Fig. 2 illustrates the overall structure proposed for the UGIX environment.

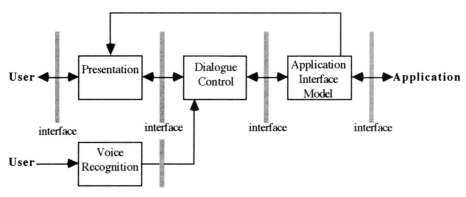

Fig. 2   Overview of UGIX(A) architecture

The presentation layer can be constructed in any standard user interface toolkit, such as Macintosh, Windows, Motif or OpenLook, a widget design facility, a screen design facility and a screen execution facility. The widget and screen design facilities operate within the constraints of the toolkit, and will be implemented as a set of executable screens. The customization environment itself and the actual resulting end user application will also simply be a set of screens with which the user may interact. The screens will be designed in terms of a set of windows, a set of widgets within each window and a set of forms. The behaviour of the windows, widgets and forms will be described in terms of the fourth generation command language.

Interaction with the screens will cause the widgets to react in the predefined manner and the execution of GIS tasks in terms of the spatial command language embedded within the fourth generation language. Equivalent commands may be issued directly via a command line interface, via widget interaction or using voice input. Each method should result in the execution of the same generic functions within the dialogue control component. The voice recognition facility issues either individual spatial command language tokens which may be used to build a complete command, or entire commands. Entire commands may be abbreviated into a spoken shorthand consisting of just a few words, rather than the user being required to speak the full command syntax for a particular task.

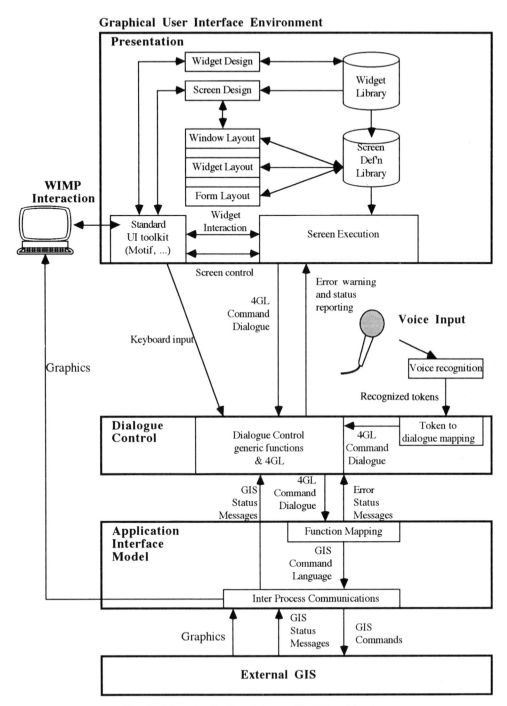

Fig. 3 More detailed breakdown of UGIX architecture

The application interface module accepts generic spatial language commands and maps them onto the command language of the underlying GIS. It then issues these commands to the GIS via an interprocess communications mechanism. The GIS responses may include alphanumeric information (which may be used to fill a form), status information (error and function status) and/or graphics. Support of a highly interactive graphics environment requires that a high speed channel be provided to display the graphical data. However, alphanumeric and status data may be routed through the dialogue control module for further processing and display. Fig. 3 illustrates the proposed architecture of UGIX(A) in more detail.

3. THE CUSTOMIZATION ENVIRONMENT FOR GIS

3.1. Background to requirements

GIS is increasingly being implemented within a corporate information system environment. It is being used as a data integration tool, providing support for spatial data management and integration with existing corporate databases. GIS provides the ability to enhance many existing corporate applications significantly, by the use of graphical displays and spatial query support. Corporate databases supporting a range of applications tend to be large and complex, with hundreds of entity types, each typically having between 10 and 50 attributes, complex relationships between entity types and complex data integrity rules. Furthermore, from the perspective of the GIS vendor, every site has its own particular requirements and requires the ability to configure the GIS to meet particular data and application requirements. Consequently, today's GIS products are becoming as generic as DBMS, able to support a wide range of applications and data models. Thus the GIS has become an enabling technology, allowing organizations to develop their own applications, rather than being the solution itself. Without customization and application development, the GIS product is not an information system; rather, it is simply some software that may potentially be used to develop an information system.

Customization is the process of adapting and extending the basic GIS product to support an organization's specific application requirements, and using the GIS to create an empty database with a suitable structure to support those requirements. We extend the scope of the customization process to include the requirements analysis and design phases.

Most of the GIS products that were available throughout the 1980s were able to support only very simple data models (e.g. layer based) with limited access to external databases. Databases were partitioned (tiles, sheets, coverages etc.) to allow simple data management facilities to be supported while maintaining reasonable performance. Applications tended to be centred on the requirements of the drafting office and the need to produce maps and charts. A common database supporting the requirements of multiple departments and applications was not considered, often because the supporting technology was not available. As the emphasis has moved from cartographic applications to corporate information management, then both application requirements and data management requirements have grown significantly more complex.

The customization facilities within many GIS products are poor, requiring technical expertise in multiple languages and subsystems. There is generally no support for analysis and design, simply a set of application modules that operate on the underlying spatial data structures, and one or more languages with which to develop the target system. Often, much of the development effort must be performed with third generation languages such as C (and even Fortran) or poor-quality macro languages. The result is that GIS customization usually requires large amounts of effort, significant levels of technical expertise and high costs. It has a relatively high risk of failure.

New product architectures specifically designed to support large, complex, multi-application customizations are significantly improving our ability to deliver corporate GIS solutions. This has primarily been achieved via tight integration of relational database management systems (RDBMS) technology and use of an integrated object-oriented language. These facilities provide a much improved development environment, increasing software development productivity by at least a factor of five over conventional third generation languages.

The ability to rapidly prototype complex applications for benchmark and demonstration purposes has also led to an increased expectation from end users. This, coupled with the increasing complexity involved in corporate GIS applications, has resulted in GIS customizations again requiring considerable effort, time and cost. The development environment might be five to ten times better, but the scale and complexity of the problems being addressed has also grown by a similar factor.

Furthermore, increased productivity during the software development phase only addresses one aspect of the systems development life cycle (SDLC) and overall customization effort. Improved development facilities can assist during the design phase by allowing rapid prototyping of user interface and critical or high risk parts of the system, but this is usually a small component of the design effort. With large, complex applications, the effort involved during the analysis and design stages is very significant.

CASE tools have been extensively used in commercial data processing applications to assist in the analysis, design and development phases of the SDLC. However, their usage within GIS projects has been extremely limited, partly due to lack of support for spatial concepts within existing CASE tools and partly due to lack of awareness of CASE within the GIS community.

If the limitations of existing commercially available CASE products can be addressed, then CASE technology may provide an answer to these problems by:

- improving productivity during the analysis and design phases;
- automatically generating code for significant portions of the target system;
- reducing the size of the knowledge barrier between the analysts and the developers.

Integration of the CASE tool directly into the GIS environment may provide additional advantages by removing the paper interface between design and development, and by providing a facility to perform ongoing system development and maintenance in a truly integrated manner.

3.2. Development of a CASE tool prototype

To evaluate the feasibility of using CASE technology for GIS customization, a prototype tool was developed. The prototype was developed to the stage of being used in a commercial GIS customization of the Smallworld GIS product for a large electrical utility organization. The design of the prototype allowed not only the descriptions but also the code for all object behaviour to be managed within the repository. Thus, the use of version management and long transaction management automatically provided the source code control environment.

The CASE environment was developed as part of a commitment to the creation of a customizable GIS environment. Unless the user is given this scope for real creative involvement in the GIS interface design it seems likely that users will simply accommodate themselves to a sub-optimal system when their knowledge could easily be incorporated through user-guided customization procedures.

4. SPATIAL EXTENSIONS TO SQL

SQL is becoming increasingly important in GIS implementations as the need to integrate GIS and other corporate applications arises. A large amount of literature has been generated discussing the pros and cons of SQL (Codd, 1988; Date, 1986, 1989) and its use or applicability within GIS (Bundock, 1991; Egenhofer, 1987; Herring, Larsen and Shivakumar, 1988). A number of national bodies in both North America and Europe (Gradwell, 1990; Dowers, 1991a) are actively involved in identifying the requirements for spatial query languages and spatial extensions to SQL, while at an international level, the special requirements of GIS, CAD/CAM and other spatial applications are being discussed by American National Standards Institute (ANSI) and International Organization for Standardization (ISO) working parties developing the SQL standard (Dowers, 1991b; Gallagher, 1991).

In response to user demand, limited SQL facilities are now being offered by a number of GIS products, including PC-based products such as GEO-SQL through to workstation/mainframe products from Intergraph, GeoVision, Smallworld and Prime (see, for example, Westwood, 1989; Charlwood, Moon and Tulip, 1987, and product descriptions from these vendors). Often, these products offer a limited SQL interface for interrogation and update of aspatial (character, numeric, date, etc.) data held in standard RDBMS products such as DB2, INGRES, ORACLE and Sybase. However, such interfaces are usually incomplete (not supporting full schema definition functionality) and are not able to define, interrogate or maintain the spatial component of the GIS database. The primary cause for this limitation is the partitioning of spatial and aspatial data into separate data management systems – the aspatial component being managed by a standard relational product, while the spatial component is managed within a proprietary non-relational data management system, such as Intergraph IGDS design files, ARC/INFO layers and coverages, GeoVision GIS and Unisys ARGIS spatial data manager, GEO-SQL Autocad. However, a number of vendors have been able to provide meaningful extensions to the query language to support spatial predicates for use in SELECT statements.

The development and integration of a query language into a GIS environment may take essentially one of two paths:

1. Adopt an existing query language and extend it to support any special application requirements.
2. Develop a new query language that directly supports the concepts underlying the application – such query languages need not be constrained to the traditional structured database languages, but rather could be based on natural language, gesture, or be purely iconic.

A number of researchers have proposed new query languages specifically designed for spatial applications (MAPQUERY, Frank, 1982; QPF, Jian-Kang, Tao and Li, 1989). However, the widespread acceptance of SQL in the general data processing community and the increasing need to integrate GIS applications into corporate information system environments have effectively removed the marketability of a non-SQL compliant query language. Organizations wishing to implement corporate GIS applications are likely to want to utilize existing data processing expertise and insist that any proposed GIS product conform to internationally accepted standards. Consequently, a large number of proposals have been made for spatial extensions to SQL (Sacks-Davis, McDonell and Ooi,1987; Abel, 1988; Herring, Larsen and Shivakumar, 1988; Pong-Chai Goh, 1989; Raper and Bundock, 1990; Bundock and Raper, 1992).

Although other query languages have been proposed and developed, for example QUEL (Stonebraker *et al.*, 1976), and Query-by-Example (QBE) (Zloof, 1977), SQL has become the major query language offered by virtually all RDBMS vendors and even non-relational DBMS (e.g. hierarchic, network and even object-oriented DBMS products offer some form of SQL). Both QUEL and QBE have been used as the basis for spatial extensions in the form of GEO-QUEL (Berman and Stonebraker, 1977) and Query-by-Pictorial-Example (Chang and Fu, 1980).

Much of the acceptance of SQL may be attributed to IBM's and Oracle Corporation's early efforts to promote SQL-based RDBMS. However, because there was no single detailed language definition available, many of the early SQL implementations were quite distinctly different. As additional RDBMS vendors started to appear on the scene, it became clear that a standard definition for the language was a necessity. The ANSI X3H2 database languages group started work on development of the standard in 1983, although this was not ratified until 1986.

Multiple organizations now provide definitions of the SQL standard, including ANSI, ISO, X/Open and the SQL/Access Group. The original American standard (ANSI X3H2, 1986) has been adopted as an international (ISO) standard, ISO 9075 (ISO, 1987), and in the USA as a Federal Information Processing Standard (FIPS). Subsequently, ANSI and ISO have defined an Integrity Enhancement Feature (IEF) (ISO 9075) (ISO, 1989) as a major extension to the original standard and incorporated it as an annex. In 1987, the X/Open consortium adopted a version of SQL for the UNIX environment, with the aim of producing a specification for a portable SQL.

SQL is a growing standard. Multiple groups are identifying requirements, often independently, and proposing new facilities for inclusion in future SQL standards. Both ANSI and ISO are working towards SQL2 and SQL3 standards (Melton, 1991). In addition, other groups are also developing SQL standards for remote database access, integration with object-oriented systems and specialist applications. In the UK, the standards committee of the Association for Geographic Information (AGI) has been active in studying GIS database requirements and the proposals for forthcoming SQL standards (Gradwell, 1990; Dowers, 1991a). Similar activities are being performed by other applications groups requiring extensions to SQL, especially for use of SQL in the areas of CAD/CAM, CASE and Information Resource Dictionary Systems (IRDS).

The work being put into development for extensions for the ISO/ANSI SQL2 and SQL3 standards is being undertaken worldwide. Working groups in many countries are putting forward proposals for adoption within SQL3. Many of the enhancements involve incorporation of object-oriented concepts and extensibility. In particular, the ability to support the definition and use of domains, user-defined data types, triggers, inheritance hierarchies and recursive relational constructs all make SQL appear much more viable as a mechanism for interrogation and manipulation of spatial data. This continuing activity in extending the language to support new capabilities acts as an assurance to user organizations that SQL will be around and useful for quite some time yet.

4.1. Current status of SQL3 draft

4.1.1. SQL3 status

SQL3 is becoming a major revision with many fundamental changes. Many of the extensions provide object-oriented facilities such as inheritance, encapsulation and modularity. However, the most significant for GIS data management are hierarchies, abstract data types (ADTs) and stored procedures.

4.1.2. Subtable/supertable hierarchies

The current SQL3 draft (Melton, 1991) includes support for subtables and supertables. A subtable is a specialization of its parent supertables, inheriting all the columns from each of its supertables, and optionally including additional columns. Any row in a subtable must correspond to one row in each of its parent supertables.

4.1.3. Subtype/supertype hierarchies

These are important in virtually all applications, including GIS. Support for multiple inheritance is an important feature and extends SQL3 beyond even languages such as ADA in this regard. A similar syntax and equivalent functionality have also been defined for ADTs, whereby one ADT may inherit from one or more parent ADTs.

4.1.4. Abstract data types

The working draft for SQL3 includes support for definition and use of (user-defined) ADTs. The definition of an ADT creates a persistent data type definition that may be used repeatedly within a database schema when defining columns. An ADT definition defines both the data structure and the behaviour of the data type (Kerridge, 1989). The data structure is specified in terms of existing data types, including both primitive data types and other ADTs, that are available within the current SQL environment (i.e. already defined in the current catalogue).

The behaviour of an ADT is defined in terms of a set of methods. Two special methods may be defined (or inherited) for an ADT:

1. An equality operator that returns a Boolean value identifying whether two instances of the ADT are equal.
2. A less-than operator that returns a Boolean value identifying whether one instance of the ADT is less than another instance.

These methods together define the semantics to be used when comparing ADTs within a comparison predicate. Additional methods may be defined to operate on the data within the ADT. These methods may return any single value of an existing data type (i.e. either a primitive data type value or an ADT value). Each method is defined in terms of a function definition. The function may be a pure SQL function, already defined within the schema, or an external function, written in a supported third generation language. Constructor and destructor methods are assumed to be supported implicitly via the INSERT and DELETE operations for a table.

4.1.5. Stored procedures

A number of RDBMS products today support the ability to store sets of SQL statements within the database and for the database server to execute those statements as a group. The primary requirement for such a facility is to improve performance in distributed database environments by reducing communications overheads between a database server and a client (Bauer, 1991). By storing the procedures within SQL at the server the client process can request a group of statements to be executed by naming the procedure, rather than sending the server one statement at a time and having the server respond to each. Furthermore, stored procedures could be shared by multiple applications, improving code reusability.

To support such a facility, the SQL3 standard allows the specification of a compound statement, which is a group of statements that are to be executed together and are bracketed by BEGIN and END statements. Within the compound statement, local variables may be declared and used.

Stored procedures may be extremely important in the context of GIS applications. Many GIS implementations in utility industries and in local and central government require distributed database management, due to the decentralized nature of the operations. Regional and district offices need rapid access to spatial data for operational reasons, while head office applications and the fuzzy nature of the boundaries between regions makes sharp segmentation of the database difficult to achieve. Spatial search, graphics retrieval and topology update/interrogation at the database server level would significantly enhance database performance and reduce network traffic.

4.2. Limitations of the SQL3 proposal for GIS applications

Egenhofer (1992) gives a comprehensive review of the limitations of SQL for use as a spatial query language, although many of the drawbacks identified are not features of any other query language currently implemented. Thus, he points out that SQL cannot handle temporal or qualitative queries, nor can it manipulate screen geometry. His most serious point is that the various extensions to SQL have not been formally defined.

4.2.1. Arrays and lists as standard data types

It has been proposed that SQL3 should provide support for lists, arrays or other types of 'generator' data types. GIS applications would immediately benefit from the ability to store structured lists of points as lines, structured sets of coordinates as n dimensional points and so on. The emerging ISO Common Language-Independent Datatype (CLID) specification (ISO/IEC, 1991) identifies a number of such generator types, including array, list, set, choice, record and range. Gallagher (1991) makes use of these data types in his discussion paper to the SQL3 working group and identifies their applicability for GIS applications. It is expected that these proposals will be integrated into the SQL3 draft.

4.2.2. Version management

The management of alternatives is a requirement common to many applications, including CASE, CAD/CAM and GIS. The design and analysis nature of many GIS applications demands the ability to create multiple alternate scenarios in the database, perform analysis or comparisons between the alternatives and possibly eventually accept one for use as the canonical version. The traditional method for managing alternatives has been to extract subsets of the database into new environments, replicating that data for each alternative. This imposes severe overheads in terms of large data volumes, poor post-extract performance and careful system management to ensure that duplicated data sets are appropriately archived and removed as necessary. Easterfield, Newell and Theriault (1990) describe new implementation techniques that may overcome many of these problems.

At present, SQL3 supports no concept of versions. Bundock (1991) suggested the syntax for the creation, deletion, posting and adoption of versions.

4.2.3. Long transaction management

Existing transaction management facilities within RDBMS have been designed and implemented to support short transactions. Most existing DBMS transaction management strategies make the assumption that a transaction will take a short time, and that other users wishing to update the same data will be prepared to wait until it is unlocked via a commit or rollback command. Traditional data processing applications generally need to lock a relatively small proportion of the database for a short period (of the order of seconds). However, in GIS, CASE and other design applications, transactions may be long, and may lock significant portions of the database. Consequently, pessimistic locking mechanisms are not suitable for long transaction management and a new technique must be used. Within SQL, the *commit* and *rollback* statements are used to indicate the end of a transaction.

In addition, SQL2 will support the concept of named checkpoints. After defining a checkpoint, subsequent rollback operations may name a particular checkpoint to roll back to. In addition to this facility, Bundock (1991) proposed explicit statements for managing long transactions.

4.3. Need for a GIS SQL3 ADT 'standard' package

The provision of generic facilities for the definition and use of ADTs will potentially make SQL3 a suitable vehicle for spatial data management. However, one can imagine the proliferation of spatial data types, variations in naming conventions, and variations in supported functions and attributes that could occur without an organized approach towards the use of ADTs. Most benefit for GIS end users will be gained if GIS vendors adopt a consistent way of using ADTs such that standard sets of ADTs are defined for different application areas. In particular, for GIS, standardization of ADT naming conventions, semantics and supported interfaces (methods and accessible attributes) will enable:

- simplified data transfer between GIS databases;
- use of the same software on multiple GIS databases;
- knowledge transportability between GIS products and applications;
- straightforward support of multiple DBMS products in conjunction with a single GIS product.

Gallagher (1991), in a discussion paper presented to SQL3 ISO and ANSI working parties, suggests that: 'it makes sense to standardize packages that have general scientific or engineering applicability. Some packages such as Geographic Information Systems, have broad appeal across different application areas and could benefit from "generic" standardization'. Stated benefits include the use of common sets of ADTs across multiple application areas, thereby promoting interoperability and the sharing of data, and encouraging performance optimization over a manageable collection of data types.

Development of a Spatial Language Interface for GIS 133

The specification of a SQL generic ADT package would not form an integral part of the SQL3 standard. Rather it might form an industry-specific standard recognized by ISO, and utilize other lower level generic ADT packages as building blocks. Gallagher (1991) proposes that all such ADT packages should be developed as a coordinated effort to avoid incompatible duplication of data types that may be used across a number of applications. He argues that now is the time to begin specification of the semantics, structures and operations for each ADT package. Such specifications would also assist development of the SQL3 standard by ensuring that all the appropriate issues are identified and resolved for the basic ADT facility.

4.4. Proposed spatial data types, operators and functions

4.4.1. Spatial data types

Using the structure outlined by Bundock (1991) the following spatial data types are proposed:

- *Point*: a position in n dimensional space
- *Node*: a *point* that provides (potential) topological connectivity
- *Link*: a linear feature terminated by two *nodes* and formed as an ordered set of *points* that define its geometry. The *link* may be any type of generalized arc, spline etc.
- *Chain*: a linear feature, being an ordered set of *links*.
- *Polygon*: a closed area delimited by an ordered set of *links*
- *Area*: an *area* is an associated set of *polygons*, some of which may form holes in the area

4.4.2. Functions for the spatial data types

For each of the data types identified above, a set of spatial operators will also be required. Bundock (1991) identified the requirement for spatial operators to be able to return not just scalar values, but also spatial data type values, lists, sets or entire relations. With the incorporation of set, list and record data types SQL3 will be able to support functions that return such values. Operator overloading will also ensure that a consistent naming convention is able to be used for operators requiring different internal behaviour with different input data types. Also note that inheritance results in subtypes possessing all the supertype functions, unless explicitly overridden.

Table 1 summarizes a list of useful operators for spatial data types. Note that with this implementation, no additional functions need be specified for set union, intersection and complement (difference) since these functions are available via the SET OF data type.

Table 1 Useful spatial operators for GIS in SQL3

Data type	Function	Type
Point	x coordinate(POINT)	Real
	y coordinate(POINT)	Real
	z coordinate(POINT)	Real
	line between(POINT,spatial)	Chain
	distance(POINT,spatial)	Real
	coordinates(POINT)	Real vector(n)
	buffer(POINT,REAL)	Area

(continues)

Table 1 (*continued*)

Data type	Function	Type
Point (*continued*)	north of(POINT,POINT) (also south, east, west)	Boolean
	coincident(POINT,POINT)	Boolean
	on(POINT,CHAIN)	Boolean
	inside(POINT,AREA)	Boolean
	near(POINT,POINT,REAL)	Boolean
Node	links attached to(NODE)	Set of link
	links ending at(NODE)	Set of link
	links beginning at(NODE)	Set of link
	connected(POINT,CHAIN)	Boolean
Link	coordinates(LINK)	List of real vector
	points(LINK)	List of point
	nodes(LINK)	List of node
	length(LINK)	Real
	line between(LINK,LINK)	Chain
	distance(LINK,LINK)	Real
	direction(LINK)	Bearing
	bounds(LINK)	Polygon
	remove point(LINK,POINT)	Link
	make current(LINK,POINT)	Point
	add point(LINK,POINT)	Link
	intersects(LINK,LINK)	Boolean
	inside(LINK,AREA)	Boolean
Chain	coordinates(CHAIN)	List of real vector
	nodes(CHAIN)	List of node
	links(CHAIN)	List of link
	length(CHAIN)	Real
	start point(CHAIN)	Point
	end point(CHAIN)	Point
	distance(CHAIN,CHAIN)	Real
	buffer(CHAIN,REAL)	Area
	line between(CHAIN,CHAIN)	Chain
	bounds(CHAIN)	Polygon
	intersects(CHAIN,CHAIN)	Boolean
	overlaps(CHAIN,AREA)	Boolean
	inside(CHAIN,AREA)	Boolean
	connected(CHAIN,CHAIN)	Boolean
Polygon	area(POLYGON)	Real
	bounds(POLYGON)	Polygon
	centroid(POLYGON)	Point
	adjacent(POLYGON,POLYGON)	Boolean
	union(POLYGON,POLYGON)	Polygon
	intersect(POLYGON,POLYGON)	Area
	difference(POLYGON,POLYGON)	Area
Area	area(AREA)	Real
	bounds(AREA)	Polygon
	centroid(AREA)	Point
	polygons(AREA)	Set of polygon
	polygon usage(AREA,POLYGON)	Polygon usage
	perimeter(AREA)	Real
	intersects(AREA,AREA)	Boolean
	overlaps(AREA,AREA)	Boolean

Table 1 (continued)

Data type	Function	Type
Area	inside(AREA,AREA)	Boolean
(continued)	adjacent(AREA,AREA)	Boolean
	union(AREA,AREA)	Area
	intersect(AREA,AREA)	Area
	difference(AREA,AREA)	Area

5. A GENERIC USER INTERFACE ENVIRONMENT

5.1. The underlying conceptual model

A major contributing factor towards the non-standardization of the GIS user interface is the lack of an underlying conceptual model for the interface. It has been suggested (Gould and McGranaghan, 1990) that the primary mechanism by which a user learns to use GIS is by metaphoric learning. Here the user is able to treat the unfamiliar environment like another, familiar, one, thus reducing the overall learning period. The general cognitive process may be partitioned into metaphoric, analogical and modelling processes. The differences between the three processes and their implications for computer systems design are reviewed by Wozney (1989). The concepts of metaphor and analogy are closely related: analogy implies that one domain behaves like another, whereas with metaphor, the target domain is more directly mapped onto the other and hence becomes the other. Consequently, the use of metaphor within the user interface is preferable since it allows users to interact with an unfamiliar system as if it is an environment with which they are familiar. This effectively reduces the learning time, reduces stress caused by unfamiliarity (i.e. makes for a happy user) and provides a conceptual framework for the new environment which may be built upon. For infrequent users, the use of metaphor may be more important, since they may never progress beyond the metaphor presented to develop a conceptual model of their own (Wozney, 1989).

Existing GUIs for non-GIS applications have often been developed using the desktop metaphor as the underlying conceptual model. The desktop metaphor is suitable for many business-related applications since the activities performed by the computer-based application have direct equivalents with the manual methods. However, it may not be readily applicable to many GIS applications, because of the lack of spatial and mapping related activities that normally occur on and around a desk.

The wide variation in GIS applications and the variation in experience of GIS users indicates that a single conceptual model is unlikely to satisfy or be applicable to all situations. If we perceive GIS to be an enabling technology for the integration of spatial and aspatial data, we must then consider it to be equivalent to a DBMS in generality, and hence not suited to a single model. In contrast, a GIS customized to suit a particular narrow application (e.g. mains fault analysis in the water industry) may provide a situation where an applicable underlying conceptual model may be utilized.

Wilson (1990) pointed out that some GIS applications may have no equivalent manual method. However, this does not imply that a conceptual model on which to base the user interface cannot be found. Rather it implies that analogy or metaphor may be suitable techniques for development of the conceptual model. Current GIS technology imposes on the user a conceptual model of geographic space that is a function of the internal structures supported by the GIS (e.g. layers, points, lines, polygons). What we should be aiming for is a user interface that permits the system customizer to present a conceptual model to the user that is relevant and applicable to both the user's background and the application in hand.

The strength of the desktop metaphor as used within the Macintosh and other PC environments for the underlying conceptual model is that it provides an organizing framework within which other operations and metaphors may exist. Gould and McGranaghan (1990) have extended this idea to suggest the need for an organizing metaphor, within which there may be other nested metaphors (which may themselves be organizing metaphors). This approach has promise since it provides a structure within which applicable and relevant metaphors may be applied, rather than trying to apply a single metaphor to all situations.

5.2. The organizing metaphor within UGIX

The UGIX GUI design is to develop an environment supporting nested metaphors. The proposed overall organizing metaphor is a building, within which there is a number of rooms, each accessible via a door (Fig. 4). It should be noted that the idea of using the room/building metaphor has been conceived previously by a number of different groups, including experimenters at the University of Waterloo (Chan and Malcolm, 1984) and built into a number of existing products (e.g. Rooms from ENVOS, and even X11 revision 4 attempts to provide a Rooms-like system).

Within UGIX, each room may possess its own organizing metaphor. Most rooms will be directly accessible from the entrance hall, although some special-purpose rooms may require access from within another room. On entering the system the user is located in the entrance hall, a neutral, public space through which the user moves to a particular environment. Doors provide access to the environments the user has access to. The door metaphor is a strong one for access into and out of different environments (Cátedra, 1990) and

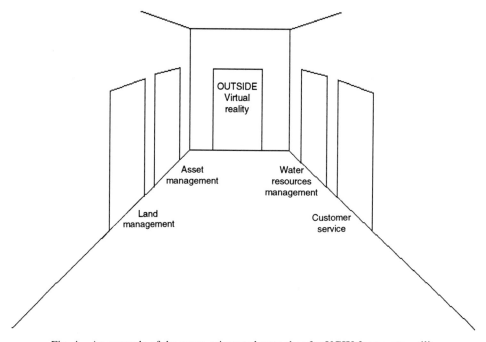

Fig. 4   An example of the room-orientated metaphor for UGIX for a water utility

provides features such as locks, opening and closing. These features may be used directly for access control, entering and leaving. Within each room, a single type of activity occurs, and a single lower-level organizing metaphor is employed. For example, one room provides an environment where the desktop metaphor is supported. Here general filing, correspondence and interfacing to external non-GIS packages (e.g. word processing, spreadsheets) takes place. Access to the aspatial part of the GIS database is available via card indexes, file folders etc. and alphanumeric reports can be created and printed.

A second room might contain the drafting office, where the map, drafting table, map cabinet and light table are the principal metaphors. Here access to the GIS database is via the map. Maps may be taken out of the filing cabinet, updated, viewed and copies taken for development proposals, etc. Eventually these may be replaced in the map cabinet following approval by the chief draftsman/engineer. Note that operations not consistent with the map metaphor may not be applied here. A third room might contain a library. Within the library, various books are stored, providing reference material, reports, archives and documentation. Updating of system documentation is performed here. A fourth room might contain the development and customization environment. A workstation metaphor is supported, which provides direct access to the fourth generation language development and customization environment.

There is one special door that leads off the entrance hall – an external door. Through the door, blue sky and clouds may be glimpsed, and on entering this environment the real-world metaphor is used for access to the GIS database. Here, there are no seams, maps or files – only a continuous world containing objects. This is where the experienced GIS user works and it is also the environment in which virtual reality may one day be accessible (Jacobson, 1990).

Users are provided with keys that are able to unlock only some doors. It is feasible to consider more specialized rooms leading off others. For example the database administrator and system manager may be in their very own room, accessible from the development and customization area. The concepts of buildings, rooms and doors are internationally and culturally neutral, providing an almost universally understood concept. A further advantage exhibited by this metaphor is that it provides an easy facility for extension. To add new environments involves simply adding another room (with door!) to the building. Within each new environment a different organizing metaphor may be used to support functions not supported elsewhere.

The possibilities of this metaphor are seemingly endless – e.g. leaving the system may simply be performed by turning off the light switch in the entrance hall, or alternatively going through the door that leads out into the night.

5.3. The iconic query language

Access to the functionality provided within each environment (room) will be predominantly via icons. The icons should fit the organizing metaphor for that environment so that they have relevance and preferably direct applicability. Consequently, the drafting office will support map cabinets from which maps may be extracted, drafting boards on which map updates and viewing may be performed and light tables on which map overlay operations may be carried out.

Consistency and simplicity are key considerations when attempting to design a user interface, whether for a GIS or for a dishwasher. A concise and simple syntax for manipulating the icons and database objects is required, and it must be both consistent and meaningful in terms of the metaphors used. Existing GUIs such as that used on the Macintosh are not fully consistent. Consistency between applications has been encouraged, because Apple provides a set of guidelines for developers to follow (Apple's human interface guidelines; Apple Computer, 1987). Consequently, most packages available for the Mac have a similar look and feel, so knowledge of one application or package provides useful knowledge on the use of others.

Certain other aspects of the Mac interface are far less consistent. In particular, the order in which the objects and the operators are selected varies from one type of operation to another. Objects to be manipulated are usually selected first, and then the operation to be applied is selected (e.g. discarding data by moving it to the wastebasket, applying a different font and ruler to a section of text). However, sometimes the operation to be performed is selected first and then the objects to which it is to be applied are identified (e.g. select the print function and then indicate which pages to print). Hence, even though operations that are common between applications are normally presented in a very similar manner the syntax for different operations may vary within an application.

A further level of inconsistency, most frequently observed by novices, is the use of the wastebasket. Why is the trash can used to eject the disk when it is normally used for deleting data? Most novices are unable to find a method for ejecting the disk, since the use of the wastebasket for deleting documents and folders implies that if the diskette is moved to the wastebasket, all folders and documents on the disk will be deleted. This latter example indicates where the use of a metaphor has been extended beyond its applicability and used in an inconsistent manner.

The impact of the English language on the syntactic ordering of operations, parameters and objects (verbs, modifiers and nouns) may not have relevance to the iconic interface. Although it is known that language structures our concept of space, it is not thought that language will adversely impact the syntactic structure of the iconic interface. In English we generally use a noun–verb–modifier ordering to state facts (e.g. Jack closed the door), but a verb–noun–modifier ordering for instructions or commands (e.g. close the door quietly, please). Most of the operations performed within the GIS tend to be instructional in that the user is commanding the system to perform some action such as modifying, deleting, or reporting, supporting the adoption of a verb–noun–modifier ordering.

However, most iconic interfaces require that the objects are selected prior to identification of the action (i.e. noun–verb–modifier, or object–action ordering). Even though this ordering is not common within the English language for instructional sentences, it does feel natural for English speaking users of the iconic interface.

Perhaps the most important aspect of this ordering is that object selection and the operations to be performed on those objects are effectively separated. They have become two discrete instructions issued by the user. Furthermore, object selection is common to virtually all operations and becomes independent of those operations, meaning that a single set of object selection techniques can be applied throughout.

One significant disadvantage to the use of object–action syntax ordering is that the selection process may select objects for which the operation may be invalid. If the operation to be performed is identified first, the object selection process can use knowledge of the operation to ensure that only appropriate (valid) objects are selected. Within existing GUIs, this problem is partially overcome by disabling functions for which the selected data is inappropriate. However, if it is not obvious which of the selected objects is causing the function to become unselectable, much operator frustration is likely to ensue.

Within the UGIX GUI, we recommend the use of object–action ordering as the basic syntactical construct for icon interaction. Object selection will be performed first.

Development of a Spatial Language Interface for GIS 139

Subsequent identification of an action will apply that action to the selected set of objects.

Fig. 5 shows an example room within the UGIX GUI environment. The furniture shown in the room is familiar, generic, storage furniture that can be used in any room to signify available data for the tasks appropriate to that room. The filing cabinets represent the store of accessible 'objects' stored in the SQL3 relational database: the drawers may be used to distinguish between tables or views. The 'objects' are rows in the tables/views where the

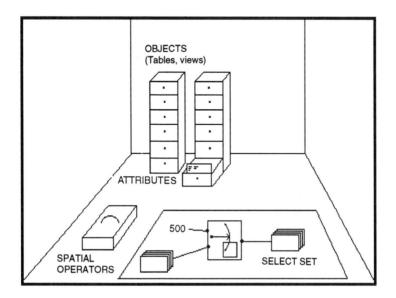

Fig. 5   An example room in the UGIX GUI environment with spatial query 'furniture'

columns may include 'normal' or abstract data types – hence the rows contain all the data, both spatial and non-spatial, describing the object. Opening the drawers allows the user to pull out sets of objects that may be queried using the spatial operators in the toolbox. The 'playpen' on the floor of the room is the area where the objects and the operators are brought together using the object–action syntax.

In the example, the objects taken from the drawers are queries using the 'within_N_of' spatial operator, where N is expressed in metres. The query as shown could be expressed as 'select all schools within 500 m of a trunk road', where the output is a 'select set' that can be reported as a table or displayed as a map. Although this design relies on some knowledge of the environment it is not considered to be any more complex than using a modern desktop publishing package, and well within the scope of the average SAP. Clearly, while a number of constraints and optimization procedures would need to be built in, the construction of such a system is practical given the availability of a stable SQL3 database environment.

6.   SUMMARY AND CONCLUSIONS

In emphasizing the need for an end-to-end design, starting at the database and moving towards the users, the UGIX project has shown exactly how a generic interface to GIS can be created by utilizing metaphorically based screen environments communicating with an underlying database using SQL3. This kind of interface can meet the requirements of many

SAPs who would like to be able to exploit GIS today, but who find the interface to current GIS an obstacle. This approach emphasizes the essential connectivity between the database and the screen metaphor: it is asserted here that experiments at either end of this channel of communication that do not consider this end-to-end linkage are unlikely to be implementable in real application environments within even the medium term. If a solution to GIS use is not found in the short term, the whole development of the discipline and practice of GIS may be harmed.

This seems to bring into question the role and practice of GIS research, since it is clear that both science and technology are interwoven in the 'user interface problem'. In this field of research the crucial medium for knowledge accumulation and sharing is the (spatial) database. However, the development of the (relational) database is now the subject of an intense technological debate, driven as much by market forces and software trends as by science. The UGIX project has aimed to contribute to this debate by submitting design statements for the SQL language dictated by the needs of interface development, in the hope that this will improve the chances of a technological step being taken which supports scientific progress. If SQL3 is approved in something like its current (1993) form then much research will be possible/required to explore its scope. This project has aimed to define the way in which the components of a communication channel between user and database can be constructed in the anticipated light of a new standard in the technological foundation of the database. If SQL3 databases become commercially available and accepted then it is likely that most research activity in the fields of query design, task definition and screen design will re-orient to this new opportunity. It seems worthwhile and appropriate to sketch out the consequences of this development and to design an approach to user interfaces that takes this into account.

The achievements of this research can be summarized as follows. The detailed specification of an approach to SQL3 design has been submitted to ISO through the AGI Standards Committee, with the aim of maximizing the suitability of this development for GIS applications. Software has been written and implemented to parse and process SQL3 queries with the aim of defining the potential and constraints on the operation of such querying procedures in a GIS environment. Software has been written and implemented to experiment with the operation of CASE tools for GIS database design and construction, as part of a customization environment that will be needed if SQL3-type databases are developed. Screen designs based upon a 'rooms' metaphor have been created to indicate the manner of access to the database that the user would have. Finally, contributions in person and in writing have been made to three international research initiatives in this field, documenting the activities of the UGIX researchers.

A number of objectives remain, which can be summarized as follows. The software has yet to be fully integrated, because the SQL3 database tools are not yet available. The scope of the SQL-SX approach to SQL3 has yet to be defined formally, and query optimization routines have yet to be formulated. Finally, UGIX has not yet been formally tested to define the scope of the interface and the nature of the feedback between the interface and the functionality. These objectives will be addressed in a future research project.

ACKNOWLEDGEMENTS

The authors acknowledge receipt of a grant from the ESRC/NERC Joint Programme in Geographical Information Handling, and assistance in the form of software tools and hardware from Smallworld Systems Ltd of Cambridge.

Valuable discussions resulted from participation in the following meetings: NATO Advanced Study Institute 'Cognitive and Linguistic Aspects of Geographic Space', organized by David Mark and Andrew Frank (Las Navas del Marqués, Spain, 8–20 July 1990); EC ESPRIT working group n.3191: 'Basic research actions for object-oriented database systems', organized by Giorgio Ausiello (Capri, Italy, 15–17 May 1991); and NCGIA Initiative 13 Discussion Meeting 'User Interfaces for GIS' (Buffalo, NY, 22–26 June 1991), organized by David Mark and Andrew Frank.

We would also like to thank Lesley Bundock for her assistance in the preparation of this paper, Tim Linsey for insights gained in the HyperArc project, Bruce Wright from the National Mapping Division of the US Geological Survey for the review he carried out of UGIX as the Senior Visiting Fellow, Barry Williams for stimulating discussions on UGIX, and David Livingstone for assistance with the maintenance of the Smallworld environment and discussions on the UGIX design.

REFERENCES

Abel, D.J. (1988) *CSIRO-DBMS User's Manual*. Technical Report TR-HA-88-1, CSIRO Division of Information Technology.

Apple Computer (1987) *Human Interface Guidelines*. Addison-Wesley, Amsterdam.

Ausiello, G. (ed.) (1991) *Int. Workshop on Database Management Systems for Geographical Applications* (Basic Research Actions for Geographical Object-Oriented Database Systems). University of L'Aquila, Rome.

Bauer, J. (1991) *Module Enhancements to Support RDA Requirements*. ANSI X3H2-91-014; ISO DBL ARL-012, February.

Berman, R.R. and Stonebraker, M. (1977) GEO-QUEL, a system for the manipulation and display of geographic data. *ACM Comput. Graphics*, **11**(2).

Bundock, M.S. (1991) SQL-SX: Spatially extended SQL – becoming a reality. *Proc. 2nd European GIS Conf. (EGIS'91)*, Brussels, 2–5 April.

Bundock, M.S. and Raper, J.F. (1991) GIS customisation: from tools to efficient working systems. *Proc. Mapping Awareness '91 Conf.*, London, 6–8 February, 101–114

Bundock, M.S. and Raper, J.F. (1992) Towards a standard for spatial extensions for SQL. *Proc. 3rd European GIS Conf.*, Munich, 23–26 March, 287–304.

Cátedra, M. (1990) Through the door: a view of space from an anthropological perspective. In Mark, D.M. and Frank, A.U. (eds.), *Proc. NATO ASI Cognitive and Linguistic Aspects of Geographic Space*, Las Navas del Marqués, Spain, 8–20 July. Kluwer, Dordrecht, 53–64.

Chan, P.P. and Malcolm, M.A. (1984) Learning considerations in the Waterloo port user interface. *Proc. IEEE 1st Int. Conf. Office Automation*. IEEE, New York.

Chang, S.K. and Fu, K.S. (1980) Query-by-pictorial-example. *IEEE Trans. Software Engng*, **6**(6), 519–524

Charlwood, G. Moon, G. and Tulip, J. (1987) Developing a DBMS for demographic information: a review. *Proc. Auto-Carto 8*, Baltimore, Maryland.

Codd, E.F. (1988) Fatal flaws in SQL. *Datamation*, **34**(16), 45–48.

Date, C.J. (1986) *Relational Database: Selected Writings*. Addison-Wesley, Reading, Massachusetts.

Date, C.J. (1989) *A Guide to the SQL Standard*, 2nd edn. Addison-Wesley, Reading, Massachusetts.

Dowers, S. (1991a) Recent developments in SQL3. Paper SCA30 presented to AGI Standards Committee Working Group A, November.

Dowers, S. (1991b) Using SQL for Geographic Data. ISO/IEC JTC1.21.3.3 SQL3 ARL-026, Discussion paper presented at ISO SQL3 Working Group meeting, Arles.

Easterfield, M.E., Newell, R.G. and Theriault, D.G. (1990) Version management in GIS – applications and techniques. *Proc. 1st European GIS Conf. (EGIS'90)*, Amsterdam, 10–13 April, 288–297.

Egenhofer, M.J. (1987) An extended SQL syntax to treat spatial objects. *Proc. 2nd Int. Semin. Trends and Concerns of Spatial Sciences*, New Brunswick.

Egenhofer, M.J. (1992) Why not SQL? *Int. J. Geogr. Inf. Sys.*, **6**(2), 71–85.

Frank, A.U. (1982) MAPQUERY: Data base query language for retrieval of geometric data and their graphical representation. *ACM Comput. Graphics*, **16**(3), 199–207.

Frank, A.U. (1991) Beyond query languages for geographic query languages: data cubes and maps. In Ausiello, G. (ed.), *Int. Workshop on Database Management Systems for Geographical Applications* (Basic Research Actions for Geographical Object-Oriented Database Systems). University of L'Aquila, Rome, 74–86.

Gallagher, L. (1991) *SQL 'Generic ADT' Packages.* ANSI X3H291-204; ISO DBL KAW-17, August.

Gould, M.D. and McGranaghan, M. (1990) Metaphor in geographic information systems. *Proc. 4th Int. Symp. Spatial Data Handling*, Zürich, 433–442.

Gradwell, D.J.L. (1990) Can SQL handle geographic data? A presentation on the work of the AGI Standards Committee's SQL Working Party. *Proc. AGI '90*, Brighton, October. AGI, London, D.3.1–D.3.12.

Green, M. (1985). Report on dialogue specification tools. In Pfaff, G.E. (ed.), *User Interface Management Systems.* Springer-Verlag, London, pp. 9–20.

Herring, J.R., Larsen, R.C. and Shivakumar, J. (1988) Extensions to the SQL query language to support spatial analysis in a topological data base. *Proc. GIS/LIS'89 Conf.*, San Antonio, Texas, November, 741–750.

ISO (1987) *Report ISO 9075 Information Processing Systems – Database Language SQL*

ISO (1989) *Report ISO 9075 Information Processing Systems – Database Language SQL*

ISO/IEC (1991) *Common Language-Independent Datatypes (CLID).* Working Draft no. 5 ISO/IEC CD 11404, JTCl/SC22/WGll N233, May.

Jacobson, R. (1990) Virtual worlds, inside and out. In Mark, D.M. and Frank, A.U. (eds), *Proc. NATO ASI Cognitive and Linguistic Aspects of Geographic Space*, Las Navas del Marqués, Spain, 8–20 July. Kluwer, Dordrecht, 507–514.

Jian-Kang Wu, Tao Chen and Li Yang (1989) QPF. A versatile query language for a knowledge-based geographical information system. *Int. J. Geogr. Inf. Sys.*, **3**(1), 51–57.

Kerridge, J.M. (1989) A proposal to add user defined data types to SQL. ANSI X3H2-89-319; ISO DBL FIR-37, September.

Kuhn, W., Willauer, L., Mark, D.M. and Frank, A.U. (1992) User interfaces for GIS: discussions at the specialist meeting. *Report on the Specialist Meeting, NCGIA Research Initiative 13*, 22–26 June. NCGIA Report 92/3. National Center for Geographic Information and Analysis, University of California at Santa Barbara.

Linsey, T.K. and Raper, J.F. (in press) A task-oriented hypertext GIS interface. *Int. J. Geogr. Inf. Sys.*, in press.

Maguire, D.J., Goodchild, M.F. and Rhind, D.W. (eds.) (1991) *Geographic Information Systems: Principles and Applications.* Longman, Harlow.

Mark, D.M. and Frank, A.U. (eds.) (1990) *Proc. NATO ASI Cognitive and Linguistic Aspects of Geographic Space*, Las Navas del Marqués, Spain, –20 July. Kluwer, Dordrecht.

Medyckyj-Scott, D., Davies, C. and Byrne, V. (1993) Discovering the trials of GIS use: the USIS study of GIS usability. *Proc. GIS Res. UK*, University of Keele, 18–20 March, 170–179.

Medyckyj-Scott, D. and Hearnshaw, H. (eds) (1993) *Human Factors in GIS.* Belhaven, London.

Melton, J. (ed.) (1991) *Database Language SQL2/SQL3.* ANSI X3H2-91-183; ISO DBL KAW-003, JTCl/SC21/WG3 N1223, July.

Openshaw, S., Cross, A., Charlton, M. and Brunsdon, C. (1990) Lessons learned from post mortem of a failed GIS. *Proc. AGI'90 Conf.*, Brighton, paper 2.3.

Pong-Chai Goh (1989) A GQL for cartographic and land information systems. *Int. J. Geogr. Inf. Sys.*, **3**(3), 245–255.

Raper, J.F. (1991) User interfaces for GIS. In Masser, I. and Blakemore, M. (eds), *Geographic Information Management: Methodology and Applications.* Longman, Harlow.

Raper, J.F. and Bundock, M.S. (1990) UGIX: A layer based model for a GIS user interface. In Mark, D.M. and Frank, A.U. (eds), *Proc. NATO ASI Cognitive and Linguistic Aspects of Geographic Space*, Las Navas del Marqués, Spain, 8–20 July. Kluwer, Dordrecht, 449–475.

Raper, J.F. and Bundock, M.S. (1991) UGIX: A GIS-independent user interface environment. *Proc. Auto-Carto 10*, Baltimore, Maryland, 25–28 March, 275–295.

Raper, J.F. and Green, N.P.A. (1992) Teaching the principles of GIS: lessons from the GISTutor project. *Int. J. Geogr. Inf. Sys.*, **6**(4), 279–290.

Raper, J.F., Linsey, T.K. and Connolly, T. (1990) UGIX – a spatial language interface for GIS: concept and reality. *Proc. 1st European GIS Conf. (EGIS'90)*, Amsterdam, 10–13 April, 876–882.

Raper, J.F. and Maguire, D. (1992) Design models and functionality in GIS. *Comput. Geosci.*, **18**(4), 387–394.

Rhind, D.W, Raper, J.F. and Green, N.P.A. (1989) First UNIX then UGIX. *Proc. Auto-Carto 9*, Baltimore, Maryland, 2–7 April. American Society of Photogrammetry and Remote Sensing, Falls Church, Virginia, 735–744.

Sacks-Davis, R., McDonell, K.J. and Ooi, B.C. (1987) GEOQL – A query language for geographic information systems. Internal Report no. 87/2, Department of Computer Science, Royal Melbourne Institute of Technology.

Smith, T.R. (1993) Integrating high-level modelling of spatial phenomena into GIS. *Proc. GIS Res. UK*, University of Keele, 18–20 March, 1–13.

Stonebraker, M., Wong, E., Kreps, P. and Held, G.D. (1976) The design and implementation of INGRES. *ACM Trans. Database Systems*, **1**(3).

Westwood, K. (1989). Toward the successful integration of relational and quadtree structures in a geographic information system. *Proc. Natl Conf. GIS – Challenge for the 1990s*, Ottawa.

Wilson, P.M. (1990) Get your desktop metaphor off my drafting table: user interface design for spatial data handling. *Proc. 4th Int. Symp. Spatial Data Handling*, Zürich, Switzerland.

Wozney, L.A. (1989) The application of metaphor, analogy and conceptual models in computer systems. *Interacting with Computers*, **1**(3), 273–283.

Zloof, M.M. (1977) Query-by-example: a database language. *IBM Sys. J.*, **16**(4), 324–343.

Section II

ENVIRONMENTAL APPLICATIONS OF GIS

Chapter Thirteen

Analytical tools to monitor urban areas

M. J. BARNSLEY AND S. L. BARR
Department of Geography, University College London

A. HAMID AND J. -P. A. L. MULLER
Department of Photogrammetry and Surveying, University College London

AND

G. J. SADLER AND J. W. SHEPHERD
Department of Geography, Birkbeck College, London

Remotely-sensed images represent an important source of spatial information on urban areas. However, while sensor technology has improved considerably over the last two decades, the techniques used to extract the desired information from the resultant images have not always developed accordingly. This chapter therefore describes the development and implementation of a set of new analytical tools designed to extract various types of spatial information relevant to urban areas from digital, remotely-sensed images. These include techniques to infer land use and land use change from an examination of the spatial arrangement of land cover types in satellite sensor images, and a system to measure the height and volume of individual buildings from stereoscopic pairs of digitized aerial photographs. The latter can be used to derive three-dimensional CAD models of buildings for dynamic visualization of urban scenes or for direct input into existing GIS databases. The utility of these new tools is demonstrated using Systeme Probatoire d'Observation de la Terre (SPOT)-HRV multispectral data of the London borough of Bromley and 1:5 000 scale aerial photography of central London.

1. INTRODUCTION

Information about urban areas is required at a variety of spatial scales and temporal frequencies, ranging from broad statistics on land use and land use change for strategic planning purposes at national and county levels, to the more detailed types of information required by local authorities. While much of this information can be derived from existing paper maps, these soon become outdated, particularly in rapidly changing environments such as urban areas. Detailed ground surveys provide another, important, source of information; however, these are usually labour-intensive, time-consuming and, as a result, are carried out relatively infrequently. Remote sensing represents a third source of information

---

Geographical Information Handling – Research and Applications. Edited by P. M. Mather
© 1993 John Wiley & Sons Ltd

on urban areas. In principle, images recorded by airborne and satellite-based sensors can be obtained at reasonably frequent intervals and at a favourable cost per unit area in comparison with traditional ground survey (Martin, Howarth and Holder, 1988; Ehlers, 1990). Remotely-sensed images can be processed to produce important basic data sets and combined with other types of spatially referenced data within an integrated GIS to generate further, value-added products. Unfortunately, while sensor technology has improved considerably over the last two decades, the techniques used to extract the desired information from the resultant images have not always developed accordingly. This chapter therefore describes the development and implementation of a set of new analytical tools designed to extract various types of spatial information relevant to urban areas from digital, remotely-sensed images.

In Section 2, discussion focuses on two closely related techniques designed to provide improved information on land use and land use change within and around urban areas from satellite sensor images. These techniques use the complex spatial mixture of spectrally distinct land cover types within urban areas as a means of inferring land use. The first is based on a simple convolution kernel which is passed across the satellite sensor image after it has been segmented into regions of uniform land cover type. The second identifies discrete land cover 'objects' within this initial segmentation and uses information on their size, shape and spatial arrangement to determine the nature of land use in different parts of the image.

Section 3 deals with information extraction at a very different spatial scale. More specifically, it discusses the development of a tool for measuring the three-dimensional structure of urban areas (i.e. building heights and volumes) from pairs of digitized, stereoscopic aerial photographs. The resultant stereo digitizing system is used to derive three-dimensional CAD models of buildings, both for dynamic visualization of urban scenes and for direct input into existing GIS databases. A prototype system has been developed which integrates these data – in the form of wireframe models and texture maps – with more conventional two-dimensional data in a three-dimensional GIS, incorporating 'point-and-click' techniques and a simple graphical user interface (GUI).

2. MONITORING URBAN LAND USE FROM SATELLITE SENSOR IMAGES

2.1. Problems in determining land use from satellite sensor images

The information that satellite sensor images might reasonably be expected to provide about urban areas includes data on their physical extent (i.e. their location and perimeter) and on the range of land cover types and land use categories present, together with their spatial distribution throughout the scene. Unfortunately, attempts to extract information of this kind have often failed to produce the levels of accuracy and detail required for town planning purposes (Barnsley, Shepherd and Sun, 1988; Sadler and Barnsley, 1990). Initially, this was attributed to the relatively low spatial resolution of early satellite sensors which, among other things, hampered the accurate delineation of the urban–rural boundary (Jackson *et al.*, 1980; Forster, 1980). However, data from current sensors with enhanced spatial resolving power have not always yielded the improvements anticipated (Toll, 1985; Forster, 1985; Martin, Howarth and Holder, 1988; Townshend, 1992). Indeed, the level of accuracy with which urban land use categories have been identified in high spatial resolution images has often been lower than in previous studies using coarser resolution data (Haack, Bryant and Adams, 1987; Martin, Howarth and Holder, 1988).

The fundamental problem involved in producing accurate land use maps from satellite sensor images is that urban areas represent complex spatial assemblages of a disparate set of land cover types – including man-made structures, numerous vegetation types, bare soil and water bodies – each of which has different spectral reflectance characteristics (Gong and Howarth, 1990; Barnsley, Barr and Sadler, 1991). Thus, while there is often a relatively simple, direct relationship between land cover and the spectral response detected by a satellite sensor, this is seldom true for land use (Gastellu-Etchegorry, 1990; Gong and Howarth, 1990; Barnsley, Barr and Sadler, 1991; Barnsley and Barr, 1992; Gong and Howarth, 1992). This problem becomes more pronounced as the spatial resolution of the sensor increases, since discrete scene elements (e.g. buildings, roads and open spaces) begin to dominate the detected reflectance of individual pixels (Woodcock and Strahler, 1987). As a result, the spectral response of urban areas becomes more heterogeneous, making consistent classification problematic (Gastellu-Etchegorry, 1990).

Although it is tempting to explain this problem in terms of inadequate or inappropriate sensor spatial resolution, it is perhaps more accurately expressed in terms of the limitations of commonly used information extraction techniques. In particular, conventional, per-pixel classification algorithms are poorly equipped to deal with the spatial variability in detected spectral response inherent in high spatial resolution images of urban areas (Woodcock and Strahler, 1987; Barnsley, Barr and Sadler, 1991; Barr, 1992). This is because they assign each pixel in the image to one of the candidate classes solely on the basis of its spectral reflectance properties, so the location of the pixel and the relationship of its spectral response to that of its neighbours are not taken into account. Similarly, it is extremely difficult to define suitable training areas for many categories of urban land use, due to the variation in the spectral response of their component land cover types (Forster, 1985; Gong and Howarth, 1990; Barnsley, Barr and Sadler, 1991). Thus, where training statistics are derived from contiguous blocks of pixels, they frequently exhibit both a multimodal distribution and a large standard deviation in each spectral waveband (Sadler, Barnsley and Barr 1991). The implication of the former is that the training statistics for urban areas violate one of the basic assumptions of the widely used maximum-likelihood decision rule, namely that the pixel values for each class follow a multivariate normal distribution. The effect of the latter is to produce a pronounced overlap between urban and non-urban land use categories in the multispectral feature space.

A variety of techniques has been used in an attempt to overcome this problem, including:

- Pre-classification image transformations and feature extraction techniques, such as median filters and various measures of image texture (Haralick, 1979; Atkinson *et al.*, 1985; Baraldi and Parmiggiani, 1990; Franklin and Peddle, 1990; Gong and Howarth, 1990; Sadler, Barnsley and Barr, 1991).
- Multi-temporal image data sets (Griffiths, 1988; Martin, Howarth and Holder, 1988; Franklin and Peddle, 1990).
- Incorporation of spatially referenced, ancillary data into the classification procedure (Barnsley, Sadler and Shepherd, 1989; Sadler and Barnsley, 1990; Ehlers *et al.*, 1991; Sadler, Barnsley and Barr, 1991).
- Spatial and contextual classification algorithms (Gurney, 1981; Gurney and Townshend, 1983; Gong and Howarth, 1989, 1992).

- Post-classification spatial processing, ranging from simple majority filters to spatial re-classification procedures (Wharton, 1982a, b; Gurney and Townshend, 1983; Gong and Howarth, 1990; Whitehouse, 1990; Guo and Moore, 1991; Barnsley, Barr and Sadler, 1991; Barnsley and Barr, 1992; Gong and Howarth, 1992).

However, few of these techniques directly address the problem of inferring land use from a complex spatial mixture of spectrally distinct land cover types. For example, pre-classification spatial filtering attempts to circumvent the problem by suppressing the spatial variability within the image (Sadler, Barnsley and Barr, 1991). This is achieved only at the expense of a reduction in the effective spatial resolution of the data set. It also produces somewhat arbitrary mean vectors for urban land use categories, by aggregating the detected spectral response of their component land cover types.

Contextual classification algorithms offer a more direct solution to the problem of inferring land use, since they employ information on the expected size, shape and location of discrete objects within the scene, as well as the spatial relationships between them (i.e. their context) to guide the classification process (Mehldau and Schowengerdt, 1990; Gong and Howarth, 1992). Use of contextual classification algorithms to map specific categories of urban land use therefore requires the accurate identification of their constituent land cover types, together with the development, formalization and encoding of rules to infer land use on the basis of particular mixtures and spatial arrangements of these cover types. Not surprisingly, this is extremely difficult to implement in a single-pass algorithm.

A simpler, but related, approach is to divide the classification process into two stages, the first involving a standard per-pixel segmentation of the scene on the basis of land cover, and the second involving some form of post-classification spatial processing of the resultant data set to infer land use. This has been termed spatial or contextual re-classification (Gurney, 1981; Wharton, 1982a, b; Gurney and Townshend, 1983). The assumption underlying this approach is that individual categories of land use have characteristic mixtures of spectrally distinct land cover types that enable their recognition in high spatial resolution images (Wharton, 1982a, b; Barnsley, Barr and Sadler, 1991; Barnsley and Barr, 1992). For example, residential districts in many western European cities are characterized by the intermixing of roofs, roads and gardens.

In the remainder of this section, two separate spatial re-classification techniques are developed and their utility for monitoring urban land use is explored. The first technique, referred to as pixel-based spatial re-classification (Gurney and Townshend, 1983; Barnsley and Barr, 1992), uses a simple convolution kernel which is passed across the satellite sensor image after it has been segmented into regions of uniform land cover type. The second technique, termed object-based spatial re-classification (Gurney and Townshend, 1983; Barr and Barnsley, 1993), identifies discrete land cover 'objects' within the initial image segmentation and uses information on their size, shape and spatial arrangement to determine the nature of land use in different parts of the image.

2.2. Development of a pixel-based spatial re-classification procedure

The work of Wharton (1982a, b) provides an early example of pixel-based spatial re-classification, in which the initial land cover segmentation is performed using a standard, unsupervised classification algorithm. The frequency of different land cover types within a 3×3 pixel block is then calculated by convolving a simple kernel with the classified image. The

Analytical Tools to Monitor Urban Areas

land use category at the location corresponding to the centre of the kernel is derived using a further unsupervised, nonparametric clustering procedure applied to these frequency data. Similar techniques have also been used more recently by Whitehouse (1990), Guo and Moore (1991) and Gong and Howarth (1992) although in these studies the frequency distribution of land cover types surrounding each pixel is compared with those of known areas of the candidate land use categories.

Although Wharton's method examines the frequency with which different class labels occur within the kernel, no account is taken of their spatial arrangement. The limitation of this approach is evident in the following example. Consider two separate 3 × 3 pixel windows, each of which has four pixels labelled as the land cover class 'Building'. In an industrial or commercial district – where these might represent a single large factory or warehouse – the pixels are likely to be clustered together in a block (Fig. 1a). By contrast, in a residential area – where the same class labels might represent individual houses – the 'Building' pixels might be arranged in a line (terraced housing) or might be physically separate (detached housing) (Fig. 1b). However, a procedure which simply calculates the frequency of different class labels within these windows has no means of distinguishing between these conditions. This example illustrates the need to find a simple method for recording both the frequency *and* the spatial arrangement of class labels within any given region of an image. One way to do this is to record the number of times that different class labels occur next to one another within a pre-defined, moving window. A simple technique to achieve this, referred to as the SPAtial Re-classification Kernel (SPARK), has been developed as part of this study (Barnsley, Barr and Sadler, 1991; Barnsley and Barr, 1992).

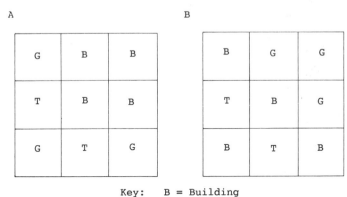

Fig. 1 Simulated 3 × 3 pixel window indicating the possible spatial arrangement of different land cover types within: (a) an industrial or commercial area, and (b) a residential district

2.2.1. The SPAtial Re-classification Kernel (SPARK)

SPARK operates by examining pairs of adjacent pixels within a rectangular kernel (i.e. those connected along an edge or by a vertex), the size of which is selected by the user. The class (land cover) label associated with each pixel is noted and is used to determine the frequency with which different cover types are adjacent to one another within the kernel.

Thus, in Fig. 1a, there are six occasions on which 'Building' pixels are adjacent to one another (hereafter referred to as a Building–Building adjacency). In the same window there are also four occurrences of Building–Tree adjacency, and so on. By comparison, although the kernel in Fig. 1b contains exactly the same number of pixels belonging to each class, there are only three occurrences of Building–Building adjacency, but six of Building–Tree. This demonstrates that, unlike the methods developed by Wharton (1982a, b) and Gong and Howarth (1992), SPARK is sensitive to the spatial arrangement of land cover types within a window, as well as to their absolute frequency. It might, therefore, be expected to distinguish subtle differences in land use.

In practice, SPARK is convolved with the land cover image to produce an adjacency vector, $A_{i,j}$, for each pixel:

$$A_{i,j}(f_{1,1}, f_{1,2}, f_{1,3}, \ldots f_{x,y}, f_{y,y}) \tag{1}$$

where the subscripts i and j denote the position of the pixel within the image. The value of each element, $f_{x,y}$, of the vector denotes the frequency with which pixels belonging to class x are adjacent to those belonging to class y, for the current position of the kernel. The length L of the vector is conditioned by the number of classes c in the image:

$$L = \sum_{m=0}^{c} (c - m) \tag{2}$$

Thus, for an eight-class image, each vector will contain 36 elements *irrespective of the kernel size*. For most studies, where the number of land cover classes is reasonably small, this represents an efficient means of storing information about the spatial arrangement of the land cover types within the image.

2.2.2. Comparison of adjacency vectors and land use 'template' vectors

The land use category of a given pixel is determined by comparing its adjacency vector with those derived from representative sample areas of the candidate land use categories; the latter will be referred to as template vectors. The sample areas used to generate each template vector are the same size as the kernel used in the spatial re-classification stage (i.e. 3×3, 5×5 or 7×7 pixels, etc.). Multiple template vectors can be defined for each land use. These may either be used independently or pooled to produce an average template vector. The advantage of using a series of independent templates for a single land use is that subtle variations in the spatial arrangement of its constituent land cover types at different locations within the image can be taken into account. However, it also results in a linear increase in computation time.

As the spatial re-classification kernel is passed over the image, the current adjacency vector is compared with each of the land use templates using the following equation:

$$\Delta_k = 1 - \left[\frac{1}{2(N^2)} \sum_{n=0}^{L} (A_{i,j}[n] - T_k[n])^2 \right]^{0.5} \tag{3}$$

$$0 \leq \Delta_k \leq 1 \qquad (4)$$

where Δ_k is a measure of similarity between the current adjacency vector and the template vector for land use category k, $A_{i,j}(n)$ is element n of the adjacency vector centred on pixel i,j within the image, $T_k(n)$ is element n of the template vector for land use category k, N is the total number of adjacency events in the kernel (determined by the kernel size, e.g. $N = 20$ for a 3×3 pixel kernel), and L is the length of the adjacency vector.

A value of 1.0 indicates a perfect match between the current adjacency vector and one of the land use templates; a value of 0.0 indicates no match. Thus, the pixel at the centre of the kernel is assigned to the land use category k, for which Δ_k is maximized:

$$P_{i,j} \leftarrow k, \text{where} \Delta_k = \max_{k=1,2,T} \Delta_k \leq \delta \qquad (5)$$

where $P_{i,j}$ is the pixel corresponding to the centre of the kernel, T is the total number of individual templates, and δ is a user-specified threshold.

The user-specified threshold, δ, can be set to prevent pixels being assigned to a land use category on the basis of a weak match between the measured adjacency vector and a land use template.

2.2.3. Test area and satellite sensor data

To test SPARK, we selected an area to the south-east of London, covering the borough of Bromley. This area encompasses various different types of urban land use, ranging from densely-occupied early 20th century housing in the north-west, through major shopping areas and inter-war industrial areas in the centre, to low-density suburbs in the south-east. Surrounding the urbanized area are very large tracts of open country, much of which is green belt land.

The data used in this investigation were extracted from a cloud-free, multispectral (XS) SPOT-1 HRV image of London (scene 32, 246; +22.46°) acquired on 30 June 1986 (Fig. 2). In particular, a 512×512 pixel subsection (approximately 10 km $\times$ 10 km) of the full image, centred on the district of Orpington, was selected for detailed study. This area exhibits a complex spatial pattern of land use, providing a stringent test for both conventional and alternative classification techniques.

2.2.4. Results – stage 1: initial land cover classification

The first stage in the spatial re-classification procedure is the production of an initial land cover map from the remotely-sensed image. A variety of techniques could be used for this purpose, including unsupervised multispectral classification, region growing, and split-and-merge procedures (Chen and Pavlidis, 1979; Mather, 1987; Rafat and Wong, 1988; Li and Muller, 1991). However, a supervised, maximum-likelihood classification algorithm was employed in this study, since it offers the greatest control over the number and nature of the classes defined. In this respect, seven broad land cover classes were identified in the Orpington subscene, namely Built, Large Structure, Tree, Crop, Grass, Soil and Water. The Built class corresponds to roads and buildings within the main urban area; no attempt has

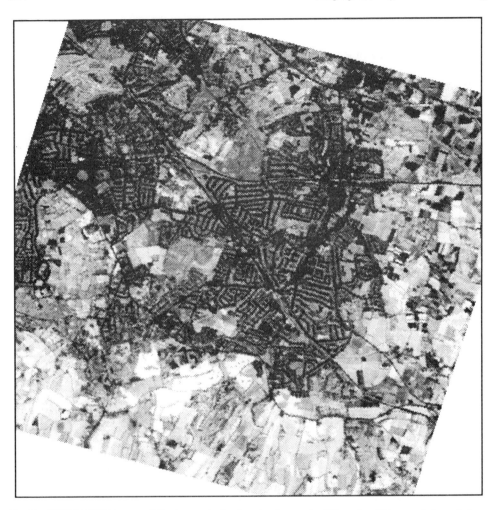

Fig. 2 SPOT-1 HRV image of Orpington in the London borough of Bromley (20 m spatial resolution; near-infrared waveband (XS3)). Note that the road pattern within the urban area is clearly evident.
© SPOT Image

been made to distinguish between these two surfaces, because of the difficulty in identifying pure pixels of either surface at this spatial resolution. Training areas for a separate class, referred to as Large Structure, have also been identified on the basis of a pronounced contrast between the spectral properties of these areas and those of the Built class. Detailed examination of the digital image and the corresponding Ordnance Survey 1:10 000 scale base maps suggest that the Large Structure class corresponds to buildings such as factories, warehouses and hospitals which often have large, flat concrete roofs. The remaining classes are reasonably self-explanatory. However, it is worth noting that the Grass class incorporates regions of open space within the urban area (for example, recreational land), as well as fields of permanent pasture lying outside it. Similarly, the Crop class incorporates, and is dominated by, areas of wheat and barley; no attempt has been made to distinguish between these crops in this particular study.

Irregularly shaped regions, sampled systematically within the image, were used to define several training areas for each of the candidate land cover classes. A second set of regions was used to define an independent test set. Some difficulty was experienced in creating the training and testing sets for the Built, Water and Large Structure classes. In the case of Water and Large Structure, this was because of their relatively limited areal extent, while for the Built class it was primarily due to the comparatively narrow, elongated regions that it forms. Consequently, the number of pixels used to train and to test these classes is quite small.

A very low rejection threshold was set for the maximum-likelihood algorithm (>5 standard deviations, i.e. <0.001% pixels rejected) to produce an image with the minimum number of unclassified pixels. This is because an adjacency event involving a Null class pixel (hereafter referred to as a Null-adjacency) presents a problem at the re-classification stage. More specifically, a Null-adjacency may obscure the true spatial pattern of land cover types present within the kernel. Thus, a Built–Null adjacency may, in reality, represent Built–Built, Built–Grass and so on. Moreover, where there is more than one Null-adjacency within the kernel, it may be impossible to determine the land use at that location. Thus, the errors introduced by using a very low rejection threshold must be balanced against the uncertainty introduced into the spatial re-classification procedure by a Null-adjacency.

The results of the initial classification are presented in Table 1 and Fig. 3. Not surprisingly, given the limited number and rather broad nature of the land cover classes identified, a very high level of classification accuracy (overall accuracy, 97.2%; kappa coefficient, 0.93) has been achieved, although the use of contiguous blocks of pixels for the test set probably means that these values represent slight overestimates of the true values.

2.2.5. Results – stage 2: spatial re-classification

Having derived a satisfactory land cover classification, SPARK has been used to re-classify the image into eight land use categories, namely medium-density residential, low-density residential, commercial/industrial, woodland, arable farmland, permanent pasture, bare soil/fallow land and open water. The distinction made here between the medium-density and low-density residential categories is somewhat subjective. However, for the purpose of

Table 1 Confusion matrix for per-pixel land cover classification

Land cover type: image	Land cover type: true						
	Built	Structure	Tree	Crop	Grass	Soil	Water
Built	25	0	0	0	0	0	0
Structure	0	20	0	0	0	0	0
Tree	0	0	156	0	0	0	0
Crop	0	0	7	73	6	0	0
Grass	0	0	0	0	191	0	0
Soil	0	3	0	0	0	91	0
Water	0	0	0	0	0	0	18
Number of test pixels	25	23	163	73	197	91	18

Average accuracy, 97.09%; overall accuracy, 97.29%; kappa coefficient, (×100) 93.0%

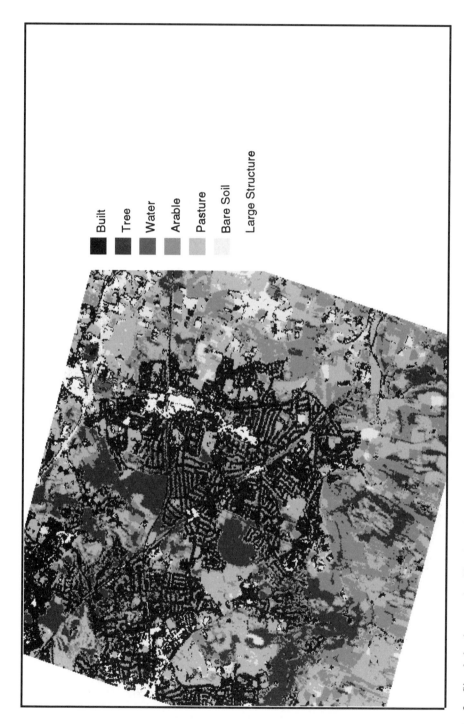

Fig. 3 Simple land cover classification of the Orpington subscene generated using a standard per-pixel, maximum-likelihood algorithm. Note that a very low rejection threshold (<0.001% of pixels) has been used to limit the number of Null class pixels passed to the spatial re-classification stage

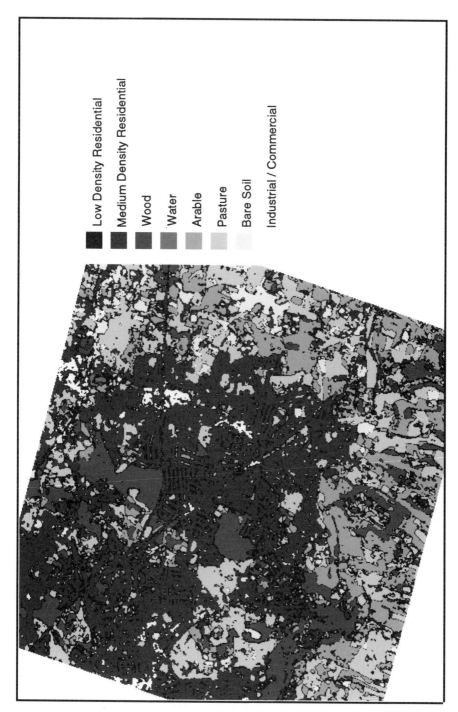

Fig. 4 Land use map produced using SPARK with a 3 × 3 pixel kernel. Note the poor performance in low-density residential districts

Table 2 Confusion matrix for land use re-classification using SPARK (3 × 3 pixel kernel)

Land use: image	Land use: true							
	Low-density residential	Medium-density residential	Commercial/industrial	Woodland	Arable	Pasture	Fallow/bare soil	Water
Low-density residential	541	13	5	0	0	5	3	5
Medium-density residential	695	625	1	0	0	0	0	0
Commercial/industrial	0	0	17	0	0	0	0	0
Woodland	41	0	0	464	0	0	0	0
Arable	13	0	0	0	73	0	0	0
Pasture	0	2	0	0	0	192	0	0
Fallow/bare soil	0	0	0	0	0	0	88	0
Water	0	0	0	0	0	0	0	7
Number of test pixels	1280	640	23	464	73	197	91	12

Average accuracy, 93.10%; overall accuracy, 85.30%; kappa coefficient (×100), 43.96%.

Analytical Tools to Monitor Urban Areas

this study medium-density housing broadly corresponds to terraced buildings with relatively small gardens, whereas low-density housing corresponds to detached and semi-detached buildings with large gardens. Template vectors have been derived for each of the land use categories using blocks of pixels selected at random from within larger sample areas. Several template vectors have been created in this way for each land use category. These have been used to define a set of independent templates for that class. The accuracy of the re-classification has been tested using an independent set of irregularly shaped sample areas. The land use in each of these regions has been determined from recent Ordnance Survey 1:10 000 scale base maps and verified through field observations.

The results obtained from SPARK using a 3 × 3 pixel kernel are given in Fig. 4 and Table 2. These indicate that SPARK performs very well for the non-urban land use categories, although this is to be expected, given that these categories generally comprise a single land cover type. By contrast, the level of accuracy achieved for the low-density residential areas is quite poor. More specifically, there is considerable confusion between this land use and the medium-density residential areas. A closer examination of Figs 2 and 4 reveals that this occurs where individual streets are widely spaced, such that the roads and some buildings (i.e. the Built class) are assigned to the medium-density residential category, while buildings which adjoin some form of open space are assigned to the low-density residential category. Consequently, the street pattern is still evident in some low-density residential districts. These results suggest that a 3 × 3 pixel kernel is too small to take into account the spatial pattern of land cover types typical of these areas.

In view of the results obtained above, tests have also been performed on SPARK using a range of other kernel sizes (Fig. 5). These indicate that the highest overall accuracy is

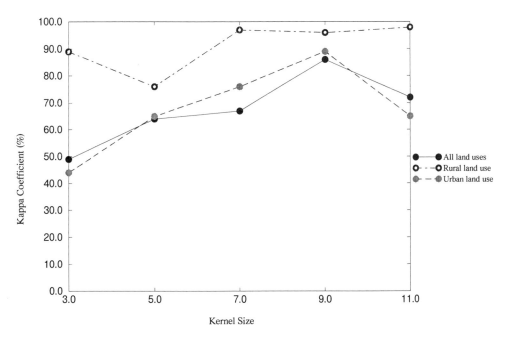

Fig. 5 Relationship between kernel size and the accuracy of spatial re-classification using SPARK, expressed in terms of the kappa coefficient (%)

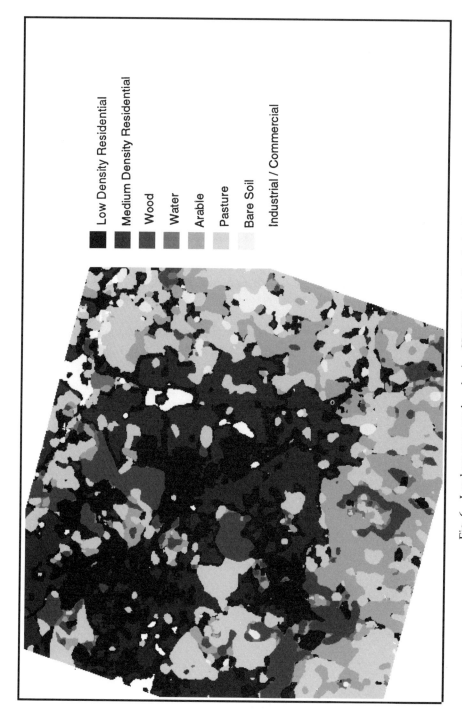

Fig. 6   Land use map produced using SPARK with a 9 × 9 pixel kernel

Table 3 Confusion matrix for land use re-classification using SPARK (9 × 9 pixel kernel)

Land use: image	Land use: true							
	Low-density residential	Medium-density residential	Commercial/industrial	Woodland	Arable	Pasture	Fallow/bare soil	Water
Low-density residential	1276	1	0	0	0	0	0	0
Medium density residential	4	639	0	0	0	0	0	0
Commercial/industrial	0	0	23	0	0	0	0	0
Woodland	0	0	0	464	0	0	1	0
Arable	0	0	0	0	73	0	0	0
Pasture	0	0	0	0	0	197	0	0
Fallow/bare soil	0	0	0	0	0	0	81	0
Water	0	0	0	0	0	0	9	12
Number of test pixels	1280	640	23	464	73	197	91	12

Average accuracy, 97.98%; overall accuracy, 96.86%; kappa coefficient (×100), 92.1%.

obtained using a 9 × 9 pixel kernel (Fig. 6 and Table 3; overall accuracy, 96.86%; kappa coefficient, 0.921), chiefly as a result of the improvement in the recognition of the urban land use categories. The use of a kernel larger than 9 × 9 pixels results in a decrease in overall classification accuracy. This is because of the smoothing effect of large kernels, particularly at the urban–rural boundary.

2.2.6. Discussion

Clearly, any technique that attempts to infer land use by examining the spatial pattern of land cover in an image will be sensitive to the accuracy of the initial land cover classification; exactly how sensitive is not known at present. One way to estimate this is to provide SPARK with several land cover images of the same scene, produced using different classification algorithms or image segmentation techniques. This can also be used to determine the sensitivity of spatial re-classification to both the number and the nature of the land cover classes identified. Further work along these lines is currently under way.

It is also evident that the accuracy achieved in the spatial re-classification stage is dependent on the size of the kernel. Different kernel sizes seem to be appropriate in different parts of the image. Thus, while a small kernel is adequate for non-urban areas, a larger kernel is required to represent the full spatial variability of land cover in urban districts. This problem is currently being addressed through the development of an adaptive SPARK, where the size of the kernel varies according to its location within the image. This, in turn, requires some means of determining the optimum kernel size at any given location. This might be inferred from some measure of the local image texture, or by using existing digital map data (for example, that indicating the urban–rural boundary), or through some iterative procedure based on a previous land use classification of the study area.

Further improvements in the basic SPARK algorithm might be obtained by using data on the 'likelihood', or confidence, with which each pixel in the land cover image is assigned to a particular class. At present, the (land cover) class label associated with each pixel is assumed to be correct. In reality, we may have only a low confidence that a pixel has been assigned the correct class. Future versions of SPARK will therefore take into account the likelihood that pairs of adjacent pixels within the kernel belong to the land cover types indicated by their class labels. The product of their respective likelihood values could then be used to establish a probability value for that adjacency event. This could be used to modify the strength of the match between the measured adjacency vectors and the land use templates.

2.3. Development of object-based spatial re-classification procedures

Although promising results have been obtained from SPARK, the use of a pre-defined kernel places an undesirable restriction on the area over which the spatial mixing of land cover types can be examined. This problem is only partly solved by using an adaptive kernel size. A more flexible approach is to examine the spatial mixing of discrete 'objects' within the image (Barr, 1992; Barr and Barnsley, 1993). The term 'object' is used here to denote a region of uniform land cover within the image (i.e. a contiguous block of pixels with the same land cover class label). Given that such objects will differ in terms of size and shape, depending on their location and context within the scene, the area over which the spatial mixing of land cover types is examined using this approach will vary throughout the image.

The identification of discrete objects such as buildings, roads and open spaces within urban areas, and the analysis of their spatial arrangement in remotely-sensed images to infer scene properties is not new. However, these techniques are more widely employed to update cartographic databases using very high spatial resolution (<1 m) images, often in the form of digitized aerial photography (Huertas and Nevatia, 1988; McKeown, 1988, 1991; McKeown, Harvey and Wixson, 1989). Using data such as these, it is possible to extract detailed information on the component elements of the scene. In urban areas these may be the node and arc primitives that form the edges of individual buildings (Huertas and Nevatia, 1988). Specific data models and associated data structures have been developed to represent these objects and their complex spatial arrangement within urban scenes (Peuquet, 1984; McKeown, Harvey and Wixson, 1989).

The nature of the data models and data structures that are used to represent scene primitives is, however, dictated by the scale (or spatial resolution) of the images under investigation (McKeown, Harvey and Wixson, 1989; Barr, 1992). In images obtained by satellite sensors, such as the SPOT-HRV, with their 20 m nominal spatial resolution in multispectral mode and 10 m in panchromatic mode, it is clearly unrealistic to extract information on building nodes or arcs for all but the largest of structures. Indeed, it may not even be possible to delineate individual buildings accurately (Barr, 1992). Instead, it would seem more appropriate to identify the general land cover 'objects' to which these features belong. For example, a closely spaced group of buildings that dominate the detected spectral response of a number of contiguous pixels in the image may be regarded as forming a 'Built' object. The size, shape and spatial arrangement of such objects might then be used to infer land use in a manner analogous to the pixel-based spatial re-classification techniques described in Section 2.2. In an object-based spatial re-classification procedure (Gurney, 1981; Gurney and Townshend, 1983), such as this, the problem becomes one of developing a data model and associated data structures that allow this information to be abstracted and encoded.

2.3.1. Object-based data structures and algorithms to abstract spatial information contained within high resolution satellite sensor images

Given an initial land cover classification of an urban scene, the resultant image may be thought of as being made up of a number of discrete regions, each comprising a contiguous block of pixels with the same land cover label (Barr, 1992; Barr and Barnsley, 1993). The size, shape and spatial arrangement of these regions, hereafter referred to as image-objects, will to a certain extent be indicative of their location and context within the scene. For example, the street patterns in many urban areas will result in a group of connected, elongated image-objects, while arable farmland will generally consist of approximately rectangular image-objects. It is therefore necessary to define a set of intrinsic (e.g. area, perimeter, shape) and extrinsic (e.g. containment, adjacency, proximity) properties that characterize these image-objects and their spatial relationships with one another (Gurney, 1981; Gurney and Townshend, 1983). This, in turn, requires the development of new data structures to represent this information and new data processing algorithms to analyse the spatial relationships between the abstracted image-objects (Barr, 1992; Barr and Barnsley, 1993).

In this study, spatial information contained within the initial land cover classification is represented at two levels of abstraction. At the lower level, information on the location and extent of individual image-objects is extracted and contained within an Object Search Map (OSM) data structure. At an intermediate level, a new data structure, XRAG (eXtended Region Adjacency Graph), has been developed to represent the intrinsic and extrinsic properties of each image-object. The following subsections describe these data structures and outline the processing stages used to derive them.

2.3.2. Deriving a low-level representation of image-objects

A low-level representation of the location and extent of individual image-objects is achieved using a contour-encoding algorithm to extract their (iconic) boundaries (Gonzalez and Wintz, 1987). The algorithm used here employs a number of simple rules to identify the starting point of each image-object and then to contour its outer boundary. During this procedure, the Freeman chain code data model (Freeman, 1961; Peuquet, 1984) is used to assign a directional vector to each boundary element according to the direction of entry from the previous image element (Fig. 7). This low-level representation of all image-objects within the scene is stored in an OSM file, which contains an identifier for each image-object, a value indicating its starting location, and a stream of characters representing the Freeman code vector for each boundary element (Fig. 8).

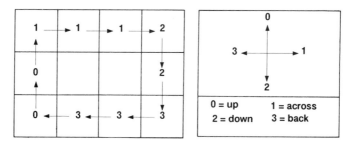

Fig. 7   Freeman chain code representation of a simple image-object

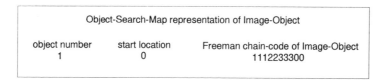

Fig. 8   Low-level OSM created from the Freeman chain code representation of image-objects

2.3.3. Intermediate-level spatial encoding of image-objects

The OSM can be processed to allow the intrinsic and extrinsic properties of each image-object to be extracted and encoded at an intermediate level of spatial abstraction (Barr, 1992; Barr and Barnsley, 1993). This is achieved by re-mapping the Freeman chain codes of each image-object into image space. During the re-mapping procedure, specific

Analytical Tools to Monitor Urban Areas   165

algorithms are invoked to extract the required properties (for example, area, perimeter, compactness, adjacency, containment and proximity) (Barr, 1992; Barr and Barnsley, 1993). Within this study, a new data structure, referred to as XRAG, has been developed to encode and represent these intermediate-level abstracted properties of the image-objects. XRAG is based upon the commonly used Region Adjacency Graph (RAG) (Schalkoff, 1989; Nichol, 1990), with extensions to allow multiple graphs to be utilized and the definition of a number of additional sub-data types to encode and represent the desired properties of the image-objects. The latter are referred to as 'externals', 'internals' and 'labels'.

2.3.3.1. Externals: graph representation of the extrinsic properties of image-objects
RAGs have previously been used to represent spatial information contained within high resolution aerial photographs in a symbolic modelling framework. The data that have been

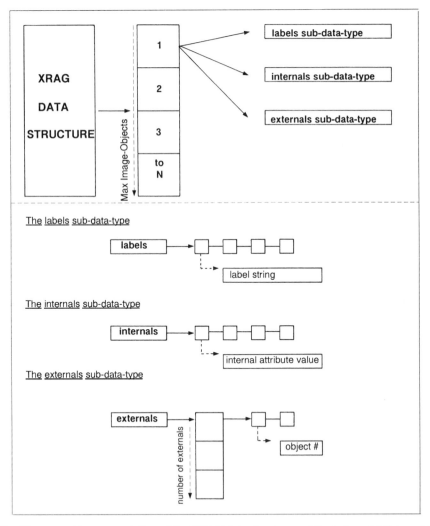

Fig. 9   Diagrammatic representation of the XRAG data structure and the 'externals', 'internals' and 'labels' sub-data types

used in this context can be characterized as having a small number of recognizable features (or objects) whose interrelationships are comparatively well defined (Schalkoff, 1989). RAGs have also been used to reduce the complexity of land cover classifications derived from images obtained by meteorological satellite sensors, such as NOAA-Advanced Very High Resolution Radiometer (AVHRR) (Nichol, 1990). By contrast, the interest here is to use graphs and graph theory as the basis for more intricate and complex spatial searching problems, which may require a number of separate spatial graphs, for example, in the representation of adjacency, containment and distance measurements, applied to very large numbers of image-objects.

The XRAG data structure developed here allows multiple graphs to be present, each holding different spatial information relating to the image-objects within the scene. This has been achieved through the implementation of an 'externals' data type. Fig. 9 shows the design of the 'externals' data type, where the graphs have been implemented in a mixed, undirected manner. Individual image-objects are represented as single elements within a contiguous list. Linked lists are used to represent the graph nodes (image-objects) that have some spatial relationship (e.g. containment, adjacency and so on) with a particular image-object. The advantage of this implementation is that, through the use of both contiguous and linked list representations, any image-object within a graph can be addressed more easily and efficiently than would be the case using a wholly linked (or contiguous) implementation. Moreover, the use of undirected graphs allows much greater flexibility in terms of the type of spatial searches that can be performed in response to user queries (Luger and Stubblefield, 1989; Kruse, Leung and Tondo, 1991).

2.3.3.2. Internals: representing the intrinsic properties of image-objects Not only can the initial land cover classification (Fig. 3) be used to derive intermediate-level information on the extrinsic properties of the component image-objects, it can also be used to generate information on a wide variety of their intrinsic (i.e. geometrical and morphological) properties. This information is encoded and represented using the 'internals' data type (Fig. 9). Internals are comprised of a linked list. Each link in the list contains the value of a specific intrinsic property of the image-object, such as its area, perimeter and shape characteristics.

2.3.3.3. Labels: representing the process-associated labels of image-objects The last data type within the XRAG data structure is referred to as 'labels'. The labels data type has been developed to permit individual image-objects to have a number of interpretations (e.g. 'man-made structure', 'residential', 'urban') that may be beneficial in the spatial segmentation of the image (Barr, 1992; Barr and Barnsley, 1993). This is akin to multiple-inheritance within object-orientated processing as, in reality, any 'object' may belong to a variety of classes. For example, in the case of urban scenes, an image-object originally labelled as Tarmac may also be considered to have a label within the groups 'Residential' and 'Urban'; alternatively, another image-object labelled Tarmac may have a label within the groups 'Runway' and 'Airport' (McKeown, 1988). The labels associated with an image-object will ultimately depend on the nature of spatial processing that has been carried out. Within the XRAG data structure each image-object may be assigned any number of different labels via the 'labels' data type. Like 'internals', the 'labels' data type is comprised of a linked list, where each element in the list contains a character string associated with a given interpretation. Fig. 9 shows a diagrammatic representation of the 'labels' data type within the overall context of the XRAG data structure.

2.3.4. Spatial processing of the XRAG data structure

A command-line query tool has been developed to assess the applicability of the XRAG data structure and object-based spatial re-classification for monitoring urban land use in high spatial resolution, multispectral satellite sensor images. This allows the user to interrogate the XRAG data structure in response to high-level spatial queries. At present, the tool uses a simple means of representing knowledge provided by the user. The user is required to supply information on the image-object labels to be processed, the intrinsic and extrinsic properties within the XRAG data structure that are to be examined, the operations to be performed and the condition (goal) to be achieved. The output from the spatial query tool is an additional label for each image-object within the XRAG data structure. This is a string, specified to be the result of the command-line query, or a null label (if the required condition is not met).

Once the spatial querying tool has parsed the high-level spatial knowledge provided by the user, it is passed to a number of intermediate-level algorithms designed to perform various types of spatial search and to make inferences based on the outcome of these searches. These algorithms analyse the XRAG data structure and search it according to the parameters specified by the user. Since many spatial segmentation tasks require operations to be performed on a variety of intrinsic and extrinsic properties of the image-objects, the intermediate-level spatial processing algorithms have been designed so that they can dynamically process the XRAG data structure and evaluate a variety of different (image-object) properties within any single routine call. For example, parameters provided by the user may require the perimeter of all image-objects contained within another particular group of image-objects to be summed. This requires an examination of both intrinsic (perimeter) and extrinsic (containment) image-object properties during a single pass.

Currently, two main algorithms have been implemented that allow this type of dynamic spatial analysis, namely Walkabout and Lookin. The first of these has been designed to perform spatial 'walks' between adjacent image-objects in response to direction from the user interface. Walkabout can also, if required, process the intrinsic properties of the image-objects visited during the spatial search. For example, Fig. 10 shows Walkabout searching

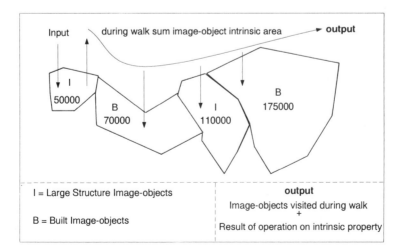

Fig. 10 Diagrammatic representation of Walkabout, an algorithm for processing the XRAG data structure. Note that Walkabout interrogates both extrinsic and intrinsic image-object properties – in this case, area and adjacency

image-objects that have the label 'Built' or 'Large Structure' and that are adjacent to one another in image space. In this case, the area of all of the image-objects visited during the 'walk' are summed to determine the total area of built-up land.

The second intermediate-level algorithm developed here is Lookin. This performs an analysis of the extrinsic XRAG property, containment. Lookin is a recursive algorithm that can analyse multiple levels of containment. As with Walkabout, Lookin can also process the intrinsic properties of image-objects visited during the spatial search. For example, Fig. 11 shows Lookin processing a series of image-objects that are contained within a larger image-object of a given land cover class and that have an area greater than or equal to some user-specified value.

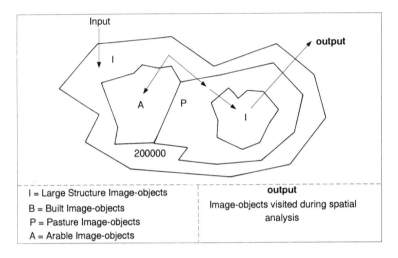

Fig. 11 Diagrammatic representation of Lookin, an algorithm for processing the XRAG data structure. Note that Lookin interrogates both internal and external image-object properties – in this case, area and containment

2.3.5. Case study: monitoring the urban–rural boundary using the XRAG data structure

In the UK the Department of the Environment (DoE) is charged with the provision of information on land use change, including data on the spatial extent of urban areas, within England and Wales (HMSO, 1984). Traditionally, this information has been derived through a combination of sample ground surveys, visual interpretation of aerial photographs and manual digitization of the appropriate paper maps. As was indicated earlier in this chapter, although satellite remote sensing offers an alternative means of acquiring this information that is both timely and cost effective, previous attempts to use satellite sensor images for this purpose have not proved wholly successful. It is suggested that this is due, in part, to the use of conventional per-pixel image classification procedures, which identify the component elements within the corresponding scene solely on the basis of their spectral reflectance properties, and which are therefore inappropriate in the context of monitoring urban areas. This is demonstrated in Fig. 12b, which illustrates the results obtained from a standard, per-pixel, maximum-likelihood classification algorithm used to generate a binary image (urban/rural) from the SPOT-HRV subscene discussed in Section 2. This should be compared directly with the urban–rural boundary that has been manually digitized from the current 1:25 000-scale Ordnance Survey base maps of this area (Fig. 12a). Not surprisingly, the standard, per-pixel, multispectral classification algorithm highlights many areas of open space (such as gardens, parks and so on) within the urban boundary. However, under the standard Office of Population Censuses and Statistics (OPCS) definition, these should form part of the urban area (HMSO, 1984). Rectifying this would require a significant amount of post-processing. A further problem with the standard multispectral classifier is that many roads and other man-made surfaces lying outside the urban area (as designated in the Ordnance Survey map) are incorporated into the urban land use category. Once again, it would be difficult to remove these outliers without considerable user-directed post-processing.

To assess the potential of the object-based techniques developed in this study, the land cover classification presented in Section 2.2 (Fig. 3) has been used to generate OSM and XRAG data structures for the Orpington subscene. As an indication of the complex spatial mixture of land cover types within this study area, the XRAG data structure compiled from the land cover image has over 15 000 discrete image-objects. Information on the intrinsic (area) and extrinsic (adjacency and containment) properties of these image-objects has been processed and encoded within the XRAG data structure.

A command-line query, based on the OPCS/DoE definition of urban areas, has been used to process the XRAG data structure. The OPCS definition considers urban areas to be comprised of regions of permanent man-made structures, including transport corridors that are built up on one or both sides, and any area completely surrounded by built-up sites (such as playing fields), provided that their areal extent exceeds 20 hectares (HMSO, 1984). The results obtained by processing the XRAG data structure using these criteria are shown in Fig. 12d.

The urban–rural segmentation obtained by processing the XRAG data structure represents a considerable improvement over that derived from the standard, per-pixel, multispectral classification algorithm. It also represents an improvement with respect to that obtained using the pixel-based spatial re-classification approach (Fig. 12c). In particular, the object-based technique has incorporated all of the intra-urban open space into the 'urban' category, and has removed most of the spurious outliers. A visual comparison of the object-based spatial re-classification results (Fig. 12d) and the manually digitized urban–rural boundary (Fig. 12a) indicates a very close correspondence between the two, although the former is not without error. The main errors appear to be ones of omission (i.e. areas incorrectly excluded from the 'urban' category). These are not spread evenly across the image, but are concentrated in two localities (circled in Fig. 13). Closer inspection of the original image and the Ordnance Survey base maps indicates that these represent districts with *very* low density housing. Despite this, the results obtained throughout the remainder of the image are encouraging and, in general, it seems that the XRAG structure offers considerable flexibility for processing and segmenting high spatial resolution multispectral images in response to user-directed queries, particularly in the realm of urban land use mapping.

2.3.6. Discussion

The results obtained from this study suggest that the XRAG data structure and the spatial command-line query tool offer considerable flexibility and potential for the segmentation of

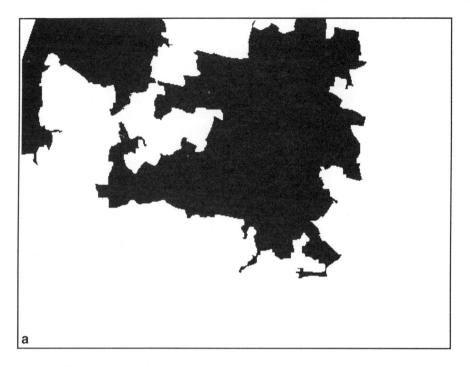

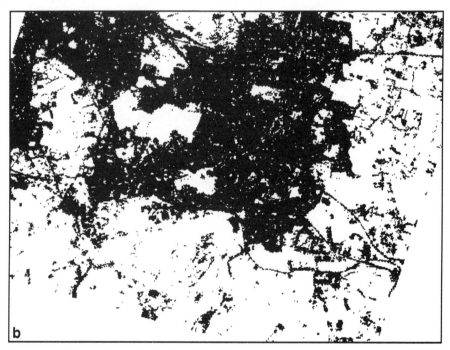

Fig. 12 Binary segmentations of the Orpington subscene (black, urban; white, rural) produced using: (a) manually digitized Ordnance Survey 1:25 000 scale base map; (b) standard per-pixel, maximum-likelihood classification of land use

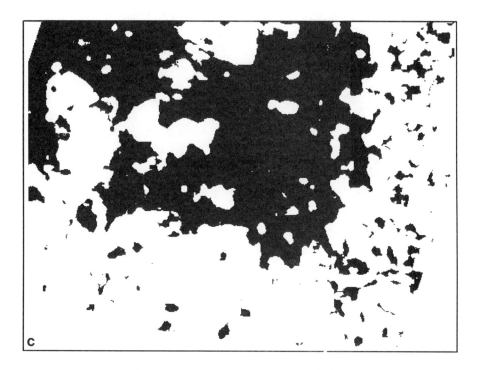

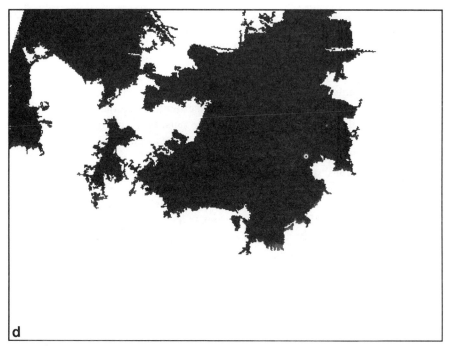

Fig. 12 (cont.) (c) pixel-based spatial re-classification (SPARK 9 × 9 pixel kernel) of simple land cover image; (d) object-based spatial re-classification (XRAG) of simple land cover image

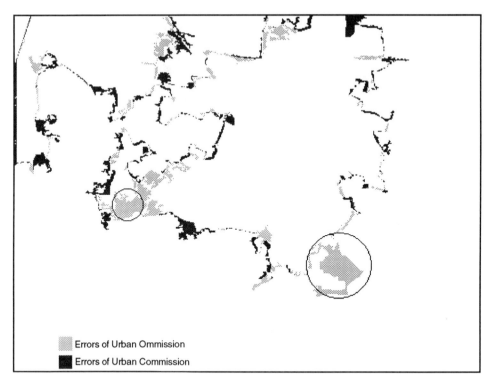

Fig. 13 Comparison between binary (urban–rural), object-based spatial re-classification of Orpington subscene and urban–rural boundary digitized from the corresponding Ordnance Survey 1:25 000 scale base map

images acquired by high spatial resolution satellite sensors, particularly for urban scenes. As with the pixel-based approach discussed in Section 2.2, the output from this object-based spatial re-classification scheme is likely to be sensitive to the initial land cover classification used to generate the XRAG data structure. Once again, the degree of sensitivity is unknown at present. It may, however, be possible to reduce the impact of any errors in the land cover classification through the use of iterative and backtracking procedures in the spatial re-classification stage.

The object-based spatial re-classification scheme described here is still at a relatively early stage of development. Most of the effort to date has been concentrated on the intermediate-level representation of the intrinsic and extrinsic properties of image-objects. The techniques to direct spatial searching and inference-making based on the resultant XRAG data structure are currently much less well developed. At present, only relatively simple spatial queries can be specified. Future work will concentrate on the development of a number of new analytical tools that allow multiple spatial queries to be handled simultaneously. Ultimately, it is hoped that, through developments such as these, intricate high-level spatial rule bases and control mechanisms can be developed. These should allow much more complex spatial segmentation tasks to be performed, compared with the relatively simple, binary segmentation discussed in the present study.

3. EXTRACTING INFORMATION ON THE THREE-DIMENSIONAL STRUCTURE OF URBAN AREAS AND ITS INCORPORATION INTO A THREE-DIMENSIONAL GIS

While Section 2 discussed techniques suitable for inferring land use at regional and national scales, this section considers data collection over much smaller areas (i.e. at a larger scale). In particular, it describes the development of an object-orientated system for extracting three-dimensional models of buildings, both for dynamic visualization of urban scenes and to provide three-dimensional data sets for input into existing GIS packages. The resultant tool, BUILD, allows surface and volume descriptions of buildings to be generated rapidly from large-scale, digital, remotely-sensed images and a pre-computed camera model. A simple conversion routine is used to port the three-dimensional data output from this system, together with the original images, into a commercially available GIS incorporating 'point-and-click' techniques and a simple GUI. These are combined for both static visualization of urban areas and to provide a three-dimensional interface for conventional GIS operations.

3.1. Deriving data on the three-dimensional structure of urban areas

Urban planning departments make decisions on the basis of spatial information from a disparate set of sources. Most of this information is provided and stored in two-dimensional (planimetric) form. Where three-dimensional data are available they are usually restricted to simple digital elevation models (DEMs) of the underlying terrain. Wider availability of accurate, detailed information on the height, shape and volume of individual buildings within urban areas would allow a number of additional analyses to be performed, ranging from telecommunications (inter-visibility) planning and urban design studies, to environmental impact assessment (EIA). Unfortunately, in the UK at least, large-scale Ordnance Survey maps have traditionally been composed almost exclusively of plan detail. Even where photogrammetric techniques are used to update these maps, the elevation values are usually discarded during data capture. Consequently, studies that require information on building heights and volumes must generate it from other sources. Stereoscopic remotely-sensed images, particularly those acquired by sensors mounted on board aircraft, provide a potential source of such data.

In this study, we examined the applicability of several techniques designed to extract information on the three-dimensional structure of urban areas from remotely-sensed images. Emphasis was placed on the use of aerial photography (1:5000 scale) digitized at a resolution of approximately 10 cm, rather than on satellite sensor data. This is because the spatial resolution of the latter is insufficient for a detailed description of urban scenes.

3.1.1. Evaluating the utility of automated stereo-matching procedures

A common way of deriving three-dimensional information from digital, remotely-sensed images is through the use of automated stereo-matching algorithms. These examine the displacement (or disparity) between conjugate points in a pair of stereoscopic images. This information can be used, in conjunction with a suitable camera orientation model, to infer the height of each point in the scene above some datum.

In this study, we examined the utility of the Otto–Chau stereo-matching algorithm for deriving information on the three-dimensional structure of urban areas (Muller *et al.*, 1988a). This algorithm has been used with considerable success in the production of small-scale DEMs from stereoscopic image pairs recorded by satellite sensors, such as the SPOT-HRV (Muller *et al.*, 1988a). It uses an adaptive least-squares approach to analyse the correlation between grey level values in small patches of the two images. This, in turn, is used to define a series of seed points from which the stereo-matcher 'grows' a sheet of disparities in a non-deterministic fashion (Muller *et al.*, 1988a).

To investigate whether stereo-matching can be applied successfully to large-scale images of urban areas, the Otto–Chau algorithm was run on several pairs of 1:5 000 scale aerial photographs, digitized at a pixel resolution of approximately 0.1 m. Stereo image pairs acquired over central London and the London borough of Bromley were selected to evaluate the utility of the stereo-matcher over different types of urban environment (i.e. a high-rise commercial area and a low-rise residential district). Unfortunately, in common with previous studies (Lowe, 1987; Horaud and Skordas, 1989; McKeown, Harvey and Wixson, 1989; Mohan and Nevatia, 1989; Mohan, Medioni and Nevatia, 1989), the results obtained were generally very disappointing, both in terms of coverage (the percentage of points matched; <40%) and blunder rate (the percentage of incorrect matches; >10%).

The poor performance of the Otto–Chau algorithm in this context is due to the fact that it operates under the constraint that disparity discontinuities are small (<2 pixels); in other words, it assumes that there are no sudden height changes across the object space. However, this is one of the characteristic properties of large-scale images of urban scenes (Lee, 1990). Other characteristics that caused problems for the stereo-matcher include building shadow, occlusions and the presence of repetitive regular patterns within the images.

Initially, it was thought that many of these problems could be solved by generating a larger number of seed points, covering the entire image. In this way, it was hoped to grow stereo-matched regions on either side of the disparity discontinuities. Two further stereo-matchers were used to generate these seed points:

1. PMF, a stereo-matcher that relies on edge detection within the images (Muller *et al.*, 1988a; Pridmore, Mayhew and Frisby, 1990).

2. PRISM, a fast edge/texture-based stereo-matcher (Muller *et al.*, 1988a).

However, while use of the PMF algorithm resulted in an improved coverage, this was achieved only at the expense of a higher blunder rate. Similarly, PRISM gave poor results because of the lack of image texture suitable for stereo-matching.

Various other techniques have been suggested for improving the results obtained from stereo-matchers applied to large-scale images of urban areas. These include texture matching, edge matching, shape-from-shading and high-level domain reasoning (Hwang, Davis and Matsuyama, 1986; Lowe, 1987; Horaud and Skordas, 1989; Irvin and McKeown, 1989; McKeown, Harvey and Wixson, 1989; Mohan and Nevatia, 1989; Hsieh, Perlant and McKeown, 1990; Hsieh, McKeown and Perlant, 1990; Liow and Pavlidis, 1990; Aviad, McKeown and Hsieh, 1991). However, although promising results have been obtained using these techniques applied to images of urban areas displaying simple block architectures, there is still some way to go before they will be ready to provide structured three-dimensional data for scene visualization or for input into GIS on a routine basis.

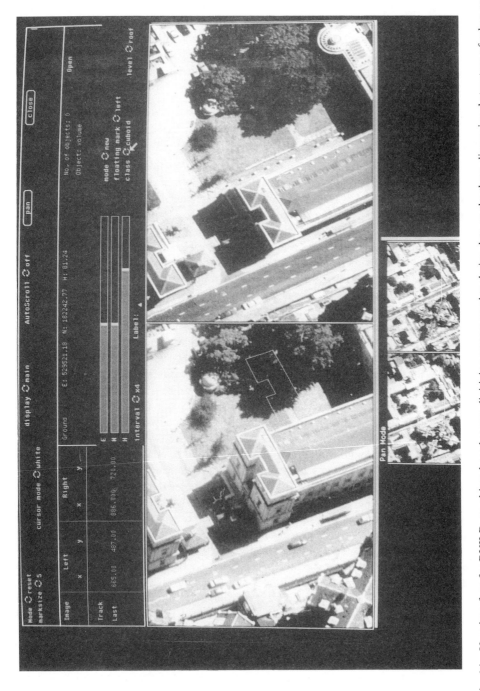

Fig. 14 User interface for BUILD, an object-based stereo digitizing system used to derive data on the three-dimensional structure of urban areas

A further problem with automated stereo-matching techniques is the need to reconstruct and recognize discrete three-dimensional objects in the output data: digital elevation models in which buildings form part of the 'terrain' are not sufficient. It is essential to be able to recognize buildings as entities that are separate from the underlying terrain; for example, when registering building footprints with digital map data within a three-dimensional GIS; when calculating building volumes; and when constructing and editing a building database.

The final problem associated with automated stereo-matching relates to the reliability of the resultant three-dimensional data. More precisely, missing and spurious values are a penalty of automated systems.

3.1.2. Developing an interactive, object-based stereo digitizing system

Given the limitations of existing automated techniques, an interactive (i.e. semi-automated) three-dimensional digitizing system has been developed to provide the required information. The system, known as BUILD, requires a stereoscopic pair of digital images and a camera orientation model. The camera orientation model is need to calculate three-dimensional coordinates from the disparity values between conjugate points in the stereo image pair.

BUILD allows the user to view a pair of stereoscopic images simultaneously on the screen and, at the same time, interactively create wireframe objects that fit the building structures currently displayed (Fig. 14). The objects that can be selected include points, lines, surfaces, cylinders and cuboids, or some combination of these. For example, the wireframes for cuboids are generated by digitizing points on a horizontal surface (usually the roof level) to form a closed polygon in one of the images. BUILD then places the corresponding wireframe model in the other image at some random height (determined by the epipolarity constraint imposed under the camera orientation model). The user then adjusts the height of this surface until the wireframe in the second image also fits the object. By altering the height of both surfaces simultaneously, a volume can be generated which describes the building as a whole. On the other hand, if a DEM is available, the user need only digitize the roof level: BUILD will automatically extrude the object down to the ground.

The drawback of having an interactive system, as opposed to a fully automated one, is that it requires a considerable amount of human resources to construct a three-dimensional representation of complex urban scenes. However, it is believed that this is largely offset by the greater reliability of the results. BUILD also has the advantage that the three-dimensional data are output in a structured, polygonal format that can be input easily into most vector-based digital mapping systems and GIS packages. Another advantage is that, by using some of the basic image processing functions inherent in BUILD, the user can enhance the images interactively to account for local variations in brightness and contrast. Furthermore, a human operator can also interpret reasonably complex scenes, inferring the structure of partially occluded objects from their surroundings and shadows.

BUILD has recently been extended to permit large-scale two-dimensional digital map data to be imported into the system. This has the advantage that the digital map data can be used to constrain the object selection process. Unfortunately, most of the available large-scale map data is supplied in a 'spaghetti' format, where polygon attribute labelling is haphazard (i.e. a line common to a building and a road could be labelled as either, causing some buildings to have incomplete polygons). Use of such data therefore requires a considerable degree of pre-processing before its incorporation into BUILD. However, once topologically structured, the digital map data can be used to generate the 'footprints' of individual buildings, with BUILD being used to add the height information.

Analytical Tools to Monitor Urban Areas 177

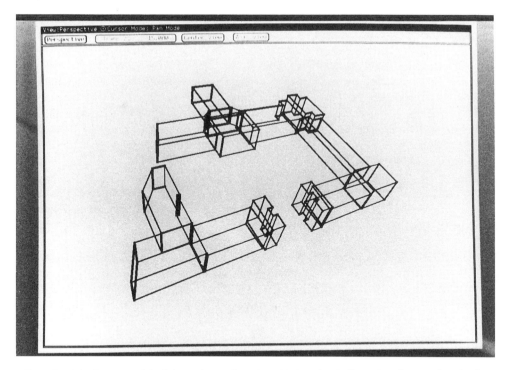

Fig. 15 Wireframe model of the main quadrangle at University College, London, produced using BUILD. Note that the model is built up from a series of simple three-dimensional objects

3.2. Scene visualization

As a post-process to BUILD, a texture map can be created from the image data and the wireframe model. The texture map is derived by projecting rays from the object space, through the camera model, into one or both of the original images. Given the texture maps and a wireframe model, scene visualizations can be generated. Figs 15 and 16 illustrate the use of a ray-tracing package (the Dynamic Visualisation Toolkit ©; Muller *et al.*, 1988b) to produce rendered views of the wireframe with roof texture maps overlaid.

3.3. Importing three-dimensional data from BUILD into a vector GIS

Having designed a practical system to extract three-dimensional information relating to urban areas from stereoscopic remotely-sensed images, the next stage is to incorporate the resultant data into a digital mapping/GIS package. This needs to be implemented so that the z-values – i.e. the heights of buildings above the underlying terrain – can be interrogated and visualized. Although many GIS packages offer a three-dimensional modelling option, the manipulation of a second set of height data (i.e. the buildings) above the general terrain surface (i.e. ground level) is often far from straightforward. For example, many systems are unable to cope with two points that differ in terms of z (i.e. height) but have the same x,y location.

In this study, work was carried out to allow data produced using BUILD to be imported directly into ARC/INFO (Revision 6). Data from BUILD are exported in WAVEFRONT (ASCII) format and imported as a terrain model using the TIN module of ARC/INFO. A

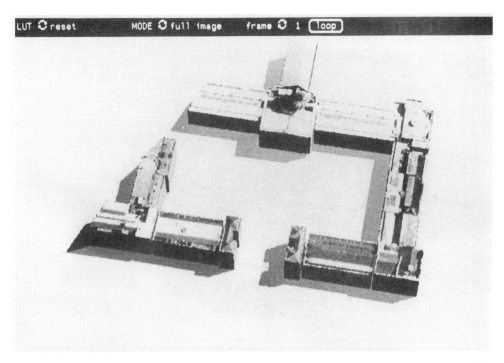

Fig. 16 Perspective view of wireframe model generated using BUILD with texture map overlaid, showing the main quadrangle of University College London

locally written Fortran routine takes the BUILD export file and generates polygons representing the 'footprints' of the building blocks, while information on the heights of the buildings is stored as attributes of the appropriate polygons. The 'building' polygons are then converted into either LATTICE or GRID format and an interpolation algorithm (TIN) is used to generate and display the building blocks as a two-and-a-half-dimensional surface (Raper and Kelk, 1991). Once the building blocks are in place on the surface they can be colour-coded or shaded to achieve a surface rendering reminiscent of the original buildings.

3.4. Developing a prototype three-dimensional GIS

Some work was also carried out to enable the building blocks to be interrogated interactively within ARC/INFO, so that information on building volume, land use and land value can be determined and displayed in response to a 'point-and-click' query. This will greatly enhance the functionality of what is currently a purely visual tool.

Two specific data analysis facilities have been developed to date. These were selected to demonstrate key features that could be expected of a full three-dimensional GIS. The first allows the user to perform an interactive query of a two-and-a-half-dimensional representation of the urban scene. The information provided in response to the user request currently consists of the dimensions and function (use) of the relevant building (Fig. 17). The second facility allows the user to perform line-of-sight analyses, given the height of the viewpoint and both the angle and direction of view (Fig. 18). Clearly, the range of analyses possible within a full three-dimensional GIS is much wider than this, and might include visual (environmental) impact assessment, building inventories and property management schemes.

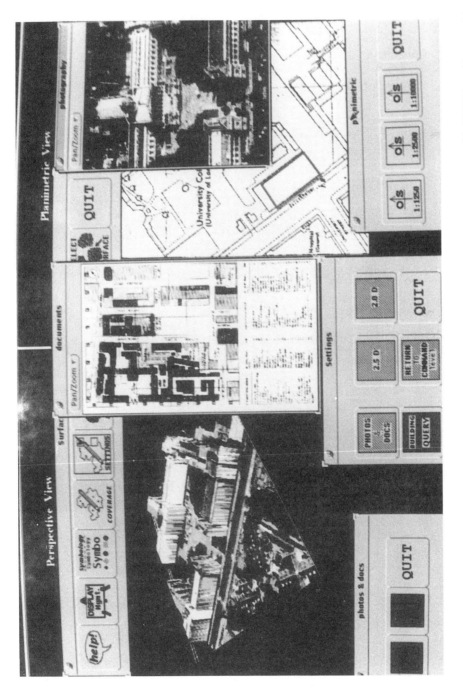

Fig. 17 Example of interactive query of a two-and-a-half-dimensional representation of an urban scene generated in BUILD and imported into ARC/INFO, showing combined use of image, map and attribute data

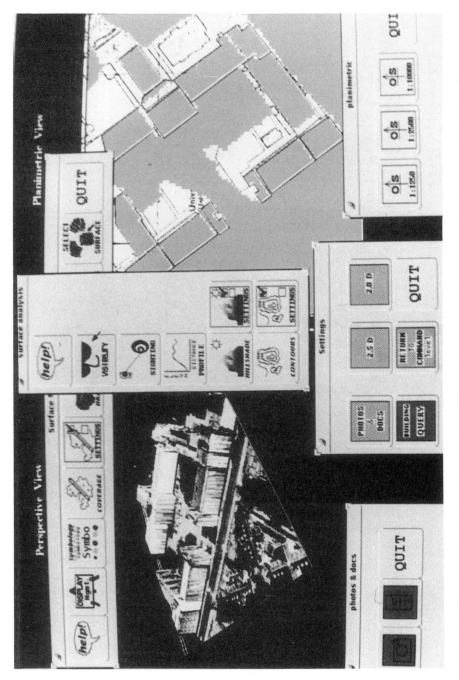

Fig. 18 Example of line-of-sight analysis performed in ARC/INFO using a two-and-a-half-dimensional representation of an urban scene generated in BUILD. Areas visible to the observer from a given viewpoint and elevation angle are shaded in the two-dimensional map on the right

3.5. Discussion

An interactive stereo digitizing system has been developed which allows the three-dimensional structure of urban areas to be determined from stereo pairs of digital, remotely-sensed images. A means of importing data generated using this system, together with the original image data, into a existing GIS package has also been implemented. This provides not only a new tool for visualizing urban environments, but also permits more meaningful interpretation of existing two-dimensional data sets, by distinguishing between buildings on the basis of differences in height and architectural style. It also opens the way for novel methods of spatial data query and building analyses, including volumetric and inter-visibility calculations. Future work in this area will focus on the development of more sophisticated functions to query and analyse three-dimensional data sets and, via a series of simple GUIs, the implementation of further application modules tailored to suit the specific needs of urban planners.

4. CONCLUSIONS

A variety of tools has been developed in this study with the aim of extracting spatial information relating to urban areas from digital, remotely-sensed images. These have been developed and implemented in response to improvements in remote sensing and general computing technology over the last few years. In particular, it has been demonstrated that enhancements in the spatial resolving power of satellite sensors offer greater possibilities for inferring land use and land use change in and around urban areas by examining the spatial mixture of spectrally distinct land cover types in the resultant multispectral images. Initial results obtained from the two spatial re-classification schemes developed in this study suggest that object-based re-classification procedures offer greater flexibility than simple, pixel-based (or kernel-based) approaches. However, the object-based techniques require further development, in terms of both the implementation of new spatial searching algorithms and the development of tools to handle complex user queries of the spatial database.

Improvements in the price–performance ratio of graphics workstations has meant that the ability to visualize and to process three-dimensional spatial data has also increased in recent years. Incorporation of such data into existing GIS packages improves their utility considerably, allowing novel methods of data query, analysis and scene visualization. Potential applications include inter-visibility studies for telecommunications planning, scene visualization for landscape architecture and urban design studies, and EIA. To take advantage of these new possibilities, an object-based stereo digitizing system has been developed. This allows the heights and volumes of buildings to be measured. The output from this system can be imported into existing GIS packages. Some work has been carried out to allow these three-dimensional data to be interrogated interactively within the ARC/INFO GIS, so that information such as building volume, land use and land value can be determined and displayed in response to a 'point-and-click' query. However, further work is required to develop these capabilities.

ACKNOWLEDGEMENTS

The authors would like to thank both the Economic and Social Research Council and the Natural Environment Research Council for support through a research grant under the ESRC/NERC Joint Programme on Geographical Information Handling. The authors would also like to acknowledge the help and advice of various colleagues in each of the participating departments, particularly David Allison, Dave Chapman, Tim Day, Duan Ming, Kevin Morris (now at the NERC Image Analysis Unit, University of Plymouth), Richard Morris, James Pearson, Peter Coppin (now at East Sussex County Council) and Dr Jamal Zemerly. Thanks are also due to Professor Paul Mather (University of Nottingham) for providing encouragement, support and advice in his role as coordinator of the ESRC/NERC Joint Programme.

REFERENCES

Atkinson, P., Cushnie, J.L., Townshend, J.R.G. and Wilson, A.K. (1985) Improving thematic mapper land cover classification using filtered data. *Int. J. Remote Sensing*, **6**, 955–961.

Aviad, Z., McKeown, D. M. and Hsieh, Y. (1991) *The Generation of Building Hypothesis from Monocular Views*. Technical Report no. 3, School of Computer Science, Carnegie – Mellon University.

Baraldi, A. and Parmiggiani, F. (1990) Urban area classification by multispectral SPOT images. *IEEE Trans. Geosci. Remote Sensing*, **28**, 674–680.

Barnsley, M.J. and Barr, S.L. (1992) Developing kernel-based spatial re-classification techniques for improved land-use monitoring using high spatial resolution images. *Proc. XXIX Conf. Int. Soc. Photogrammetry and Remote Sensing (ISPRS'92)*, 2–14 August. Int. Archs Photogrammetry and Remote Sensing: Commission 7, Washington DC, 646–654.

Barnsley, M.J., Barr, S.L. and Sadler, G.J. (1991) Spatial re-classification of remotely-sensed images for urban land use monitoring. *Proc. Spatial Data 2000*, Oxford, 17–20 September. Remote Sensing Society, Nottingham, 106–117.

Barnsley, M.J., Sadler, G.J. and Shepherd, J.S. (1989) Integrating remotely-sensed images and digital map data in the context of urban planning. *Proc. 15th Annual Conf. Remote Sensing Soc.*, Bristol, 13–15 September. Remote Sensing Society, Nottingham, 25–32.

Barnsley, M.J., Shepherd, J. and Sun, Y. (1988) Conversion and evaluation of remotely-sensed imagery for town planning purposes. In Muller, J.-P. (ed.), *Environmental Applications of Digital Mapping, Proc. Eurocarto Seven*. ITC Publication no. 8, Enschede, The Netherlands, 134–143.

Barr, S.L. (1992) Object-based re-classification of high resolution digital imagery for urban land use monitoring. *Proc. XXIX Conf. Int. Soc. Photogrammetry and Remote Sensing (ISPRS'92)*, 2–14 August. Int. Archs Photogrammetry and Remote Sensing: Commission 7, Washington DC, 969–976.

Barr, S.L. and Barnsley, M.J. (1993) Object-based spatial analytical tools for urban land-use monitoring in a raster processing environment. *Proc. 4th European GIS Conf. (EGIS'93)*, Genoa, Italy, April, 810–822.

Chen, P.C. and Pavlidis, T. (1979) Segmentation by texture using a co-occurrence matrix and a split-and-merge algorithm. *Comput. Vision, Graphics and Image Processing*, **10**, 172–182.

Ehlers, M. (1990) Remote sensing and geographic information systems: Towards integrated spatial information processing. *IEEE Trans. Geosci. Remote Sensing*, **28**, 763–766.

Ehlers, M., Greenlee, D., Smith, T. and Star, J. (1991) Integration of remote sensing and GIS: Data and data access. *Photogrammetric Engng Remote Sensing*, **57**, 669–675.

Forster, B.C. (1980) Urban residential ground cover using Landsat digital data. *Photogrammetric Engng Remote Sensing*, **46**, 547–558.

Forster, B.C. (1985) An examination of some problems and solutions in monitoring urban areas from satellite platforms. *Int. J. Remote Sensing*, **6**, 139–151.

Franklin, S.E. and Peddle, D.R. (1990) Classification of SPOT-HRV imagery and texture features. *Int. J. Remote Sensing*, **11**, 551–556.

Freeman, H. (1961) On the encoding of arbitrary geometric configurations. *IEEE Trans. Electronic Comput.*, **10**, 260–268.

Gastellu-Etchegorry, J.P. (1990) An assessment of SPOT XS and Landsat MSS data for digital classification of near-urban landcover. *Int. J. Remote Sensing*, **11**, 225–235.

Gong, P. and Howarth, P.J. (1989) Performance analyses of probabilistic relaxation methods for land-cover classification. *Remote Sensing Environ.*, **30**, 33–42.

Gong, P. and Howarth, P.J. (1990) The use of structural information for improving land-cover classification accuracies at the rural–urban fringe. *Photogrammetric Engng Remote Sensing*, **56**, 67–73.

Gong, P. and Howarth, P.J. (1992) Frequency-based contextual classification and gray-level vector reduction for land-use identification. *Photogrammetric Engng Remote Sensing*, **58**, 423–437.

Gonzalez, R.C and Wintz, P. (1987) *Digital Image Processing*. Addison-Wesley, New York, 275–287.

Griffiths, G.H. (1988) Monitoring urban change from Landsat TM and SPOT satellite imagery by image differencing. *Proc. Int. Conf. IEEE Geosci. Remote Sensing Soc.*, Edinburgh, 13–16 September. Remote Sensing Society, Nottingham, 493–497.

Guo, Liu Jian and McM. Moore, J. (1991) Post-classification processing for thematic mapping based on remotely-sensed image data. *Proc. Int. Conf. IEEE Geosci. Remote Sensing Soc.*, Espoo, Finland, 3–7 June. IEEE, New York, 2203–2206.

Gurney, C.M. (1981) The use of contextual information to improve land cover classification of digital remotely-sensed data. *Int. J. Remote Sensing*, **2**, 379–388.

Gurney, C.M. and Townshend, J.R.G. (1983) The use of contextual information in the classification of remotely-sensed data. *Photogrammetric Engng Remote Sensing*, **49**, 55–64.

Haack, B., Bryant, N. and Adams, S. (1987) An assessment of Landsat MSS and TM data for urban and near-urban land-cover digital classification. *Remote Sensing Environ.*, **21**, 201–213.

Haralick, R.M. (1979) Statistical and structural approaches to texture. *Proc. IEEE*, **67**, 786–804.

HMSO (1984) *Key Statistics for Urban Areas*. Great Britain Office of Population Censuses and Surveys Registrar General, Census 1981. HMSO, London.

Horaud, R. and Skordas, T. (1989) Stereo correspondence through feature grouping and maximal cliques *IEEE Trans. Pattern Analysis and Machine Intelligence*, **11**, 1168–1180.

Hsieh, Y.C., McKeown, D. and Perlant, F. (1990) *Performance Evaluation of Scene Registration and Stereo-Matching for Cartographic Feature Extraction*, Technical Report, CMU-CS-90-193, Carnegie–Mellon University, November.

Hsieh, Y.C., Perlant, F. and McKeown, D. (1990) Recovering 3D information from complex aerial imagery. *Proc. Image Understanding Workshop*, London, September, 670–691.

Huertas, A. and Nevatia, R. (1988) Detecting buildings in aerial images. *Comput. Vision, Graphics and Image Processing*, **41**, 131–152.

Hwang, V.S., Davis, L.S. and Matsuyama, T. (1986) Hypothesis integration in image understanding systems. *Comput. Vision, Graphics and Image Processing*, **36**, 321–371.

Irvin, R.B. and McKeown, D. M. (1989) Methods for exploiting the relationship between buildings and their shadows in aerial imagery. *IEEE Trans. Systems, Man and Cybernetics*, **19**, 1564–1575.

Jackson, M.J., Carter, P., Smith, T.F. and Gardner, W. (1980) Urban land mapping from remotely-sensed data. *Photogrammetric Engng Remote Sensing*, **46**, 1041–1050.

Kruse, R.L., Leung, B.P. and Tondo, C.L. (1991) *Data Structures and Program Design in C*. Prentice-Hall, Englewood Cliffs, New Jersey, pp. 382–403.

Lee, D. (1990) Coping with discontinuities in computer vision: their detection, classification, and measurement. *IEEE Trans. Pattern Analysis and Machine Intelligence*, **12**, 321–344.

Li, K. and Muller, J.-P. (1991) Segmenting satellite imagery: A region growing scheme. *Proc. Int. Conf. IEEE Geosci. Remote Sensing Soc.*, Espoo, Finland, 3–7 June. IEEE, New York, 1075–1078

Liow Yuh-Tay and Pavlidis, T. (1990) Use of shadows for extracting buildings in aerial images. *Comput. Vision, Graphics and Image Processing*, **49**, 242–277.

Lowe, D. (1987) Three-dimensional object recognition from single two-dimensional images. *Artif. Intell.*, **31**, 355–395.

Luger, G.F. and Stubblefield, W.A. (1989) *Artificial Intelligence and the Design of Expert Systems*. Benjamin Cummings, New York, 77–88.

Martin, L.R.G., Howarth, P.J. and Holder, G. (1988) Multispectral classification of land use at the rural–urban fringe using SPOT data. *Can. J. Remote Sensing*, **14**, 72–79.

Mather, P.M. (1987) *Computer Processing of Remotely-Sensed Images*. John Wiley, Chichester, 289–309.

McKeown, D.M. (1988) Building knowledge-based systems for detecting man-made structures from remotely-sensed imagery. *Phil. Trans. R. Soc. Lond., Ser. A*, **324**, 423–435.

McKeown, D.M. (1991) Feature extraction and image data for GIS. *Spatial Data 2000, Proceedings of the Remote Sensing Society Annual Conference*, Christ Church, Oxford, 17–20 September. Remote Sensing Society, Nottingham, 3–11.

McKeown, D.M., Harvey, W.A. and L. E. Wixson (1989) Automated knowledge acquisition for aerial image interpretation, *Comput. Vision, Graphics and Image Processing*, **46**, 37–81.

Mehldau, G. and Schowengerdt, R.A. (1990) A C-extension for rule-based image classification systems. *Photogrammetric Engng Remote Sensing*, **56**, 887–892.

Mohan, R., Medioni, G. and Nevatia, R. (1989) Stereo error detection, correction, and evaluation. *IEEE Trans. Pattern Analysis and Machine Intelligence*, **11**, 113–120.

Mohan R. and Nevatia, R. (1989) Using perceptual organisation to extract 3-D structure. *IEEE Trans. Pattern Analysis and Machine Intelligence*, **11**, 1121–1139.

Muller, J.-P., Collins, K.A., Otto, G.P. and Roberts, J.B.G. (1988a) Stereo matching using transputer arrays. *Int. Archs Photogrammetry and Remote Sensing*, **27**, 559–586.

Muller, J.-P., Day, T., Kolbusz, J., Dalton, M., Pearson, J.C. and Richards, S. (1988b) Visualisation of topographic data using video animation. *Int. Archs Photogrammetry and Remote Sensing*, **27**, 602–616.

Nichol, D.G. (1990) Region adjacency analysis of remotely-sensed imagery. *Int. J. Remote Sensing*, **11**, 2089–2101.

Peuquet, D.J. (1984) A conceptual framework and comparison of spatial data models. *Cartographica*, **21**, 66–113.

Pridmore, T., Mayhew, J.E.W. and Frisby, J. P. (1990) Exploiting image-plane data in the interpretation of edge-based binocular disparity. *Comput. Vision, Graphics and Image Processing*, **52**, 1–25.

Rafat, H.M. and Wong, A.K.C. (1988) A texture information-directed region growing algorithm for image segmentation and region classification. *Comput. Vision, Graphics and Image Processing*, **43**, 1–21.

Raper, J. F. and Kelk, B. (1991) Three-dimensional GIS. In Maguire, D.J., Goodchild, M.F. and Rhind, D.W. (eds), *Geographical Information Systems: Principles and Applications*, vol. 1. Longman, Harlow, 299–317.

Sadler, G.J. and Barnsley, M.J. (1990) Use of population density data to improve classification accuracies in remotely-sensed images of urban areas. *Proc. 1st European GIS Conf. (EGIS'90)*, Amsterdam, 10–13 April, 968–977.

Sadler, G.J., Barnsley, M.J. and Barr, S.L. (1991) Information extraction from remotely-sensed images for urban land analysis. *Proc. 2nd European GIS Conf. (EGIS'91)*, Brussels, 2–5 April, 955–964.

Schalkoff, R.J. (1989) *Digital Image Processing and Computer Vision*. John Wiley, New York, pp. 349–364.

Toll, D.L. (1985) Effect of Landsat Thematic Mapper sensor parameters on land cover classification. *Remote Sensing Environ.*, **17**, 129–140.

Townshend, J.R.G. (1992) Land cover. *Int. J. Remote Sensing*, **13**, 1319–1328.

Wharton, S.W. (1982a) A contextual classification method for recognizing land use patterns in high resolution remotely-sensed data. *Pattern Recognition*, **15**, 317–324.

Wharton, S.W. (1982b) A context-based land use classification algorithm for high resolution remotely-sensed data. *J. Appl. Photographic Engng*, **8**, 46–50.

Whitehouse, S. (1990) A spatial land-use classification of an urban environment using high-resolution multispectral satellite data. *Proc. 16th Annual Conf. Remote Sensing Soc., Remote Sensing and Global Change*, University College, Swansea, 19–21 September. Remote Sensing Society, Nottingham, 433–437.

Woodcock, C.E. and Strahler, A.H. (1987) The factor of scale in remote sensing. *Remote Sensing Environ.*, **21**, 311–332.

Chapter Fourteen

The application of GIS to the monitoring and modelling of land cover and use in the United Kingdom

R. W. GOODING, D. C. MASON AND J. J. SETTLE
NERC Unit for Thematic Information Systems (NUTIS), University of Reading

AND

N. VEITCH AND B. K. WYATT
The Environmental Information Centre, Monks Wood, Huntingdon

The overall purpose of the project described in this chapter was to develop algorithms and operational procedures, for use with geographic information systems (GIS), that would permit integrated handling of remotely-sensed imagery and other data for the purposes of land cover mapping and for ecological modelling applications. To this end, we have developed a suite of routines that operate between a conventional image processing system and a conventional GIS. This package takes as its input a satellite image, with or without ancillary cartographic information, and breaks this down into the constituent scene elements by a process of segmentation with optional clustering and classification. The process has an almost wholly automatic path, or may be used as a kind of computer-aided scene segmenter. In addition, a land use information system has been built up, and is being added to continually, for the ecologically sensitive area of the Dorset heathlands. This area has received an increasing amount of attention and there is interest from a variety of bodies in preserving what remains of the heath and, in some cases, restoring old heathland. This is a problem that an appropriate marriage of ecological modelling and GIS should be able to take on. The prototype system has been used as a general demonstrator and to identify areas where further research should be directed.

1. INTRODUCTION

The collaborating groups have a history of research in remote sensing, and of participation in large projects on land cover mapping. The NERC Unit for Thematic Information Systems (NUTIS) has mostly been interested in novel techniques of information extraction, while the Institute of Terrestrial Ecology's (ITE) interest lies in obtaining accurate land

Geographical Information Handling – Research and Applications. Edited by P. M. Mather
© 1993 John Wiley & Sons Ltd

cover data as input to real ecological applications. NUTIS has taken part in two previous projects in which the techniques of computer vision have been adapted to the study of remotely-sensed scenes. The first of these was funded under the Alvey programme (Corr *et al.*, 1989), and from this an image segmenter was developed which combined standard image processing operations with an elementary expert system. The resulting segmenter performed very well on agricultural scenes, where the set of expert rules was most highly developed; however, the need to extend these methods to cope with the less regular and more poorly defined patchworks of semi-natural vegetation was appreciated. The second large undertaking, the MuSIP project (Sawyer *et al.*, 1992) was funded under the EC ESPRIT programme; the aim of this was to provide a generic system for image analysis and understanding, based on a so-called picture understanding database (PUD) which acted as a crude GIS. The Environmental Information Centre, in partnership with the British National Space Centre and the Department of the Environment, has undertaken the compilation of a digital map of land cover by automatic classification of multi-seasonal data from the Landsat Thematic Mapper instrument. The resultant database provides information on land cover in 25 broad classes at a field-by-field scale, which extends across the whole of Great Britain; it is the first national land use survey since the 1930s. The data are increasingly being used for national land use planning, as a baseline for predicting ecological consequences of environmental change and as a means of estimating other environmental variables (such as levels of chemical fertilizers, infiltration rates, botanical diversity, etc.) which, while not directly accessible from remote sensing, nevertheless can be expressed in terms of land cover. The land cover database underpins much current and planned work in the ITE aimed at the establishment of quantitative models of ecological change (Wyatt and Fuller, 1992).

The principal thrust of the research of this joint project was to extend work done on image segmentation in previous studies to the more difficult area of semi-natural vegetation. The work proceeded in two parallel streams, with effort devoted on the one hand to providing a robust and efficient set of routines for getting satellite data into GIS form, and on the other to setting up a testbed in which the outputs from the first process will eventually be used to initialize ecological models. The particular application here was a study of the Dorset heathlands and neighbouring countryside, with the GIS being used to identify areas where restoration of the heathlands might be possible.

2. REMOTE SENSING, GIS AND ECOLOGICAL MODELS

The links between ecological modelling and GIS are self-evident, given the essentially spatial nature of many ecological models. Indeed, the spatial constituents of landscape ecology are effectively the lines and polygons of a conventional two-dimensional vector GIS. The problems that the relatively new discipline of landscape ecology attempts to tackle concern the flow of materials, energy and species between landscape elements, and this is the sort of model that GIS ought to be able to handle quite effectively.

The arguments for the coupling of GIS and remotely-sensed imagery are well rehearsed. Satellite imagery is a cheap source of synoptic data for GIS; on the other hand, to obtain the full benefit of high resolution satellite images we need to make use of co-registered data from other sources, and GIS are a natural vehicle to bring these different data sources together. So far, the remotely-sensed data have been seen principally as a source of information on current land cover, which can be determined with some confidence from high resolution images. This can then be used as a surrogate for a number of ecological and micrometeorological variables.

The most usual way of eliciting the information content of the satellite scene is the classification of the image (allocation of pixels to one of a number of land cover classes), followed by a connected component operation to form regions, followed by the extraction of the boundaries; the boundaries are passed into a vector GIS and the class information incorporated as an attribute of the enclosed polygon (raster GIS have a slightly different data path, but the principle is not truly different).

The approach taken here follows the experience gained under earlier projects into land use mapping, and is based upon reversing the order of data extraction: the regions are determined first, a process known as segmentation, and a subsequent operation determines the ground cover classes of the extracted regions. A large body of literature exists to show that this approach to processing gives a more accurate final class map. This is basically because the uncertainty when allocating a region is much less than that when allocating a pixel, and no amount of post-classification processing can make up for that fact. Just as important, from the point of view of inputs to a GIS, is that the boundary information from segmentation is much more accurate than if we start with a per-pixel classification; errors in classification occur mostly around boundaries. The Alvey project (Corr *et al.*, 1989) on satellite image processing showed the practical advantages to be gained by the fusion of imagery with cartographic data, which could be used as a model to guide the segmentation, and also showed the potential of knowledge-based segmentation, using decision rules and a simple expert system. The system was sensitive, however, to the initial model segmentation and worked much less well in the complete absence of cartographic information. However, cartographic information is of limited value in areas of semi-natural vegetation where boundaries between regions may follow no obvious pattern and obey no simple rules. One of the main conclusions to emerge from the Alvey study, therefore, was the need for a robust, reasonably accurate first segmentation derived from the imagery alone, and this was identified as one of the principal objectives of the project.

3. MAPPING POTENTIAL RESTORATION

There is currently a widely based interest in increasing the area of scarce habitats in the UK for conservation purposes. This entails a need for methods to assess rapidly the existing resources and the potential for expansion. In this study we concentrated on one such sensitive environment, lowland heath, and in particular the Dorset heaths, drawing upon the research in this area being carried out by staff at ITE Furzebrook. Groups that have expressed interest in the future of these habitats include English Nature, whose concern is with the potential for re-creating heathland on conifer plantations, and the Royal Society for the Protection of Birds (RSPB), which owns land in Dorset and is interested in heathland fragments for bird habitat. The government has also recognized the importance of these habitats and is encouraging landowners to re-establish areas of heath under the Countryside Stewardship scheme (funded by the Countryside Commission). In the longer term we hope to see the development of quite sophisticated land use change models, based on landscape ecology, for assessing the viability of present heathland patches and for analysis of the likely landscape evolution under different restoration scenarios. However, given that the appropriate models do not yet exist to the required level of spatial detail, and the lack of any tradition in obtaining ecological model parameters from remotely-sensed imagery, we have started with a simpler approach. This is to map the existing heathland fragments from satellite imagery, and to determine the suitability or otherwise of other land use parcels, according to their current land use type, their size and their location with respect to heathland fragments.

3.1. The study area

The Dorset heaths are located on the south coast of England, close to Bournemouth and Poole harbour (Fig. 1). The heaths developed on well drained acid soil derived from tertiary deposits of the Poole basin (Moore, 1962). Over the past 200 years their extent has decreased to around 14% of their former range (around 30 000 hectares in 1811). The losses have been almost entirely caused by human activity – agriculture, forestry and urban growth – and much of what remains is under pressure. The fragmentation and isolation of the remaining heaths has significant consequences for the remaining rare species, and it has become important to assess the viability of the remaining patches as functional heathland ecosystems and identify those sites that can best be managed for conservation. In order to

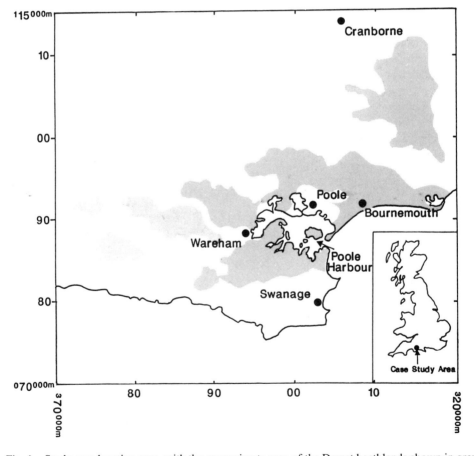

Fig. 1 Study area location map, with the approximate area of the Dorset heathlands shown in grey

do so, it is necessary to characterize the status of the patches, their shape, size and the land use matrix in which they exist. These factors determine the susceptibility of these heathland enclaves to invasion by other species, and their potential to remain as significant reservoirs of heathland species.

The strong dependence of heathland on soil type means that it should be possible to map the maximum former or potential distribution by mapping the soil distribution; this approach has recently been adopted by ITE, at a resolution of 1 km square. Soil information is, however, unavailable at larger scales, so our estimates of former heathland extent are based on cartographic sources dating back to 1811. Though not explicitly mapped as heath, areas of rough unenclosed land shown on the 1811 and 1896 Ordnance Survey map sheets may be considered heath if they occur on the appropriate soil types (Moore, 1962); the 1934 Land Utilization survey provides further information, as do field surveys carried out in 1960 and 1974 (Moore, 1962; Webb and Haskins, 1980). The maximum extent is taken from a composite of the 1811 and 1896 Ordnance Survey maps; a composite was used because some of the land taken into agriculture during the Napoleonic blockade by 1811 had reverted to heath by 1896.

Because the two parts of the study were proceeding in parallel, the outputs from the automatic segmenting software were not available. To begin with, some 1984 Landsat imagery was studied and classified in traditional image processing fashion; later, the ITE land cover map, derived from pixel-by-pixel classification of combined summer and winter imagery from the Landsat Thematic Mapper, became available and was used to map current land use on former heathlands.

The main cover types on the former heaths are arable, grassland, urban land and evergreen forestry. Urban land has no real restoration potential, while evergreen forestry has much; heathland species are found to survive within these areas and the soil remains acid. Further work at ITE has shown that, on former heaths, less soil alteration is evident on managed grassland than on arable fields. The reason for this is the increased fertility of the soil and the destruction of the heath seed bank under arable agriculture. The increased fertility allows non-heath species to out-compete characteristic heathland flora, and the depletion of the seed bank means that heathland regeneration is unlikely to occur naturally. However, the closer a fragment is to existing heathland, the less of a problem this is, and distance from existing heathland is an important variable in determining viability of restoration.

3.2. Data processing

The Dorset Heathland Information System is being developed in a commercial raster GIS, the initial digitization of historic heathland boundaries having been done on a variety of systems and converted. In addition to the historic boundaries, contours, forestry, coasts, drains, rivers, field boundaries and roads have also been digitized. Digital boundary information for some parts of the heath were derived from aerial photography to permit the incorporation of very fine detail. Heathland boundaries have been included for the 19th century, and for the years 1934, 1960, 1974, 1984 and 1990. Land use information for various historic periods is continually being added to the database.

The ITE land use map consists at present of a sequence of raster images of tokens denoting cover type. The land use map was processed to eliminate obvious errors, certain classes were amalgamated and the thematic image was subjected to a post-classification majority filter. The boundaries were automatically digitized and the regions imported into the GIS. As a result of this we estimate the present area of heathland (1990) to be 4700 hectares, a decline of 20% on the 1978 estimate.

The suitability for restoration of heathlands can be mapped from the land use image, and the following figures have been obtained. On former heathlands we find 5400 hectares of evergreen woodland (3400 of these within 100 m of surviving heath); 1500 hectares of post-1960 grassland, and 4400 hectares of longer-established grass. In addition, 6900 hectares of arable land are now located on former heaths. If landscape models prove all of these to be viable then it should be possible to effect a significant increase in heathland area.

It should be emphasized that the indicated areas are not those where regeneration will occur naturally; rather, they show the potential for restoration and offer a means for focusing subsequent field survey and model-based research. The ITE land cover map will not itself be the best source of land cover information for this further work because landscape models need information on boundaries. Adjacent fields of the same cover type might be separated by a hedge, or parcels of woodland separated by firebreaks, but are likely to have been merged into single units by the pixel-based processing described above. These boundaries, however, have significant ecological importance. We should aim, therefore, to extract as much of this information as possible from the original satellite imagery, which impels us towards segmenting the image before attempting to allot land cover categories to picture elements.

4. IMAGE ANALYSIS SOFTWARE

A special suite of image processing procedures, consisting of about 15 000 lines of code, was developed for this project, mostly for use on an image of regions. This subsystem is intended to sit between an everyday image processing system at one end and a GIS at the other. We take from the image processing system a suitably geo-referenced satellite image and generate from this a vectorized set of regions, together with various attributes of each region. These could be quite simply the averaged radiances, or some measure of the texture as derived from those digital values; land cover labelling is another option, which we have treated at more length. The heart of the system is a region editing module, which is intimately bound up with clustering and classification routines. Other elements have been written but may be bypassed. The system is designed to run with any degree of user interaction: a scene can be segmented completely interactively, completely automatically, or at various levels in between. Only the allocation of land cover labels to regions does not yet have an automatic version. The most important elements of this are:

1. An initial image segmentation procedure.

2. An interactive region editor.

3. Fuzzy clustering and allocation routines.

4. A routine for incorporating historic land cover information over larger areas to provide information on likely current cover, using the method of belief functions.

The region editor is closely coupled to the clustering algorithm; segmentation performance is improved by feeding back the results of classification (Corr *et al.*, 1989). The reason for this is probably that the classification stage introduces a certain amount of

The Application of GIS to Monitoring and Modelling

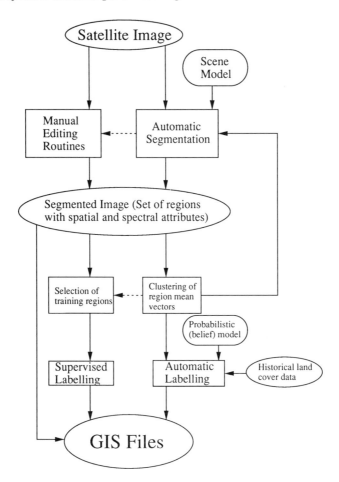

Fig. 2 Main pathways (manual and automatic) by which GIS files may be extracted from satellite images

information about the global properties of the scene. The initial, automatic, segmentation is in fact an optional element in the overall system, since the region editor can act as a fully functional, interactive segmenter in its own right. The system is also designed so that this initial segmentation is essentially an external routine, interchangeable at will. The same applies to the algorithm used to allocate classes to regions, although the fuzzy clustering, which can act as an unsupervised classifier, is part of the core. The end product of these various routines is a set of boundaries, regions and class attributes, which can be transferred to a GIS. In addition, uncertainties for the integrity of boundaries and regions can be generated if the initial segmentation, defined below, is used. Fig. 2 shows the relationships of the various parts of the system.

4.1. Segmentation

There are two standard approaches to the problem of dividing a scene into its natural picture elements. The first of these is to identify homogeneous regions directly, by lumping together groups of contiguous pixels with similar grey levels, and expanding these groups until no more pixels are added to a region. This is known as region growing. The other principal strategy is edge-finding, which aims to identify the uniform areas by drawing around them, that is, by finding the boundaries between them. Most segmentation algorithms contain elements of both, and that is the case with the automatic segmentation procedure developed in this project. It is designed for images containing lots of small, fairly homogeneous areas, although – unlike the approach taken in the Alvey or MuSIP projects – we make no assumptions about the shapes of areas. The method begins with an edge-finding and tracing algorithm, based on Martellia (1976), which is described in greater detail elsewhere (Gooding *et al.*, 1991). Each band of the satellite image gives rise to its own edge image, and these are then combined (logical OR). A standard connected component finder then derives a list of regions from the list of edges. Because fairly relaxed thresholds are set for this stage, the first product of the process is highly oversegmented; that is, it has far more edges present than actually exist in the scene. Few real boundaries that are detectable at the resolution of the image are likely to have been missed; however, a later stage of processing attempts to trap any that might have fallen through. The following stage is to merge the small regions, taking out the false edges. This is done by applying a series of statistical tests to each boundary segment to see if it is real. The nature of the evidence looked for is:

1. Are the regions on either side of the boundary significantly different in mean grey level?
2. Is the average difference in grey level across the edge significant compared to the variation in either region?
3. Is the mean-squared difference across the edge more than would be found on a random line through the union of the two separated regions?

Criteria (2) and (3) appear to be similar, but different types of edge are in fact being tested for, and experience has shown that both tests need to be kept if some real edges are not to be lost at this stage. If any of these three tests, for any band, is satisfied, then no merge will take place. The final list of regions depends on the order in which these merges are carried out, and so, to remove any arbitrariness, the order is determined by the significance of test (1); the first boundary to be tested is between the pair of regions with the most similar mean grey levels. Once these merges have been accomplished the segmentation is almost complete, but a final loop is included; each region is tested to see if it should be split using a histogram thresholding technique on the grey levels of the satellite band with the greatest variability for that region. If an edge is found in this way, it is tested for significance and the original region could be reconstituted. The purpose of this last round of splitting is to detect and test any real edges that were not strong enough to be found by the edge following procedure, which has to use a threshold based on global statistics; however, the existence of this last loop in the processing means that the final segmentation is not very sensitive to the precise value of the threshold, nor indeed to the fine detail of the initial edge following.

4.2. Region editing

A segmentation that has been created by completely automatic means is bound to contain errors, and it is desirable to be able to eliminate obvious blunders after the event, or simply to override the decisions of the automatic segmenter. For this reason an interactive editor for regions was developed. Basic features of the region editor are:

1. Merging of adjacent regions (this is hardly ever used).
2. Splitting regions by one of a number of methods.
3. Deleting pixels from a region (to form either new very small regions or isolates).

Only the second of these needs to be described further; the first and third are very straightforward to implement. The splitting option is a little more difficult to realize in a satisfactory way; we wish to avoid the detailed drawing over the scene of break lines. A number of ways of splitting a region has therefore been implemented within this system. The first of these is a simple method based on histogram splitting of the grey levels of a single band; this may divide the region cleanly but is more likely to create a number of islands within the original region. It works best when such small inclusions genuinely do exist – clumps of trees in open grassland, for example. Another aims to split a region at an internal neck by a combination of mathematical morphological operations; the region is eroded until it just splits into two; the two regions thus formed are then dilated back to fill the original space. Any odd pixels left over can be merged with the most similar of the two new regions, or left as isolates. Breaks along a boundary that has been missed can be put into effect by pointing near the boundary required and selecting the option; the most likely edge near the point is found and extended in each direction, using the edge following procedure, until the boundary of the region is reached. The user does not need to position a cursor precisely, nor to draw the line, although this option is also available. In fact, it is possible to segment small scenes perfectly adequately with just this option, when the operation becomes a computer assisted, rather than an automatic, segmentation.

4.3. Clustering

The main purpose of this step in the process is to identify cluster centres in the feature space whose points are the means of individual regions. From these cluster centres the class specific information needed for classification (namely, the means and covariance matrices of different classes) can be generated. Only the larger regions are used in this step because smaller regions have a higher proportion of mixed pixels and so a greater variability; if many relatively small regions are used then it is better to work with the means of the interiors of the regions. The distribution of region means in the feature space produces much better clusters than are exhibited by the clustering of individual pixel values. The actual method used is a variant of fuzzy clustering (Kent and Mardia, 1988; Fisher and Pathirana, 1990). While the results are generally very good, errors do still occur and the clustering and classification steps have been lumped together with the region editor to provide a fast and efficient means of interactive classification. A measure of error is provided by the dissimilarity of a region's mean vector to that of the cluster centre to which it has been allocated; the average error per cluster can be used to decide whether the number of clusters needs to be increased, although it is more difficult to ascertain whether too many clusters have been specified. Some simple point-and-click methods of identifying the cover type of regions, when known, have been included; these can serve either as an element of a supervised region classifier, or as a means of identifying the cover type of a cluster.

4.4. Integration with ecological databases

A very large amount of historical information exists at different scales for the Dorset heaths. ITE's Heathland Survey for Dorset records the presence of different species within 200-m squares, on a scale of 0 (missing) to 3 (greater than 50% coverage). However, the resulting numbers may be apparently inconsistent. If we imagine a woodland with a grass understorey covering one of these squares, then both cover types would be deemed to be 100% present. A simple means of interfacing the satellite imagery with this database was constructed, which allows the interrogation of the database via a display of the scene. A species can be specified and the squares where it occurs are overlaid on the image; alternatively, point-and-click operations will give the relevant survey information, should the pixel selected lie within one of the squares. This sort of operation is a useful aid to interpreting the remotely-sensed imagery and can form part of the interactive classification element of the core. In addition, a prototype was developed for the processing of the historical database in such a way as to produce estimated probabilities for land use and land cover types of immediate interest. This uses a simple rule base and the calculus of belief functions to manipulate the uncertain evidence of the survey data. One unfinished area of work was the planned use of these data with the cluster information to attempt automatic labelling of the clusters.

The relationship of the different parts of the system are set out in Fig. 2, which shows the two main pathways (manual and automatic) by which GIS region files may be extracted from satellite images, and indicates how some of the automatic procedures may be used to enhance the manual operations.

4.5. Results

The segmentation procedures were tested using the method and test image described in Mason *et al.* (1988). The test image was an Airborne Thematic Mapper image of part of the agricultural countryside around Blewbury, in Oxfordshire. The output segmentation is compared to a reference segmentation, produced by matching digital map boundaries to the remotely-sensed image. There are two types of error we might wish to consider when appraising a segmentation, which loosely correspond to errors of commission and omission and are known as 'under-merge' and 'over-merge' errors, respectively (Levine and Nazif, 1982). An over-merge error occurs when more than one region on the ground corresponds to a single region in the segmentation, and under-merging happens when a region in the reference segmentation overlaps more than one region in the test segmentation. In Mason *et al.* (1988) it was shown that a traditional segmentation gave a combined error of 46.5% on the test image, whereas the best knowledge-based method, without map data to guide the process, had an error value of 33.8%. The methods described in this chapter, which take no advantage of the likely shape of the fields, gave an error value of 30%, most of these occurring in a highly textured built-up area. Some problems with the reference segmentation were in fact noted, particularly some obvious mis-registration of field boundaries, and so a freehand reference segmentation was generated. The error measure for this, more realistic, comparison fell to 15%. This matches the best results attained under the Alvey project, which had the benefit of digitized map data to provide an initial guess at segmentation, and the use of a rule base on shape and texture. The results of supervised classifications on a number of other test images were compared, on both a per-pixel and a region basis. As expected, more accurate results were invariably obtained by classifying the segmented image. In one case the error rate was nearly halved by using a segmentation based on just two bands out of eleven available.

5. AREAS OF FUTURE RESEARCH

The single most important piece of unfinished business in the image processing package centres on the labelling of an unsupervised classification. We feel that, with a reasonably complete data set of historical information on land cover and a simple set of rules relating to the spectral properties and textural characteristics of broad cover types (woodland, grassland, etc.) this should be feasible using existing tools and algorithms, such as those developed under the Alvey programme. One of the planned products from the ITE land cover map is a database of cover types at 1-km scale over the United Kingdom; proportions of each cover type within these larger squares are being prepared to assist with the evaluation and validation of the land cover map. This data set will form an invaluable source for an automatic labeller, such as has been suggested.

Another area where further study needs to be carried out is the integration of GIS with ecological models. In fact, the incorporation of environmental models within GIS needs general attention if GIS are to realize their full potential and become more than organizers of environmental data. With regard to the specific problem of the Dorset heath study, landscape ecology models would appear to be the direction to take, testing further the suitability of different strategies for heathland restoration. The philosophy and, so far as they exist, the techniques of landscape ecology, constitute the most appropriate theoretical vehicle for such study, which can then address such questions as whether it is best to try to expand existing patches, or to re-create patches in the spaces between the existing ones. Segmentation of remotely-sensed imagery of high spatial resolution, as developed under this Special Topic, will provide an accurate patchwork of landscape elements, and these boundaries are in a form ready for inclusion in a GIS. Land cover for this area can also be obtained to a fairly high degree of accuracy from the satellite imagery. The identification of land use then (nearly) completes the definition of the landscape structure. However, land use and land cover are not the same and the generation of a land use map may involve a certain amount of modelling on the land cover map; different forms of grassland have very different ecological roles, according to whether they are able to set seed. Additional effort may be required to characterize further the nature of boundaries between elements, and simple spectral models should enable some information on boundary type to be obtained from the remotely-sensed imagery. Experimental work is needed to refine further, and in some cases simply to define for the first time, ecological constants such as dispersal rates and the effectiveness of certain types of boundary as barriers to nutrient and species flow.

REFERENCES

Chapman, S.B., Clarke, R.T. and Webb, N.R. (1989) The survey and assessment of heathland in Dorset, England, for conservation. *Biol. Conservation*, **47**, 137–152.

Corr, D.G., Tailor, A.M., Cross, A.M., Hogg, D.C., Lawrence, D.H., Mason, D.C. and Petrou, M. (1989) Progress in automatic analysis of multi-temporal remotely-sensed data. *Int. J. Remote Sensing*, **10**(7),1175–1195.

Fisher P.F. and Pathirana, S. (1990) The evaluation of fuzzy membership of land cover classes in the suburban zone. *Remote Sensing Environ.*, **34**, 121–132.

Gooding, R.W., Settle, J.J., Veitch, N. and Radford, G.L. (1991) Image segmentation for areas of semi-natural vegetation: The Dorset Heaths. In Dowman, I. (ed.), *Spatial Data 2000. Proceedings of the Remote Sensing Society Annual Conference*, Christ Church, Oxford, 17–20 September. Remote Sensing Society, Nottingham, 232–241.

Kent, J.K. and Mardia, K.V. (1988) Spatial classification using fuzzy membership models. *IEEE Trans. Pattern Analysis and Machine Intelligence*, **10**(5), 659–670.

Levine, M.D. and Nazif, A.M. (1982) An experimental rule-based system for testing low-level segmentation strategies. In Preston, K. and Uhr, L. (eds), *Multicomputers and Image Processing: Algorithms and Programs*. Academic Press, New York, 149–158.

Martellia, A. (1976) An application of heuristic search methods to edge and contour detection. *Communs ACM*, **19**(2), 73–83.

Mason, D.C., Corr, D.G., Cross, A., Hogg, D.C, Lawrence, D.H., Petrou, M. and Tailor, A.M. (1988) The use of digital map data in the segmentation and classification of remotely-sensed images. *Int. J. Geogr. Inf. Sys.*, **2**(3), 195–215.

Moore, N.W. (1962) The heaths of Dorset and their conservation. *J. Ecol.*, **50**, 369–391.

Sawyer, G., Mason, D.C., Hindley, N., Johnson, D.G., Jones-Parry, I.H., Oddy, C.V.J., Pike, T.K., Rye, A.J., de Salabert, A., Serpico, B. and Wielogorski, A.L. (1992) MuSIP Multi-Sensor Image Processing system. *Image and Vision Comput.*, **10**(9), 589–609.

Webb, N.R. and Haskins, L.E. (1980) An ecological survey of the heathlands in the Poole Basin, Dorset, England, in 1978. *Biol. Conservation*, **17**, 281–296.

Wyatt, B.K. and Fuller R.M. (1992) European applications of space-borne Earth observation for land cover mapping. In *Environmental Observation and Climate Modelling through International Space Projects, vol. II: Remote Sensing for Environmental Monitoring and Resource Management*, European International Space Year Conference, Munich, Germany, 30 March–4 April. European Space Agency, Report ISY-1, Paris, 655–659.

Chapter Fifteen

GIS and distributed hydrological models

R. ROMANOWICZ AND K. BEVEN
Centre for Research on Environmental Systems and Statistics, Lancaster University

AND

R. MOORE
Institute of Hydrology, Wallingford

The objectives of the research reported in this chapter were:

1. To develop modelling procedures within the TOPMODEL framework that exploit the properties of spatial data available from a geographic information system (GIS).

2. To carry out these developments within a GIS framework capable of handling data that varies spatially and temporally.

3. To assemble a data set of both fixed and time varying spatial data for one or more catchments in north-western England within the area covered by the North West Water Hambledon Hill rainfall radar.

The Water Information System (WIS), developed by the Institute of Hydrology as a hydrology-orientated GIS, was used as the spatial and time variable input, user interface and graphic output of the results for the hydrological modelling. A physically-based, quasi-distributed hydrological model (TOPMODEL) was linked with WIS as one of its application modules. Special techniques within the WIS framework allowing on-line data manipulation (data retrieving and updating) were developed. In particular, the database design was modified to handle time-series grids and the user interface was upgraded. The modelling procedures within the TOPMODEL framework were also enhanced to make use of the new types of spatial data available from WIS. Monte-Carlo simulation within a generalized likelihood uncertainty estimation (GLUE) framework, performed on a transputer system and initialized from WIS, was used for the calibration of TOPMODEL. The results of the uncertainty estimation can be visualized using the graphical display facilities of WIS. The results of this research demonstrate the feasibility of incorporating modelling capabilities within a GIS.

Geographical Information Handling – Research and Applications. Edited by P. M. Mather
© 1993 John Wiley & Sons Ltd

1. INTRODUCTION

The purpose of this chapter is to discuss and illustrate the procedures involved in linking a hydrological model (TOPMODEL) and a GIS. WIS is a hydrology-orientated GIS developed by the UK Institute of Hydrology. It provides methods and tools for the storage, retrieval and graphic display of both spatial and non-spatial data. Thus, it can serve as a data manager and as distributed input for distributed hydrological models. Hydrological models are dynamic in nature, which poses additional requirements for on-line data manipulation, retrieval and updating.

WIS was developed by the Institute of Hydrology in collaboration with ICL. It is based upon a four-dimensional database designed to capture, store, retrieve and visualize hydrological data that varies in both time and space. It is a unified database, containing both spatial and non-spatial data. Based on C, Fortran and embedded SQL data retrieval and storage, WIS can serve as the user interface, data manager and graphic output for the distributed hydrological models.

The distributed hydrological model TOPMODEL uses digital elevation data as well as spatial information on soil, rain and vegetation for the prediction of soil moisture distribution in the catchment. Until now, the preparation of data for each TOPMODEL application required a long and demanding period of data analysis. The display of TOPMODEL results – hydrographs and a variety of different types of maps – required considerable preparation. The linkage of TOPMODEL with WIS had the aim of overcoming these difficulties. The developed system has the novel feature of fully automatic internal manipulation of data. WIS serves as the only source of spatially and temporally variable hydrological data. Users, via the interface, specify the site (catchment) and time periods to be modelled together with the range of calibration parameters. The model then takes its data from the database and subsequently returns the results there for display and evaluation.

Previous use of WIS had concentrated on the storage, display and analysis of static catchment characteristic data (topography, river networks, water quality classifications and so on). In this project these data sets were used as the basis for fully dynamic modelling using the distributed TOPMODEL, including procedures for choice of model structure, model calibration and uncertainty estimation and the display of both hydrographs and spatial predictions within WIS.

The application of TOPMODEL within the WIS framework for the prediction of flows and soil moisture distributions on the Salwarpe catchment is presented. The sensitivity analysis and calibration of the model is performed using GLUE on a transputer system and initialized from within WIS. The results of the uncertainty estimation are visualized on the WIS graphical facilities.

2. A BRIEF DESCRIPTION OF TOPMODEL

2.1. Main objectives of the TOPMODEL work

The integration of TOPMODEL within the WIS structure gives rise to two important problems. The more crucial is the problem of generalization of the model structure to allow for its use in varying topographic and hydrologic conditions with different data availability. This problem can be called the 'calibration problem', as it can be solved by a proper sensitivity–identification–validation algorithm. In order to avoid over-parameterization (e.g. Blackie and Eeles, 1985; Beven, 1989) the smallest possible number of parameters should be chosen for calibration.

The other significant problem involves the restructuring of the model to allow it to be run interactively, thus permitting the user to view intermediate results in graphical form and so assess the performance of the model and the validity of the initial parameter values.

The possible applications of TOPMODEL are:

- Catchment runoff forecasting and evaluation of the spatial distribution of runoff production.
- Evaluation of spatial distribution of evapotranspiration and soil moisture content on a catchment scale or at the nodes of a mesoscale grid of a general circulation model.
- Distributed prediction of flow in different river reaches for flood or pollution protection.

2.2. Basic assumptions

TOPMODEL is a set of simple concepts for modelling catchment response involving primarily subsurface flows and dynamic saturated contributing areas (although versions allowing spatially-variable infiltration excess runoff production also exist). It uses physically-based parameters and a spatially-distributed topographic index computed from a digital terrain model (DTM). The TOPMODEL package can take into consideration the individual circumstances of a particular catchment. This is possible because of the way in which TOPMODEL produces spatial patterns of hydrological variables that may be compared, quantitatively or qualitatively, with responses observed in the field. The requirement of building this functionality into the WIS framework thus presents a challenge that will depend crucially on the graphical presentation of the modelling results within WIS. The theory underlying the model is presented in a number of papers (e.g. Beven and Kirkby, 1979; Beven and Wood, 1983; Beven et al., 1984; Beven, 1986; Sivapalan, Beven and Wood, 1987). Similar ideas have been developed by O'Loughlin (1986). The model takes into account spatial heterogeneity of topography and soil. As a basic assumption it separates the calculation of spatially-averaged dynamics of the catchment and local soil moisture accounting. Both parts of the model are linked together with the help of catchment distribution functions, which can be evaluated on the basis of topographic and soil data (Quinn et al., 1991).

2.3. Model structure

The schematic structure of the model is presented in Fig. 1. The model consists of five basic components, the first of which refers to the initialization of the model based on the initial value of the observed flow from the catchment. The second component describes the root zone storage S_r, with maximum storage S_{rmax}, which should be filled before the infiltration to the unsaturated zone takes place.

In the third component the distribution of soil moisture deficits over the catchment is evaluated. Rather than operate on a spatial grid, TOPMODEL evaluates soil moisture deficits at any point i, S_i, for the discrete values of the frequency distribution of the soils/topographic index, $\ln(a_i/T_i\tan\beta_i)$, where a_i is the local upslope drainage area, β_i is the local slope angle and T_i denotes local saturated transmissivity. The static relation between the local values of deficits and averaged over the catchment deficit, S_t has the form:

$$S_i = S_t + m[\gamma - \ln(a_i/T_i\tan(\beta_i))] \tag{1}$$

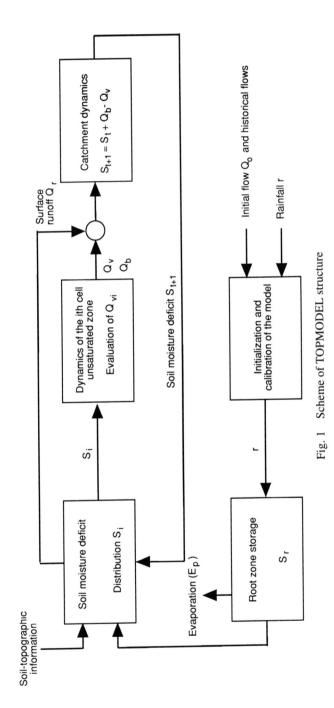

Fig. 1 Scheme of TOPMODEL structure

where

$$\gamma = (1/A) \int_A \ln(a_i/\tan\beta_i)da - (1/A) \int_A \ln T_i da \qquad (2)$$

m is a recession parameter and A is the total catchment area.

Equation (1) assumes an exponential relation between transmissivity and storage deficit and applies for any steady recharge rate. With its help we can obtain maps of the soil moisture deficit for each time step of a TOPMODEL run. The fourth component calculates a vertical flux rate recharging the saturated zone. For each value i of the soil topographic index, the dynamics of the ith cell of the unsaturated zone is evaluated and the recharge to the saturated zone Q_{vi} is found. The fifth component describes the dynamic changes of the catchment average soil moisture deficit S_t:

$$dS_t/dt = Q_b - Q_v \qquad (3)$$

where Q_b (mh^{-1}) is the total hill slope drainage to the stream channels and Q_v (mh^{-1}) is the total amount of recharge from the unsaturated zone to the saturated zone in the catchment. It can be shown that under these assumptions Q_b should fulfil the relation:

$$Q_b = Q_0 \exp(-S_t/m) \qquad (4)$$

where the coefficient Q_0 is given by:

$$Q_0 = \exp(-\gamma) \qquad (5)$$

The drainage rate from the unsaturated zone into the groundwater zone over the whole catchment Q_v is given by:

$$Q_V = \sum_{i=1}^{n} Q_{vi} \qquad (6)$$

where n denotes the number of soil topographic index increments $a_i/T_i\tan\beta_i$.

The surface layer of the soil may become saturated in two ways: through excess of rainfall rate over the available storage (saturation from above) or through the rise of the water table level due to drainage from upslope (saturation from below). In both cases the excess of water at the surface will form the surface runoff Q_r. The initial value of average saturation over the catchment is evaluated from the relation (4), assuming knowledge of the initial flow from the catchment in the first time period of the modelled process. In order to account for the time delays in water flow, a flow routing model is introduced. It utilizes the map of distances along the river network from the catchment outlet formed in the course of digital terrain data analysis.

3. WIS–TOPMODEL IMPLEMENTATION

The structure of the flow of information within the WIS–TOPMODEL package is shown in Fig. 2. It requires fast, on-line data management for inserting, updating and retrieval of the database data in real time, while allowing for the visual effects of the choice of the model structure and parameters to be compared. As the scheme suggest it is a closed system – all the data for a chosen catchment are retrieved from the WIS database and inserted back to the database after the transformation in TOPMODEL. For any external application such as parallel computing the flow of information will be the same, which makes the system very flexible. A TOPMODEL application includes five main procedures:

- Catchment characteristics
- Generalized likelihood uncertainty estimation (GLUE)
- Flow prediction
- Soil moisture distribution
- Map display

The first option allows the choice of the catchment from the map, and gives its description and the list of available data. When a particular catchment is chosen, all data required

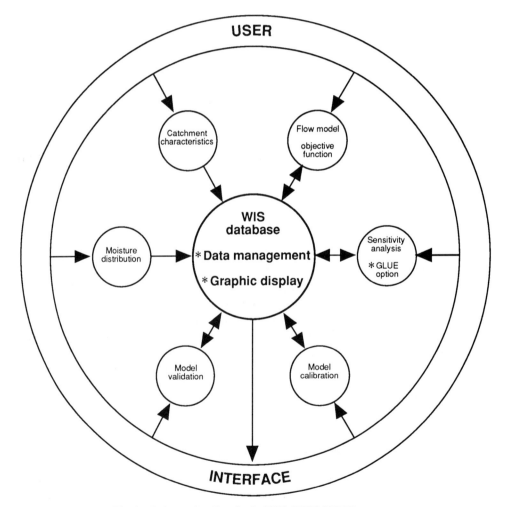

Fig. 2   Information flow in the WIS–TOPMODEL system

for the TOPMODEL run are retrieved from the database and passed to the general TOPMODEL input files. The second application (GLUE) switches WIS to the Meiko transputer system to perform the sensitivity analysis of the catchment model. The third option is used to model the catchment runoff production, and the fourth evaluates the spatial distribution of soil moisture and evaporation fluxes in the catchment, while the fifth is used to display the spatial predictions.

The GLUE procedure is described fully by Beven and Binley (1992). It allows the user to choose the set and value range of parameter values for the sensitivity analysis, and then starts to run Monte-Carlo simulations on the Lancaster University Meiko transputer system, with up to 80 processors. The results of the simulations are sent to the WIS directory, where the proper likelihood maps are formed. Confidence limits for the predicted flows are evaluated and loaded into the database. Graphic display of sensitivity maps and hydrographs in a WIS window allows comparison of the results and the choice of the parameter values.

When the transputer system is not available, the sensitivity analysis of the model may be performed in the flow prediction option. That option also allows the calibration and validation of the model. The structure of the model and its performance will depend on the catchment characteristics and available measurements. The goal of the modelling helps us to choose a proper criterion (goodness of fit function) for the comparison of the model performance. Calibration is carried out using an automatic non-gradient optimization routine. The model structure should then be validated on a different time period of the rainfall and discharge record. Comparison of the hydrographs of simulated and observed flows adds further information, allowing the user to choose the model structure and parameter values.

In the soil moisture distribution option, soil moisture maps are computed on the basis of the TOPMODEL soil moisture deficit evaluations and a simple infiltration model. The changes of subsurface soil moisture content and evaporation with time can be visualized dynamically on-line in the map display option, taking advantage of the static spatial relationships in TOPMODEL by the use of appropriate changes of the colour look-up table. The radar rainfall data, where available, can be analysed within the catchment choice option.

4. APPLICATIONS FOR CATCHMENT MODELLING

To illustrate the use of TOPMODEL within the WIS framework, data from the River Salwarpe catchment were used. The catchment area above the gauging station at Harford Mill is 184 km^2. The Salwarpe catchment is located in the lower Severn basin and the area is devoted mainly to mixed agricultural land use. Only event-based hourly flow and rainfall data were available for the catchment, with the longest event being 92 hours in duration. The DTM applied has a 50-m grid. For flow prediction in a small catchment, with an area of less than 10 km^2 the results of simulation may be accurate enough using the simplest version of TOPMODEL without a flow-routing procedure and with only two calibration parameters (Romanowicz et al., 1993). Also, where continuous flow data over a period of at least a few months are available, the initialization of TOPMODEL does not present problems, as the influence of the initial conditions on the model performance quickly ceases. However, for the Salwarpe catchment the initial conditions influenced the choice of parameters because of the lack of continuous data. Because the length of the Salwarpe catchment is about 30 km, the flow-routing algorithm was added to the TOPMODEL routine to take into account the delays of water flow. The sensitivity analysis showed that three parameters (the recession parameter m, the catchment average transmissivity T and channel flow velocity, CHV) influenced the model performance significantly. The sensitivity map evaluated using the GLUE option is shown in Plate IX. The sensitivity analysis allowed the choice of suitable feasibility regions for the parameters. Later, the model was calibrated using the Powell non-gradient optimization (Powell, 1970) with logarithmic transformation of variables to constrain the space domain. Because of the large catchment area and the lack of long, continuous series of data the efficiencies of the model performance for the validation period are not very great (about 60%). Relative soil moisture content and evaporation rate distribution maps for the beginning of the simulation period for the Salwarpe catchment are shown in Plates X and XI, respectively.

5. DISCUSSION

The work reported in this chapter has achieved two aims:

1. Preparing the user–WIS interface to allow for the various TOPMODEL options.
2. Adjusting the TOPMODEL application to a GIS environment and thus taking advantage of the new possibilities provided by WIS.

The user–WIS interface allows for the choice of the catchment from the map and presentation of all available data for the given site (including radar estimates of spatial rainfall patterns). In the next step the user can perform the sensitivity analysis of the chosen flow model to its parameters using the parallel Meiko transputer system external to WIS. The results of the sensitivity analysis can be displayed using the WIS graphic facilities. Sensitivity analysis allows evaluation of the parameter region for which the performance of the model is the best under the given series of rainfall–runoff data. The flow prediction option allows adjustment of the TOPMODEL parameters in calibration mode and subsequent validation of its performance for new sets of rainfall–runoff data. The hydrographs illustrating the model performance can be displayed on the WIS screen. According to the results acquired, the structure of TOPMODEL as well as the objective function used for the evaluation of model performance can be adjusted on-line by the user. The soil moisture distribution option gives the possibility of evaluation and display of subsurface soil moisture content and evaporation maps for the catchment for different rainfall patterns. The change of subsurface soil moisture content and evaporation with time can be animated during simulation in the map display option in a quasi on-line mode, by appropriate changes of the colour look-up table. An example of the use of the WIS database structure together with the Meiko parallel computing system indicates that further external applications can be added to the WIS structure. That option can be of a vital importance when using any satellite or radar images, which require specific data processing and calibration using software specially designed for that purpose (e.g. MATLAB, Magiscan System GENIAS).

ACKNOWLEDGEMENTS

This work was supported by NERC grant GST/02/491 and the ENCORE program of the EC. Colleagues from the hydrology section of Lancaster University, particularly Paul

Quinn and Jim Freer, as well as Nick Bonvoisin, Andrew Howes, Robert Flavin and David Morris from the Institute of Hydrology are thanked for their help and support in implementing WIS at Lancaster University.

REFERENCES

Beven, K.J. (1986) Runoff production and flood frequency in catchments of order n: An alternative approach. In Gupta, V.K., Rodriguez-Iturbe, I. and Wood, E.F. (eds.), *Scale Problems in Hydrology*. Reidel, Dordrecht, 107–131.

Beven, K.J. (1989) Changing ideas in hydrology – the case of physically based models. *J. Hydrol.*, **105**, 157–172.

Beven, K.J. and Binley, A.B. (1992) The future of distributed models: model calibration and uncertainty prediction. *Hydrol. Processes*, **6**, 279–298.

Beven, K.J. and Kirkby, M.J. (1979) A physically based variable contributing area model of basin hydrology, *Hydrol. Sci. Bull.*, **24**(1), 43–69.

Beven, K.J., Kirkby, M.J., Schoffield, N. and Tagg, A. (1984) Testing a physically-based flood forecasting model TOPMODEL for three UK catchments. *J. Hydrol.*, **69**, 119–143.

Beven, K.J. and Wood, E.F. (1983) Catchment geomorphology and the dynamics of runoff contributing areas. *J. Hydrol.*, **65**, 139–158.

Blackie, J.R. and Eeles, C.W.O. (1985) Lumped catchment models. In Anderson, M.G. and Burt, T.P. (eds.), *Hydrological Forecasting*. John Wiley, Chichester, 311–346.

O'Loughlin, E.M. (1986) Prediction of surface saturation zones in natural catchments by topographic analysis. *Water Resources Res.*, **22**, 794–804.

Powell, M. (1970) A survey of numerical methods for unconstrained optimisation. *SIAM Rev.*, **12**, 79–97.

Quinn, P., Beven, K., Chevallier, P. and Planchon, O. (1991) The prediction of hillslope flow paths for distributed hydrological modelling using digital terrain models, *Hydrol. Processes*, **5**, 59–79.

Romanowicz, R., Beven, K., Freer, J. and Moore, R. (1993) TOPMODEL as an application module within WIS. *HydroGIS'93 Proc.*, Vienna, April.

Sivapalan, M., Beven, K. and Wood, E.F. (1987) On hydrological similarity. II: A scaled model of storm runoff production. *Water Resources Res.*, **23**(12), 2266–2278.

Chapter Sixteen

Automated derivation of stream-channel networks and selected catchment characteristics from digital elevation models

J. HOGG
School of Geography, University of Leeds

J. E. McCORMACK, S. A. ROBERTS AND M. N. GAHEGAN
School of Computer Studies, University of Leeds

AND

B. S. HOYLE
Department of Electronic and Electrical Engineering, University of Leeds

A method for automated derivation of stream-channel networks and selected catchment characteristics from digital elevation models (DEMs) is described in this chapter. Entities such as depressions and plateaux, on which drainage directions are undefined, are handled as special features and are identified using operations on connected regions. Unlike traditional methods based on pixels, the method described here facilitates construction of a consistent stream-channel network, provides a solution for network and spatial analysis and may be applied using DEMs at a broad range of scales. To illustrate the method, selected stable catchment characteristics are derived automatically using a square grid DEM with 1152 × 1152 elevations at 50 m spacing of the Peak District, England. The derived catchment characteristics include stream-channel networks, delineation of catchment boundaries, catchment area, catchment length, stream-channel long profiles, stream order, height ratios, slope and aspect. Other important characteristics of river catchments, such as the stream-channel density, stream-channel bifurcation ratios, stream-channel order and number, and length, gradient and area of streams of that order, may also be derived. The method may be applied at a broad range of scales. When applied to medium- and small-scale DEMs covering extensive areas, such as those derived from stereoscopic SPOT (Systeme Probatoire d'Observation de la Terre) images or from synthetic aperture radar on the ERS-1 satellite using radar interferometry, results facilitate comparative analysis of hydrological and geomorphological characteristics of river catchments and stream-channel networks, and

---

Geographical Information Handling – Research and Applications. Edited by P. M. Mather
© 1993 John Wiley & Sons Ltd

collection of parameters for process–response models of erosion, solutes and water flows over extensive regions. The method may be extended to include geology, soils, vegetation and other data in a single, integrated GIS for regional studies of hydrological, geomorphological and climatic processes.

1. INTRODUCTION

Runoff or stream flow over the earth's surface comprises movement of water under the influence of gravity. Each stream-channel receives runoff from its catchment area and delivers it typically to a larger stream, river, lake or the sea. River catchments and their stream-channel networks form well-defined topographic units, separated from each other by divides and organized in a hierarchy of scales on the basis of stream ordering. They can be dissected and their components analysed on the basis of a number of topographic characteristics, as demonstrated in the pioneering work of Horton (1932, 1945). Such topographic characteristics can be studied to determine how the morphometry of a river catchment influences hydrologic processes, or used as parameters in process–response models of erosion, sediment and solute transport by water.

Increasingly sophisticated techniques for measurement and analysis of the topographic characteristics of river catchments and their stream-channel networks have begun to emerge. New and improved technology for measuring topography over the earth's surface allows the creation of DEMs over extensive areas. DEMs, which are rectangular grids of elevations over part of the earth's surface, create significant new opportunities for automated analysis of river catchments and their characteristics, including more refined methods of identifying spatial variations in hydrological parameters. As a major factor in determining the spatial distribution of hydrological properties, topography plays a crucial role. It is important in quantifying surface drainage and infiltration processes and in linking local and global climatic models to regional hydrology, erosion and sedimentation. Technological advances have increased the speed of collecting DEMs, improved their position and height accuracy, and extended their geographical coverage from local to regional, continental and global scales. For these reasons, there has been growing interest in recent years in exploring the potential of DEMs for modelling the flow of water through river catchments.

Such water flow is influenced by many factors other than topography, including geology, soils, vegetation and climate. Because these factors can be represented to some extent in digital form, the facilities of a geographical information system (GIS) will be used more and more to store and process them for visualization, for spatial analysis of interrelationships and for distributed physically-based modelling of hydrological processes (Bathurst and O'Connell, 1992). The need to devise new methods of representing river networks and spatial distributions of features within a single integrated GIS formed a major stimulus for this research. The focus of this chapter is restricted, however, to automated derivation of stable characteristics of river catchments and stream-channel networks. These characteristics provide essential parameters for many hydrological, geomorphological and ecological models, and their derivation at a range of scales posed a challenging – yet realistic – research goal for the resources and time available.

The broad aims of the project were to study:

- ways of structuring individual terrain elevations and topographic features such as river networks and spatial distribution of gradient, aspect and river catchment divides in a GIS;

- the spatial distribution of surface gradients and contributing area to each elevation point;

- algorithms for automated derivation of river catchments and their stream-channel networks with a view to modelling hydrological surface processes over a broad range of spatial and temporal scales.

We review briefly the characteristics of the river catchment and its stream-channel network, the technological advances in methods of collecting digital terrain data and recent research into the use of such data in hydrology and geomorphology. We outline the method that we have developed and present results to illustrate its potential, using DEM data of the Peak District, England. Finally, we discuss the results and their implications for modelling terrain and surface processes over river catchments, and indicate how the work might be extended.

2. BACKGROUND

2.1. Characteristics of river catchments and stream-channel networks

The need to understand the relationship between precipitation over a river catchment and the resulting flow in a river is a fundamental problem in geomorphology and hydrology. The relationship depends critically on the time scale being considered and on the area of the river catchment. However, numerous other factors are also involved. These include the nature and spatial distribution of rainfall and the characteristics of the river catchment and its stream-channel network, including topography, geology, soils and vegetation, which all need to be described quantitatively (Scheidegger, 1991).

Current understanding of the river catchment and its stream-channel network rests firmly on the work of Horton (1932, 1945), who demonstrated a nested hierarchy of stream-channels at different scales, with different hydrological and morphological features associated with each. Horton's work on the morphometry of river catchments and how this links to hydrological processes has since been extended by many workers, notably Langbein (1947), Schumm (1956), Strahler (1964), Scheidegger (1965), Shreve (1966) and Smart (1972). His proposals for a theory of sheetflow, sheetwash erosion and rill formation have similarly been extended (Kirkby, 1971; Smith and Bretherton, 1972). Horton (1945) stimulated the use of numerical methods in water flow, sediment transport and erosional processes which lead to the formation of stream-channel networks within river catchments. Reviewing the development of ideas on drainage morphometry, Gardiner and Park (1978) noted that the Horton (1945) formulation of the morphometry of river catchments was carried out with a view to deriving methods by which discharge events could be predicted for ungauged rivers. The application of morphometric methods to hydrological and geomorphological analysis has become an important part of contemporary research, with the use of multiple regression techniques to derive equations between geometric and seasonal characteristics of a river catchment and selected measurements of water flow over different periods.

Gardiner and Park (1978) stress that Horton (1945) focused on a small number of basic characteristics of river catchments and their stream-channel networks, but that a wealth of further characteristics or indices have been proposed since. Many have been devised on the discovery of new relationships, but a considerable amount of redundancy exists among these characteristics. From the numerous quantitative measurements that have been proposed, the most important geometrical measurements of river catchments are based on: (i) area, (ii) length, (iii) shape, (iv) relief and (v) stream-channel network (Chorley, Schumm and Sugden, 1984).

2.1.1. Area

The definition of the river catchment can be straightforward, although there may be difficulties in some situations in interpreting the divide from topographic maps, aerial photographs or DEM data, especially in relatively flat terrain. The total area of a river catchment, and the area of a stream-channel network of a given order, are important characteristics because they affect the size of the peak hydrograph and the magnitude of total runoff. When related to stream-channel length, area is one of the most sensitive and important parameters. When linked to stream-channel ordering and scale, it provides an index of the amount of stream-channel flow that can be produced by a particular network. Drainage density, the length of stream-channel per unit area, is a critical determinant of stream-channel runoff, and sediment and solute transport rates. Stream-channel frequency, the number of streams of all orders per unit area, is also a critical determinant of processes, and may be easier to measure than drainage density. As area influences the amount of water discharge, it is one of the most important characteristics of a river catchment. Consequently, accurate and automated methods of measuring the area of river catchments and sub-catchments form a crucial part of the analysis of drainage basins.

2.1.2. Length

Several characteristics of a river catchment and its stream-channels are based on measurements of length. Catchment length, the maximum length of a river catchment from its mouth to the divide, is somewhat subjective but it provides a characteristic that is relatively easy to measure manually compared to the length of the main stream-channel. Measurement of the number and lengths of stream-channels forms an important part of the analysis of the hierarchy or composition of networks (Ferguson, 1977). The length of stream-channels of a given order and the total length of stream-channels within a river catchment are important. The character and extent of stream-channels affects the availability of sediment and the rate of water yield from a catchment (Gregory and Walling, 1973).

2.1.3. Shape

The shape of river catchments may be expressed in various ways, such as the form factor, which is the ratio of catchment area over the square of catchment length. Measurements of the shape of river catchments have not generally fulfilled initial expectations for predicting processes, although they do have some value for general studies. With three-dimensional models of river catchments, surface form may prove to be more valuable for correlation with discharge than other measurements, but is more complicated to derive (Anderson, 1973). The total length of the catchment boundary is used in several characteristics, such as catchment circularity, which is the ratio of catchment area over area of a circle with the same perimeter length. Most measurements of length, such as that of stream-channels, can be computed from DEM data, but difficulties can arise where characteristics are not well defined.

2.1.4. Relief

Other characteristics of river catchments and their stream-channel networks are based on measurements of relief and gradients. The difference in elevation between high and low points is used to derive the relief ratio and ruggedness number. These provide measurements of available potential energy. The greater the relief, the greater the erosional forces acting on the river catchment. Slope has two components, gradient and aspect (Skidmore, 1989). The maximal angle of slope at a point may be related to water flow and erosional, transport and sedimentary processes. The direction of maximal slope or aspect at any point may be related to solar heating of soils, soil moisture regime and ecological factors. Slope may also be related to other catchment characteristics, such as the long profile of rivers and stream-channels of a given order. Numerous methods have been proposed for expressing relief and gradient characteristics of a river catchment, including the mean slope, maximum catchment relief and relief ratio (Gregory and Walling, 1973). While these characteristics are relatively straightforward to measure from topographic maps, they do not provide information about the spatial distribution of relief and slopes throughout river catchments. With increasing interest in distributed process-response modelling within river catchments, there is a growing need to identify contributing areas and quantify their characteristics. Most measurements of relief, gradient and aspect can be derived automatically from DEM data, including hypsometric curves for a given river catchment, although the accuracy depends on the quality and spacing of the original elevations.

2.1.5. Stream-channel network

The pattern of a stream-channel network in a river catchment has long been recognized as an important catchment characteristic. Patterns may be related to different lithologies at various scales. They may also be related to water flow, erosion and sediment deposition. Horton (1932, 1945) proposed an ordering system by which stream-channel networks may be subdivided into constituent sub-catchments and by which the hierarchical structure may be analysed. Numerous relations have been established between stream-channel order and other characteristics, such as the lengths, gradients and drainage areas of stream-channels of each order, and with water discharge, sediment and solute processes (Smart, 1978). Methods of ordering stream-channels are not necessarily based on the same theories or intended for the same purpose, and no single method appears best for all applications (Dunkerley, 1977). The methods of Horton and Strahler contain elements of subjectivity but tend to be used most frequently by geographers.

A river network can be regarded mathematically as a tree, with the river mouth as its trunk and the stream-channel network as its branches. The labelling of the branches within the tree has been done in various ways. In Strahler's (1964) method, every fingertip stream-channel is order 1. Where two stream-channels of equal orders meet, they produce a stream-channel of one order higher, whereas where two stream-channels of unequal order meet they produce a stream-channel of order equal to the higher of the two. A weakness of Strahler's system is that the order in which three stream-channels combine affects the outcome. Horton's (1945) system is similar to Strahler's but maintains the order of the main stream-channel from its mouth to its source. Scheidegger (1965) proposed a method by which every junction in a stream-channel network was associated with an increase in stream-channel order, according to a mathematical formula using logarithms to base 2.

Shreve (1966) introduced a topological method of link magnitudes, which is directly related to Scheidegger's method, while Graf (1975, 1977) proposed a cumulative stream-ordering system. While there have been other proposals, they are not as widely used as the Strahler and Shreve methods (Gardiner, 1975). Each method of stream-ordering can be implemented for specific applications. In this study, the method of Shreve was adopted.

2.2. Understanding relationships between characteristics of river catchments and physical processes

Relationships between precipitation and runoff are influenced by numerous characteristics of river catchments and their stream-channel networks (Jarvis, 1977). Understanding the relationships between these characteristics is important, for both scientific and practical reasons. For any given period, predicting the movement of water through a river catchment forms an important part of hydrological research (Maidment, 1993). It has practical implications for water management and for applications in diverse fields of study, including soils, agriculture, forestry, civil engineering and terrain trafficability. Various hydrological parameters derived from the river catchment and its stream-channel network are used in models of hydrological systems at a variety of spatial and temporal scales (Walling, 1987). Explanation of past and contemporary land forms, and prediction of changes under different climatic conditions, rests to some extent on research into the physical processes of water flow, erosion, sediment transport and deposition within river catchments. As a process-response system, the river catchment is in dynamic equilibrium, because process can influence form and form can influence process, with numerous feedback relationships between the two (Gregory and Walling, 1973).

The focus of much recent research has been on methods of integrating hydrological modelling and GIS to provide a spatial dimension to the study of river catchments (Anderson, 1989). Kemp (1992) stressed that hydrological models include two distinct geographic models. Water originates as a distributed input to the river catchment, so modelling runoff of water as a distributed process is an important component. Water quickly concentrates, however, in stream-channel networks, which requires network analysis models. Digital terrain data is extremely important as input to both types of model. For that reason, interest in efficient methods of collecting high quality and accurate DEM data has grown.

2.3. Technological advances in methods of collecting digital terrain data

Over the past decade, there have been vast improvements in the speed, accuracy and geographical extent over which data about the terrain surface can be acquired (Doyle, 1978; Li, 1992). The improvements are mainly due to developments in information technology, which have brought major changes to all areas of surveying and mapping and have played a key role in automating the collection of digital terrain data (Gugan, 1992). To illustrate this point, consider four areas where automation is having a major impact: (i) positioning and navigation systems; (ii) fully digital photogrammetric workstations; (iii) digital methods of extracting information from existing maps; and (iv) derivation of elevation data from synthetic aperture radar on the ERS-1 satellite using radar interferometry.

2.3.1. Positioning and navigation systems

The NAVSTAR Global Positioning System (GPS) is a satellite-based positioning system for determining position on the earth's surface at any time (Cross, 1991). It is having a major impact on methods of ground and aerial surveying and is expected to have an even greater impact over the next few years after completion of the experimental phase in 1994 (Ackermann, 1992). GPS observations can be made to locate positions on the earth's surface using pseudo-ranges or the so-called carrier frequency phase measurement (Ashkenazi and Dodson, 1992). With pseudo-ranges, the accuracies achieved were of the order of 15 m in horizontal position and 25 m in height but, since 1990, the introduction of Selective Availability has downgraded the system for civilian users to about 100 m and 150 m, respectively. According to Ashkenazi and Dodson (1992) the phase observation is intrinsically precise to a few millimetres but currently has inaccuracies of tens of metres. These are gradually being reduced by various methods so that positions can be determined to submetre accuracy using static, semi-kinematic and differential GPS. Hence, it offers a way of precise positioning for a wide range of applications, including the determination of ground control points, survey flight navigation and related applications involving generation of digital terrain heights over extensive areas. When allied to developments in electronic ground surveying, the speed and accuracy of measuring position and elevation has increased significantly in recent years (Kennie and Petrie, 1990).

2.3.2. Fully digital photogrammetric workstations

Most digital terrain data is currently collected using conventional methods of aerial photography and analogue photogrammetry but analytical and automated methods of aerial photogrammetry have become more important in recent years (Petrie, 1990). The combination of an analytical plotter and a digital correlator offers an automated way of acquiring dense, digital elevations over extensive areas from stereoscopic pairs of overlapping photographs. The impact of such analytical plotters on the collection of DEM data has already been considerable but human operators are still required.

Analytical plotters are evolving rapidly, however, into fully digital photogrammetric workstations (Bonjour and Newby, 1990; Gruen, 1992). These carry out photogrammetric tasks in an iterative and automated way using digital image data as input (Dowman, Ebner and Heipke, 1992). They can be designed to operate on one or many different types of image data, such as synthetic aperture radar, raster scanned aerial photographs and satellite images. While high performance systems use special hardware, low-cost systems running on personal computers are available for generating DEM data (Welch, 1990). All the evidence from the literature suggests that fully digital photogrammetric workstations will eventually replace the current generation of analytical plotters and that there will be some convergence of photogrammetric systems and GIS. This is partly because digital photogrammetric workstations offer greater speed and lower costs for producing DEM data and partly because of the need for integration of information from different sources (Bonjour and Newby, 1990).

The first of a new generation of satellite imaging systems, the Systeme Probatoire d'Observation de la Terre (SPOT) produces digital images with 10 m resolution and stereo imaging capabilities. These lend themselves to computer processing in digital photogrammetric workstations and to automatic generation of DEM data by stereo-matching (Muller, 1989; Day and Muller, 1989). Numerous investigators have developed and evaluated algorithms for creating DEMs from SPOT images (Gugan and Dowman, 1988; Ley, 1988; Theodossiou and Dowman, 1990). However, earth scientists have been slow to investigate potential applications of such DEMs for regional studies of the spatial distribution of hydrological and geomorphological processes over a landscape.

2.3.3. Digital methods of extracting information from existing maps

Techniques for automated conversion of topographic paper maps to digital data have similarly been transformed by developments in computer technology (Jackson and Woodsford, 1991). While manual methods are still used, raster-scan and line-following technologies are being introduced to expedite the whole operation. Commercial systems for raster scanning and line following of topographic maps have evolved from scanning to automated data conversion and have impressive speed, reliability and quality of output. Since they use special hardware and software, they tend to be beyond the means of research workers.

Low-cost scanners connected to desktop computers offer an alternative solution and have advanced considerably in recent years (Drummond and Bosma, 1989). For example, Ansoult and Soille (1990) reported on a method of image processing to derive digital data for GIS from scanned thematic colour maps. The method used mathematical morphology to segment a raster-scanned thematic map into its different areas and extract separate geometrical and colour patterns on the basis of radiometric values. The paper map can essentially be disaggregated into its separate colour layers, objects can be labelled automatically, text strings read and features edited to create raster files for a GIS.

2.3.4. Derivation of elevation data from SAR on ERS-1 using radar interferometry

Early results from the synthetic aperture radar (SAR) on the ERS-1 satellite show that it has considerable potential for derivation of DEM data over extensive areas of the earth's surface. Using methods of radar interferometry, detailed DEMs could be produced (Clery, 1992). The SAR on ERS-1 beams microwave signals downwards and detects the reflected signal. After intensive computer processing, this signal is converted into a detailed image of the earth's surface with a pixel resolution of about 25 m. The SAR system detects the phase of the return wavelength, which may be precisely related to distance of the surface from the antenna.

Radar interferometry relies on taking two images of the same area of the earth's surface from closely spaced orbits of the ERS-1 satellite (Kingsley and Quegan, 1992). The phase information of one image is then related to that of the other. Phase values for each corresponding pixel of the two images are subtracted to produce an image known as an interferogram, which records only the differences in phase between the two images. These phase differences give variations in altitude of each pixel and can be used to construct a map of topography of the imaged area. From these, DEMs can be derived. Initial research results show heights are accurate to about 4 m (Clery, 1992). Given the all-weather capability of SAR and its extensive coverage, this and other work on processing of SAR images shows that rapid progress is being made in this area.

The contrast between the rapid developments in computer hardware and software, positioning and navigation systems, digital photogrammetric workstations and automatic raster-scanning systems for converting paper maps to digital form and traditional methods of gathering data in hydrology and geomorphology could scarcely be greater. The characteristics of river catchments and stream-channel networks have been derived in various ways in the past, but a basic distinction can be made between manual and automated methods. Manual methods of analysis of paper topographic maps involve making measurements on maps to derive characteristics such as length of river catchment, mean slopes and stream-channel density. Such manual methods are somewhat cumbersome, slow and prone to error for all but single, small catchments. With large catchments, or with many small catchments (often used in comparative studies), manual methods have increasingly severe limitations. Consequently, many researchers have investigated ways of automating the process of deriving some or all of the characteristics.

2.4. How terrain data has been analysed in hydrology

Approaches to automating the process of quantifying characteristics of river catchments and their stream-channel networks fall into three broad categories, discussed in the following subsections.

2.4.1. Extraction of physiographic features from digitized contours

Many topographic features such as ridges, breaks of slope, saddles, pits and peaks are portrayed in contours on topographic maps. This is especially the case if the latter have been drawn directly from a three-dimensional stereoscopic model of the terrain in a photogrammetric plotter. Several studies have been made to extract topographic data from contours in general and from the distinctive characteristics of river catchments and their stream-channel networks in particular (Mark, 1975; Evans, 1980; Roy, Gravel and Gauthier, 1987; Weibel, 1989). A method for interpolation of contours into arrays of randomly distributed points was proposed by Yoeli (1984). Aumann, Ebner and Tang (1991) reported on two methods of automatically extracting ridge and drainage lines from digitized contours, one based on vector and one on raster sets of contour data. Tang (1992) extracted geomorphological elements from digitized contours and recombined them with contours to provide high quality digital terrain models.

2.4.2. Extraction of physiographic features from remote sensing images

Remote sensing techniques have been studied to obtain physiographic information about the characteristics of river catchments and their stream-channel networks for applications in hydrology, geomorphology and landscape ecology (Astaras and Soulakellis, 1992). Aerial photographs continue to have major benefits for identification and measurement of topographic features such as slopes, breaks of slope, ridges, depressions and stream-channel delineation, especially when stereoscopic pairs of vertical aerial photographs are used with height as one of the major determinants (Stewart, 1968; Speight, 1968). Parry and Turner (1971) showed the value of using infrared aerial photographs in channel detection and delineation when water was present in rivers. A major difficulty with aerial photographs is that vegetation often obscures stream-channels, particularly first-order stream-channels in heavily vegetated regions. Another difficulty relates to the interpretation of ephemeral stream-channels, which are sometimes difficult to detect.

The potential of satellite remote sensors for obtaining information about river catchments and their stream-channel networks has been studied by many researchers. Engman and Gurney (1991) provide an excellent review of recent advances in remote sensing in hydrology, including derivation of watershed geometry and related physiographic characteristics. France and Hedges (1986) compared Multispectral Scanner (MSS) and Thematic Mapper (TM) images from Landsat for their application in hydrology and found that the TM data provided much more information, but not nearly as much as aerial photographs or a 1:50 000 scale map. Lopez-Garcia and Thornes (1992) studied the use of remote sensing techniques to obtain physiographic information about river catchments and their stream-channel networks. In a study to explore the use of satellite images for estimating drainage density in semi-arid areas of Spain, they found a strong linear relationship between the image and actual drainage density but results depended on scale. Below a certain area, errors increased rapidly due to the coarse resolution of satellite images. Results were affected by different images, environments and solar angles but the resolution of satellite images was found to be insufficient for delineation of many smaller stream-channels. As these are often the most important in geomorphological and hydrological studies of environmental processes, the limitations of remote sensing from Landsat satellites are apparent.

2.4.3. Extraction of physiographic features from digital terrain data

Methods of extracting characteristics of river catchments and stream-channel networks from digital terrain models have been studied extensively over the past three decades (Moore, Grayson and Ladson, 1991). Two different types of elevation data are typically used:

- Rectangular grids of regularly-spaced elevations, known as DEMs.
- Triangulated irregular networks (TINs), which consist of a network of elevations of irregular size, shape and orientation, often using randomly located elevation points over a surface.

2.4.3.1. The DEM approach

This approach is in many ways the simplest and most widely used in landform studies. Elevations are regularly spaced and held as an array of elevations (Petrie and Kennie, 1990). The choice of grid spacing is typically related to the purpose for which the data has been collected and to the size of the total area of study. Each elevation covers an elemental area or cell, the size of which depends on the grid spacing. For example, for large-scale projects, DEM with a grid spacing of 1 m may be used to give a cell of 1 m^2. In the case of the 7.5-minute quadrangle maps at a scale of 1:24 000 produced by the United States Geological Survey, DEM data has a spacing of 30 m to give a cell of 30 m × 30 m.

While DEMs have merit in terms of ease of computer implementation and computational efficiency, they have several shortcomings for modelling terrain. The distribution of elevation points in not closely related to the characteristics of the terrain. Thus, the density of elevation points must theoretically be high enough to portray the smallest terrain features present in the area being studied. If this is done, then the density of elevations collected will be too high in most other areas of the DEM (Peucker et al., 1978; Petrie and Kennie, 1990). One solution that would avoid such redundancy and inefficiency would be to vary the density of elevations according to relief (Hutchinson, 1988). This makes collection of DEM data and computations based upon it more complex but provides more accurate representations of topography. While the choice of grid size affects the results obtained and the computational efficiency, in practice it is often necessary to make use of DEM data that is already available. Such data may not be entirely appropriate in terms of grid spacing, position and elevation accuracy.

Methods for extracting terrain parameters and watersheds from DEM have been proposed by several researchers (Collins, 1975; Peucker and Douglas, 1975; Marks, Dozier and Frew, 1984). The problem has usually focused on the definition of river-channel flow paths. These are derived by examining the direction of maximal slope at each elevation point. Slope is typically derived from DEM using local interpolation methods because they are the simplest to implement. Where there are depressions or plateaux, the definition of river-channel flow paths is undefined (Mark and Goodchild, 1982). Depressions and plateaux therefore create difficulties when attempting to extract river-channel flow paths. Mark (1984) and O'Callaghan and Mark (1984) proposed an algorithm for automated detection of drainage networks from DEM, in which they created depressionless DEMs and assigned flow directions. The method has been extended by Band (1986), Jenson and Domingue (1988) and Tarboton, Bras and Rodriguez-Iturbe (1991). Hutchinson (1988) suggested an interactive finite difference method for use with irregularly distributed data for removing spurious depressions.

Several methods for defining catchment boundaries have been proposed. Such boundaries are often highly convoluted and difficult to define consistently. Many solutions have been adapted, including that proposed by O'Callaghan and Mark (1984), in which the dimensionless DEM is used to derive a direction of drainage matrix and a weighting matrix, from which a drainage accumulation matrix is derived (Jensen, 1985; Jensen and Domingue, 1988; Band, 1989). Soille and Ansoult (1990) and Vincent and Soille (1991) reported a new approach based on mathematical morphology, pioneered by Serra (1982, 1988), which provides a formal solution to extracting watersheds.

Various other methods have been proposed for extracting topographic features from DEM. Douglas (1986) conducted a series of experiments to locate ridges and channels to create a new type of DEM. Tribe (1991) conducted a detailed review of various algorithms in a study to recognize valley heads from DEM.

One of the most important characteristics of a river catchment is slope, which is defined by a plane tangent to the surface, as modelled in a DEM at a point. As indicated earlier, slope has two components: gradient and aspect. Derivation of slopes from DEM has been studied by many workers because slope and aspect are used extensively in hydrological and geomorphological research (Moore, Grayson and Ladson 1991; Franklin, 1987) and in remote sensing for improving accuracies of thematic mapping (Franklin *et al.*, 1986). Skidmore (1989) reported on a comparison of six different algorithms for calculating gradient and aspect from DEM in order to determine which were most accurate.

2.4.3.2. The TIN approach The TIN approach is popular because of its efficiency in storing data and its capacity for faithful surface representation of the terrain (Lee, 1991). It has been used extensively, particularly in large-scale engineering applications, and numerous methods of generating the TIN have been reported (Gold, Charters and Ramsden, 1977; McCullagh, 1988). All data points in a TIN form vertices of the triangles that are used to approximate the terrain. The use of TIN offers a relatively simple way of including significant terrain features such as peaks, pits, ridges, stream-channels and breaks of slope (Palmer, 1984). Many authors agree that the TIN approach gives a better definition of the surface representation of part of the earth (Mark, 1979; Weibel, 1989; Petrie and Kennie, 1990). For this reason, TIN is used extensively in civil engineering for the design of highways, calculating earthwork volumes and visual analysis.

Palacios-Velez and Cuevas-Renaud (1986) developed an algorithm for automated delineation of river networks, ridge and river catchment perimeters from TIN data and extended this work for dynamic hydrologic modelling (Palacios-Velez and Cuevas-Renaud, 1989). Jones, Wright and Maidment (1990) reported on a method of delineating river catchments from TIN data. While other workers have used TIN data for hydrological research, one of the difficulties lies in determining the upslope connection of a slope facet. Another lies in the irregularity of the TIN, which makes computation of characteristics of river catchments more difficult than for DEMs.

Many other approaches have been studied. For example, Wolf (1992) used weighted surface networks to represent special types of graphs with the vertex sets consisting of the critical points such as pits, passes and peaks and the edge sets consisting of the critical surface-specific lines such as ridges and stream-channels of corresponding topographic surfaces. The interlocking ridge and stream-channel networks are formalized by weighted surface networks. These are used to extract stream-channel networks and delineate river catchments from digital elevation data.

3. LIMITATIONS TO PREVIOUS STUDIES AND AN ALTERNATIVE METHOD

Whether using the DEM, TIN or other data, studies have demonstrated different ways in which river catchments and their stream-channel networks can be derived either wholly or in part (Mark, 1988; Jenson, 1991). In many cases, however, there have been difficulties in dealing with exceptional situations, such as plateaux and depressions (Moore, Grayson and Ladson, 1991). In some cases, studies appear to have been limited to some extent by the choice of data structures, computational methods, or both. There has often been a compromise between spatial resolution and computational expense: either the size of data sets has been limited to a relatively small number of elevations, or substantial computer resources have been applied to deal with extensive geographic areas. The method described below represents an attempt to overcome these limitations and to derive automatically, quickly and efficiently a broad range of topographic characteristics for hydrological and geomorphological applications.

3.1. Feature-based analysis

A stream-channel network is relatively easily derived from a DEM, provided that every cell (excluding cells on the boundaries) possesses a neighbour at a lower elevation. A suitable rule is applied at each cell in order to determine the direction of stream flow and throughflow on the hillside. A simple model might assume a single flow in the direction of steepest descent; a more sophisticated one may assume flow into more than one neighbour (Quinn *et al.*, 1991).

A DEM will generally contain plateaux and depressions, so some cells will not possess lower neighbours. We refer to plateaux and depressions here as 'features' because they are handled explicitly. They may be the result of errors in elevation data (or limitations in the level of resolution) or may correctly represent the underlying terrain. We handle such features, firstly, by identifying them and, secondly, by excluding them from the normal analysis. Thus, we treat them as special entities. A stream-channel network is derived for the remaining points of the DEM using the chosen flow rule (either single or multiple flow). Once this is done, we can then link flows into and out of such features in whatever manner is deemed appropriate. Overall control of the process is such that there is no necessity to check for closed paths in the network.

3.2. Handling of large data sets

The volumes of DEM data now becoming available are often much larger than can be handled efficiently on desktop computers and workstations due to limitations on the amount of random access memory (often of the order of a few megabytes of data). The size of hard disks, however, is not nearly so limiting. Even on personal computers, hard disks with gigabytes of space are becoming increasingly widely available.

One solution is to divide physically a large set of DEM data into smaller sections and to perform an analysis on each section of flow directions on hillslopes and of stream-channel networks. This is an inefficient and somewhat untidy approach, since results in one section depend critically on those in other sections, and the rejoining of sections is not a trivial problem because they need to be placed in order. Streams will usually meander through several sections during their descent, often crossing from one section to another and back again as they progress. Dependencies are critical when, for example, a depression overlaps two or more sections. In this case, the outflow point cannot be determined without reference to all of the sections involved. In addition, an algorithm for climbing a stream-channel network to perform an analysis of the catchment divide or accumulation of flows into a point does not appear to be feasible with such a solution.

A second possible solution is to rely on the computer's virtual memory. This gives the impression of more memory than is physically present, by storing some of the data on disk, moving it back and forth between disk and memory as it is needed. These operations are performed 'behind the scenes' and do not concern the user. Virtual memory is not optimized, however, for handling the kind of two-dimensional operations encountered during stream-channel network analysis and the result will usually be excessive (and often prohibitively expensive) movement of data between disk and memory. In any case, in practical terms, virtual memory is usually somewhat limited in capacity on most machines.

Our approach to handling large sets of DEM data is to use the principle of virtual memory and to divide the two-dimensional array of elevation values into square sections (pages) of data. During the analysis, only a few of these pages are typically held in computer memory at any one time, the remainder being stored on hard disk. The page handling is 'behind the scenes' and does not concern the user of the software during analyses of stream-channel networks. If we can ensure that manipulations on DEM data are predominantly local in nature, then the overhead in moving pages from memory to disk will be relatively unimportant compared to other processing times. This is indeed the case, as outlined below. The net result is that the usual memory limitation on size of DEM data is removed (McCormack, 1993).

3.3. Description of the method

The analysis of river catchments and their stream-channel networks involves a number of stages. Firstly, we identify the features in the DEM – the depressions and plateaux – using ideas based on mathematical morphology (Serra, 1982, 1988; Vincent and Soille, 1991). Secondly, drainage directions are determined for all other areas of the DEM. Thirdly, flow paths into and out of depressions can be determined to form a full stream-channel network of flow links. Fourthly, river catchment divides can be derived and a flow accumulation performed by climbing the stream-channel network of flow links from the end points and calculating the upslope area contributing to every point. Many other operations can then be performed using the flow pathway.

3.4. Depressions and plateaux

Plateaux are identified using a recursive neighbour algorithm. Neighbours of a cell are examined to see if they are at the same elevation. If so, then the neighbours of each neighbour are examined in turn until all the cells comprising the plateaux have been collected (McCormack *et al.*, 1993).

Depressions are identified by simulated filling until overflow. The perimeter of a simple depression expands during filling until an outflow point is found. Such a point will have a lower neighbour, which is not inside the perimeter and which defines the direction of water flow. Dependencies between depressions are important; for example, two depressions with a common boundary may fuse to form a composite depression with a single outflow cell. Also, in practice, a larger depression will often enclose a number of smaller depressions. A consistent approach is therefore required in order to handle such dependencies (McCormack *et al.*, 1993). This is accomplished by the simple rule that if a depression in the process of being filled touches a previously identified feature (depression or plateau), then the presently filling depression takes over all the cells of the touching feature.

3.5. Flow network

Flow directions are determined firstly for cells that do not appear in features. Since these cells have at least one lower neighbour, this is a relatively simple computation. As noted above, it may be based on single or multiple flow directions.

Secondly, the stream-channel network can be completed by joining feature inflows to outflows. This is done by drawing a straight line path from an inflow to the nearest outflow point. Although there are many other possibilities, the straight line approach works well.

3.6. Watersheds and accumulation

Completion of the stream-channel network means that it is possible to traverse stream-channels from their lowest points up to their sources and thus perform a watershed analysis or an accumulation.

This process is implemented efficiently with a recursive algorithm for climbing the stream-channel network. Starting at a given point of a stream-channel network, every path flowing into that point is followed in turn. At each new point that is accessed, every path flowing into it is also followed in turn. The algorithm is efficient in that every cell need be accessed only twice, once to determine its value (flow directions) on the ascent and once more to deposit a value (for example, the accumulated upslope area at that point or the watershed identifier) on the subsequent descent.

3.7. Local operations and size of DEM

If computer processing is confined to a relatively small area of the DEM, then it is only necessary to have access to a small portion of it at any one time. This can be achieved by careful algorithmic design. Computer processing of features – the depressions and plateaux – can be implemented so as to be based predominantly on local operations. Algorithms can be designed that ensure that all the cells in a locality are processed before moving to another. The assignment of flow directions is free to be performed one block at a time for the whole DEM and so is very efficient.

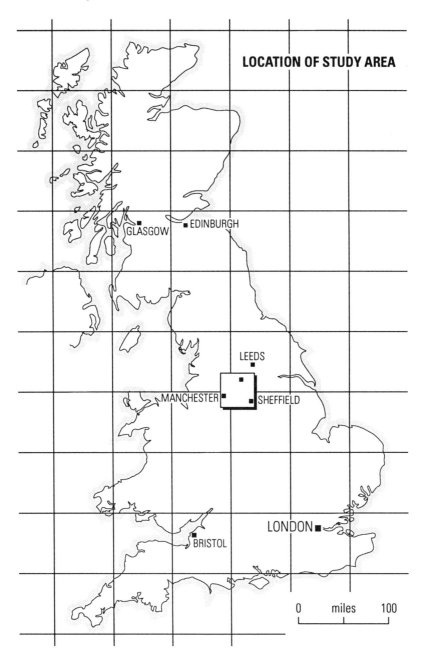

Fig. 1   Map showing the location of the study area

The algorithm for climbing stream-channel networks can also be based predominantly on local operations. It decomposes a river catchment into its component watersheds, with each component being decomposed in turn. The natural order of processing such watersheds is to process one area and then to move to an adjacent area.

3.8. Including other data

The treatment of features can be extended to include other classes of geographical data; for example, information about obstacles such as reservoirs, dams, and canals, or about land use, and so on. The essential property of a feature is that it will usually override normal processing in some way. In the case of depressions or plateaux, the assignment of direction of flows does not follow the normal rules. Similarly, if the position of a river is known, then normal processing might be overridden in order to constrain the network to follow certain paths. This is typically the case where, for example, rivers cross wide flat floodplains.

3.9. The set of DEM data

To illustrate the potential of the method, a DEM supplied by the Ordnance Survey was used. This was produced at a scale of 1:50 000 by the Directorate of Military Survey from the contours on the Ordnance Survey Landranger maps (Ley, 1990). The contour interval on these maps is 10 m. Height values at each intersection of a 50 m horizontal grid were interpolated from the contours. Heights were rounded to the nearest metre and have an accuracy that depends on the nature of the terrain but is about 3 m. A DEM of 1152×1152 elevations was taken from nine tiles, each containing 400×400 elevations.

The study area covers part of the southern Pennines, England (Fig. 1). Extending from Manchester and Stockport in the west to Sheffield in the east and from Huddersfield in the north to Castleton in the south, it covers some 3300 km^2. It is an area with varied topography, geology, vegetation and land use. Elevations range from about 12 m to 636 m. The gritstones, sandstones, shales and limestones have been affected by minor volcanic activity and by recent glacial history to form a complex landscape. The vegetation and land use reflect the geology, topography and climate, with vast expanses of peatbogs and heather moorlands at higher levels and grasslands for sheep grazing in the valleys. Much of the area provides resources for water supply, leisure, recreation and hill farming.

4. RESULTS

The DEM is shown in Fig. 2 as a two-dimensional image on a grey level scale. The elevation values are integer numbers. Fig. 3 shows a three-dimensional representation of part of the DEM, covering the Ladybower and Derwent reservoirs (square (3,0) in Fig. 2, taking the bottom left-hand corner as square (0,0)).

The stream-channel network is derived without user intervention from the raw DEM elevation values (Fig. 4). In the example, we use the direction of steepest descent for those cells that are not on features. In this set of DEM data there are several large plateaux, caused by reservoirs, and numerous smaller depressions in the valleys. The disk storage space required for computer processing is of the order of 4 megabytes. About 100 kilobytes of random access memory are needed for efficient processing. This figure is not significantly affected by increasing the size of the DEM. The processing time on a standard 386 PC, with a 25 MHz processor, for deriving the stream-channel network of flow links from the raw DEM of elevation values in this study was about 8 minutes.

An accumulation map is shown in Fig. 5. This is derived by climbing the stream-channel network of flow links from the end points and computing the upslope area contributing to every cell. The accumulated area is displayed on a logarithmic scale, with the dark areas representing a large upslope area. The stream-channel network may be classified on the basis of stream-channel order, as illustrated in Fig. 6.

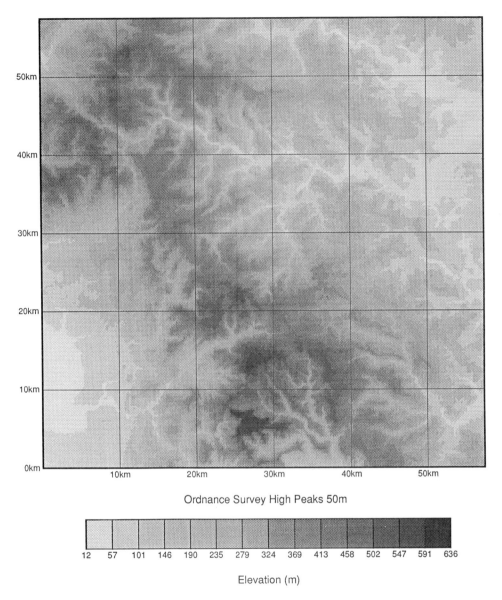

Fig. 2 DEM of 1152 × 1152 elevation values at 50 m grid spacing shown as a two-dimensional image on a grey-level scale

Fig. 7 shows the perimeter of the river catchment superimposed on the elevation map. This is derived by climbing the stream-channel network and attaching a label to every point in the river catchment. Edge detection of the labelled river catchments is used to extract the perimeters.

Slope and aspect form an intrinsic part of the automated derivation of the river-channel network. They may be represented in maps, as shown in Figs 9 and 10, with the user choosing the categories of slope and aspect to suit a particular application.

Fig. 3 A three-dimensional view from the west of part of the DEM, showing Ladybower and Derwent reservoirs

The total area of a river catchment, or sub-catchment, is obtained directly by counting the elevation values with attributes such as labels of a particular value. Each elevation covers an area of 50 m × 50 m (except along the outside edge of the DEM) and hence area can be derived, each elevation in this case representing an area of 0.25 hectares. The density of stream-channels is obtained from the total length of stream-channels and the area they cover. Similarly, density of stream channels of a given order can be derived.

The long profiles of stream-channels are derived by traversing the stream-channel network and recording distance and elevation in a table (Fig. 8). Various length and height ratios may be derived. Hypsometric curves and other relations between hypsometric integral and drainage density may be derived using a similar approach.

5. DISCUSSION

The results show that a relatively large DEM containing plateaux and depressions can be processed quickly and without user intervention on a standard desktop computer, in this case running Microsoft Windows 3.1. Only a small amount of computer memory is needed for efficient processing, regardless of the size of the DEM. Processing times are short enough to make it practical to handle relatively large DEMs. For example, a DEM of 4096 × 4096 elevations requires about 1.5 hours of processing time on a personal computer with a 80386 processor running at 25 MHz. Times for larger DEM are given in McCormack *et al.* (1993).

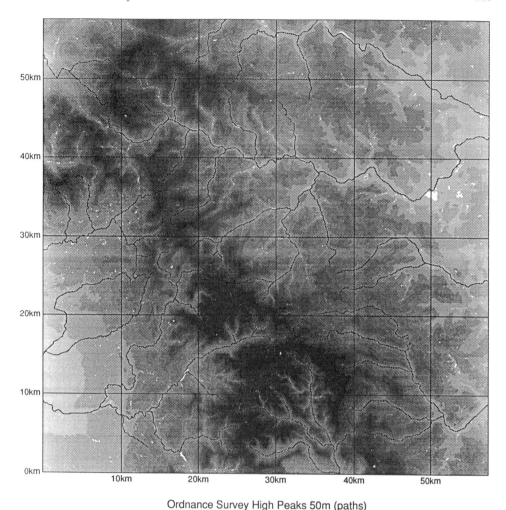

Ordnance Survey High Peaks 50m (paths)

Fig. 4 DEM showing major stream channels, with 'features' on which slope is undefined shown as white areas

The results illustrate the potential of the method but need to be qualified. Firstly, the choice of quality and grid spacing of DEM data usually needs to be made with reference to the nature of the terrain and to the specific objectives for geomorphological or hydrological modelling of river catchments and their stream-channel networks. In this case, the choice was controlled more by economics and expediency than by specific modelling objectives or applications. The DEM was selected to illustrate the method because it was available.

Secondly, the DEM data was processed without manual interference. The results show the stream-channel network derived from the DEM by the computer. No account was taken of canals, water abstractions, underground pipes for water transfer between river catchments, water discharges into rivers or diversions of stream-channels in urban areas. An option is available to constrain stream-channel paths, to modify the results manually in various ways and to take account of local factors, but this was not used in the study.

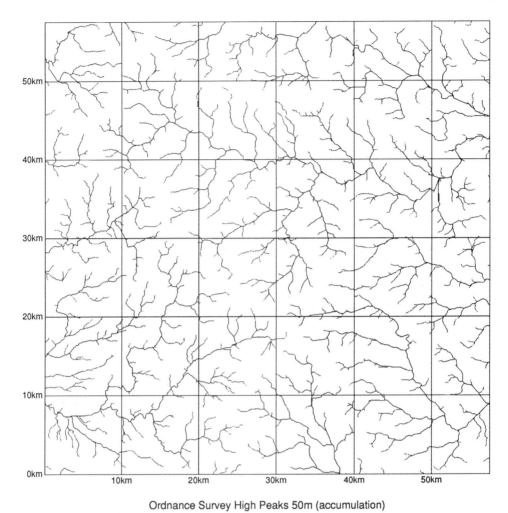

Ordnance Survey High Peaks 50m (accumulation)

Fig. 5 A flow path accumulation map derived by climbing the stream-channel network of flow links from the end points and computing the upslope area contributing to every elevation point. The accumulated area is displayed on a logarithmic scale, with dark areas representing elevation points with large upslope areas

6. CONCLUSIONS

This study has examined methods of structuring data from individual terrain elevations and algorithms for modelling processes over the earth's surface. Results have shown that selected stable characteristics of river catchments and their stream-channel networks can be derived for relatively large DEMs using a standard personal computer. Results have also shown that the method provides an avenue for linking spatial and network data derived from DEM to GIS for environmental modelling at a range of scales.

The technology for gathering DEM data has been examined briefly. It was shown that the availability, accuracy and geographical coverage of DEM data at various scales will

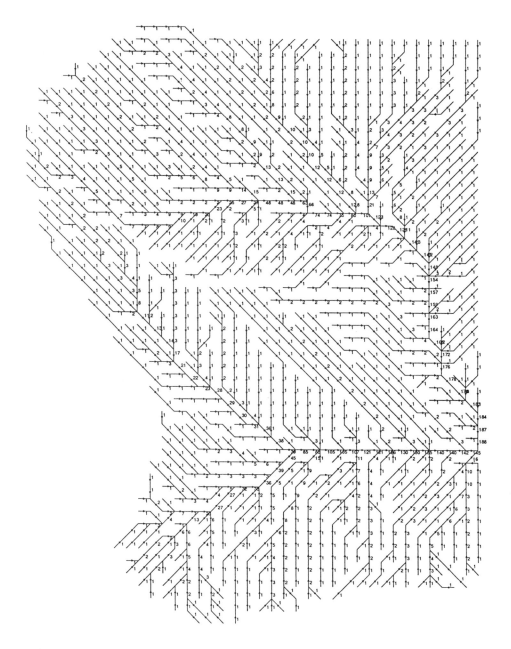

Fig. 6 Stream-channel network classified on the basis of Shreve's method of stream ordering

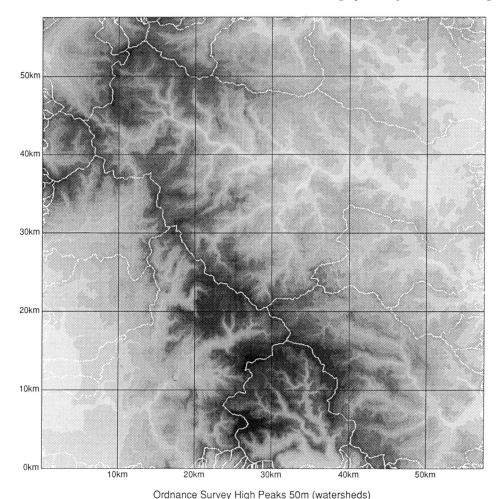

Ordnance Survey High Peaks 50m (watersheds)

Fig. 7 The boundaries of river catchments superimposed on the elevation map

increase as the technology for gathering it continues to advance. What is required from the scientific community is improved methods of making use of such data to model earth surface processes and to develop our understanding of environmental relationships. The increasing sophistication of global climate models will require, for example, the inclusion of accurate and contemporary land surface parameters from local, regional and continental areas. By focusing on the kind of problems geomorphologists, hydrologists and landscape ecologists wish to solve, and by devising a new data structure and algorithm to suit, a method has been demonstrated that shows potential and has implications for use of DEM and GIS. While there is a need to evaluate the method further in specific hydrological and geomorphological applications, the results have provided an indication of what can be achieved.

The results have shown one method for solving the problem of representing terrain and surface processes in GIS. They have shown how the choice of data structure and algorithm can provide an insight into ways in which DEM data may be used in future to derive spatial

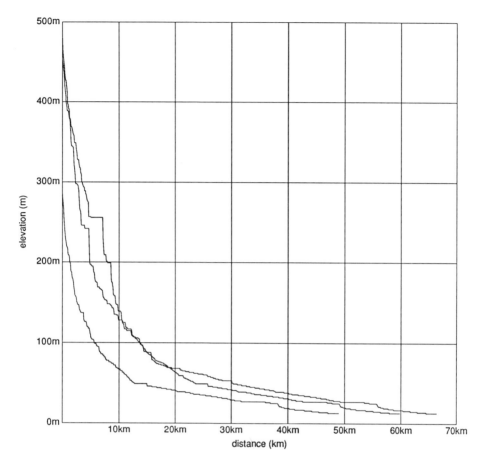

Fig. 8 The long profiles of selected stream-channels within the study area, with reservoirs indicated by horizontal steps in the profile

and network data. Such data may be studied further to devise ways of solving a broad range of problems in physical geography and environmental science within GIS. For example, the method may be applied to examine regional drainage patterns and to compare derived catchment characteristics from different physiographic regions at various scales and over different rock, soil, climatic and vegetative conditions.

While the focus of this chapter has been on an approach to automated derivation of established characteristics of river catchments and their stream-channel networks from DEMs, further investigation ought to focus on new or hybrid measurements that might usefully be derived from DEM data for this purpose. The traditional approach to describing catchment characteristics has been by sampling diagnostic geometric variables, which could be derived manually from topographic maps or aerial photographs. This approach might no longer be appropriate in a technological environment that is changing rapidly and creating new opportunities. Given the growing potential to collect DEM and to derive the spatial distribution of parameters and features such as slopes and river channel network in digital form, the question of how best to make use of such data for hydrological,

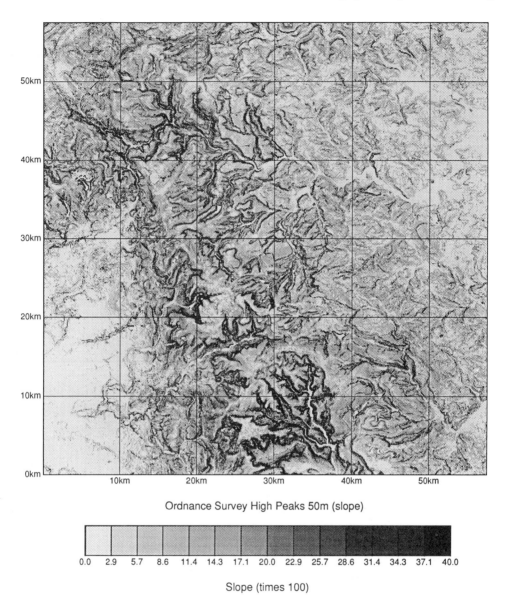

Fig. 9 Slope classification derived from DEM

geomorphological or other geographical research might fairly be raised as a topic for further investigation.

ACKNOWLEDGEMENTS

This work was carried out under the ESRC/NERC Joint Programme on Geographical Information Handling, grant no. GR GST/02/493.

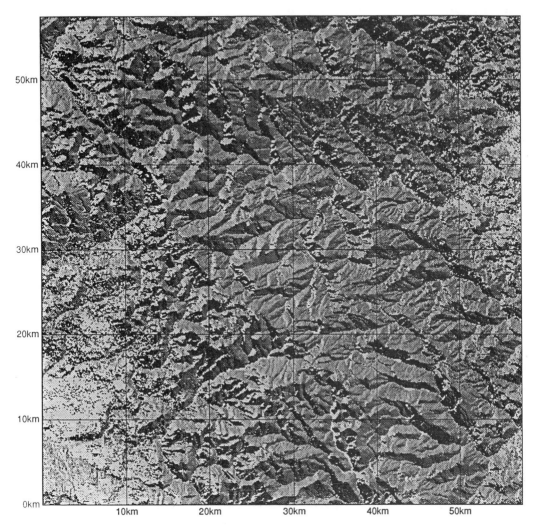

Ordnance Survey High Peaks 50m (aspect)

180   154   129   103   77   51   26   0   -26   -51   -77   -103   -129   -154   -180
West-North-East-South-West
Aspect (degrees)

Fig. 10 Aspect classification derived from DEM

REFERENCES

Ackermann, F. (1992) Kinematic GPS control for photogrammetry. *Photogrammetric Record*, **14**(80), 261–276.

Anderson, M.G. (1973) Measure of three dimensional drainage basin form. *Water Resources Res.*, **9**(2), 378–383.

Anderson, M.G. (1989) Physical hydrology. *Prog. Phys. Geogr.*, **13**(1), 93–102.

Ansoult, M.M. and Soille, P.J. (1990) Mathematical morphology: a tool for automated GIS data acquisition from scanned thematic maps. *Photogrammetric Engng Remote Sensing*, **56**(9), 1263–1271.

Ashkenazi, V and Dodson, A. (1992) Positioning with GPS. *Mapping Awareness and GIS in Europe*, **6**(7), 7–9.

Astaras, Th.A. and Soulakellis, N.A. (1992) Contribution of digital image analysis techniques on Landsat-5 TM imageries for drainage network delineation: a case study from the Olympus Mountain, W. Macedonia, Greece. *Proc. 18th Annual Conf. Remote Sensing Soc.*, University of Dundee, 163–172.

Aumann, G., Ebner, H. and Tang, L. (1991) Automatic derivation of skeleton lines from digitised contours. *ISPRS J. Photogrammetry and Remote Sensing*, **46**, 259–268.

Band, L.E. (1986) Topographic partition of watersheds with digital elevation models. *Water Resources Res.*, **22**(1), 15–24.

Band, L.E. (1989) A terrain-based watershed information system. *Hydrological Processes*, **3**, 151–162.

Bathurst, J.C. and O'Connell, P.E. (1992) Future of distributed modelling: the Systeme Hydrologique Europeen. *Hydrological Processes*, **6**, 265–277.

Bonjour, J.D. and Newby, P.R.T. (1990) Architecture and components of photogrammetric workstations. *Photogrammetric Record*, **13**(75), 389–405.

Chorley, R.J., Schumm, S.A. and Sugden, D.E. (1984) *Geomorphology*. Methuen, London.

Clery, D. (1992) Radar traps Earth's shifting surface. *New Sci.*, **136**(1848), 19.

Collins, S.H. (1975) Terrain parameters directly from a digital elevation model. *Can. Surveyor*, **29**, 507–518

Cross, P A. (1991) GPS for GIS. *Mapping Awareness and GIS in Europe*, **5**(10), 30–34.

Day, T. and Muller, J.-P. (1989) Digital elevation model production by stereo-matching SPOT imagepairs: a comparison of algorithms. *Image and Vision Comput.*, **7**(2), 95–101.

Douglas, D.H. (1986) Experiments to locate ridges and channels to create a new type of digital elevation model. *Cartographica*, **23**, 29–61.

Dowman, I.J., Ebner, H. and Heipke, C. (1992) Overview of European developments in digital photogrammetric workstations. *Photogrammetric Engng Remote Sensing*, **58**(1), 51–56.

Doyle, F.J. (1978) Digital terrain models: an overview. *Photogrammetric Engng Remote Sensing*, **44**(2), 1481–1485.

Drummond, J. and Bosma, M. (1989) A review of low cost scanners. *Int. J. Geogr. Inf. Sys.*, **3**(1), 83–95.

Dunkerley, D.L. (1977) Some comments on stream ordering schemes. *Geogr. Analysis*, **9**(4), 429–431.

Engman, E.T. and Gurney, R.J. (1991) *Remote Sensing in Hydrology*. Chapman and Hall, London.

Evans, I.S. (1980) An integrated system of terrain analysis and slope mapping. *Z. Geomorph., Neue Floge, Suppl.*, **36**, 274–295.

Ferguson, R.I. (1977) On determining distances through stream networks. *Water Resources Res.*, **13**, 672–674.

France, M.J. and Hedges, P.D. (1986) A hydrological comparison of Landsat TM, Landsat MSS and black and white aerial photography. *Symp. Remote Sensing for Resources Development and Environmental Management*, Enschede, The Netherlands, 717–720.

Franklin, J., Logan, T.L., Woodcock, C.E. and Strahler, A.H. (1986) Coniferous forest classification and inventory using Landsat and digital terrain data. *IEEE Trans. Geosci. Remote Sensing*, **24**, 139.

Franklin, S.E. (1987) Geomorphic processing of digital elevation models. *Comput. Geosci.*, **13**(6), 603–609.

Gardiner, V. (1975) Drainage Basin Morphometry. BGRG Technical Bulletin no. 14, Norwich, GeoAbstracts.

Gardiner, V. and Park, C.C. (1978) Drainage basin morphometry: review and assessment. *Prog. Phys. Geogr.*, **2**(1), 1–35.

Gold, C.M., Charters, T.D. and Ramsden, J. (1977) Automated contour mapping using triangular element data structures and an interpolant over irregular triangular domain. *Comput. Graphics*, **11**, 170–175.

Graf, W.L. (1975) A cumulative stream-ordering system. *Geogr. Analysis*, **7**, 335–40.
Graf, W.L. (1977) Measuring stream order: a reply. *Geogr. Analysis*, **9**(4), 431–433.
Gregory, K.J. and Walling, D.E. (1973) *Drainage Basin Form and Process: A Geomorphological Approach*. Edward Arnold, London.
Gruen, A. (1992) Tracking moving objects with digital photogrammetric workstations. *Photogrammetric Record*, **14**(80), 171–185.
Gugan, D.J. (1992) Operational mapping from satellite imagery. *ISPRS J. Photogrammetry Remote Sensing*, **47**(1), 43–50.
Gugan, D.J. and Dowman, I.J. (1988) Accuracy and completeness of topographic mapping from SPOT imagery. *Photogrammetric Record*, **12**(72), 787–796.
Horton, R.E. (1932) Drainage basin characteristics. *Trans. Amer. Geophys. Un.*, **13**, 350–361.
Horton, R.E. (1945) Erosional development of streams and their drainage basins: hydrophysical approach to quantitative morphology. *Bull. Geol. Soc. Amer.*, **56**, 275–370.
Hutchinson, M.F. (1988) Calculation of hydrologically sound digital elevation models. *Proc. 3rd Int. Symp. Spatial Data Handling*, Sydney, Australia, 117–133.
Jackson, M. and Woodsford, P. (1991) GIS data capture hardware and software. In Maguire, D.J., Goodchild, M.F. and Rhind, D.W. (eds.), *Geographic Information Systems: Principles and Applications*. Longman, Harlow.
Jarvis R.S. (1977) Drainage network analysis. *Prog. Phys. Geogr.*, **1**, 271–295.
Jenson, S.K. (1985) Automated derivation of hydrologic basin characteristics from digital elevation model data. *Proc. Auto-Carto 7, Digital Representations of Spatial Knowledge*. American Society of Photogrammetry and Remote Sensing, Washington DC, 301–310.
Jenson, S.K. (1991) Applications of hydrological information automatically extracted from digital elevation models. *Hydrol. Processes*, **5**, 31–44.
Jenson, S.K. and Domingue, J.O. (1988) Extracting topographic structure from digital elevation data for geographic information system analysis. *Photogrammetric Engng Remote Sensing*, **54**(11), 1593–1600.
Jones, N.L., Wright, S.G. and Maidment, D.R. (1990) Watershed delineation with triangle-based terrain models. *J. Hydraul. Engng*, **116**, 1232–1251.
Kemp, K.K. (1992) Spatial models for environmental modeling with GIS. *Proc. 5th Int. Symp. Spatial Data Handling*, Charleston, South Carolina, 524–533.
Kennie, T.J.M. and Petrie, G. (1990) *Surveying Engineering Technology*. Blakie, Glasgow and London.
Kingsley, S. and Quegan, S. (1992) *Understanding Radar Systems*. McGraw-Hill, London.
Kirkby, M.J. (1971) *Hillslope Process–Response Models Based on the Continuity Equation*. Institute of British Geographers Special Publication no. 3, 15–30.
Langbein, W.B. (1947) *Topographic Characteristics of Drainage Basins*. US Geological Survey Water Supply Paper no. 968C, 125–157.
Lee, J. (1991) Comparison of existing methods for building triangular irregular network models of terrain from grid digital elevation models. *Int. J. Geogr. Inf. Sys.*, **5**(3), 267–285.
Ley, R.G. (1988) Some aspects of height extraction from SPOT imagery. *Photogrammetric Record*, **12**(72), 823–832.
Ley, R.G. (1990) A DTM of Great Britain based on the 1:50 000 scale Ordnance Survey Series – a production viewpoint. Paper presented at the Institute of British Geographers Annual Conference, Glasgow, January 1990.
Li, Z. (1992) Variation in the accuracy of digital terrain models with sampling interval. *Photogrammetric Record*, **14**(79), 113–128.
Lopez-Garcia, M.J. and Thornes, J.B. (1992) Drainage density estimation from satellite imagery in semi-arid environments. *Proc. 18th Annual Conf. Remote Sensing Soc.*, University of Dundee, 378–387.
Maidment, D.R. (1993) GIS and hydrologic modeling. In Goodchild, M.F., Parks, B.O. and Steyaert, L.T. (eds), *Geographic Information Systems and Environmental Modeling*. Oxford University Press, Oxford.
Mark, D.M. (1975) Geomorphic parameters:a review and evaluation. *Geogr. Annr*, **57A**(3/4), 165–177.
Mark, D.M. (1979) Phenomenon-based data structuring and digital terrain modelling. *Geo-Processing*, **1**, 27–36.

Mark, D.M. (1984) Automated detection of drainage networks from digital elevation models. *Cartographica*, **21**, 168–178.
Mark, D.M. (1988) Network models in geomorphology. In Anderson, M.G. (ed.), *Modelling Geomorphological Systems*. John Wiley, Chichester, 73–97.
Mark, D.M. and Goodchild, M.F. (1982) Topographic model for drainage networks with lakes. *Water Resources Res.*, **18**, 275–280.
Marks, D., Dozier, J. and Frew, J. (1984) Automated basin delineation from digital elevation data. *Geo-Processing*, **2**, 299–311.
McCormack, J.E. (1993) An efficient approach to processing spatial data for environmental applications. Research Report 93.34, School of Computer Studies, Univ. of Leeds.
McCormack, J.E., Gahegan M.N., Roberts, S.A., Hogg, J. and Hoyle B.S. (1993) Feature-based derivation of drainage networks. *Int. J. Geogr. Inf. Sys.*, in press.
McCullagh, M.J. (1988) Terrain and surface modelling: theory and practice. *Photogrammetric Record*, **12**(72), 747–779.
Moore, I.D, Grayson, R.B. and Ladson, A.R. (1991) Digital terrain modelling: a review of hydrological, geomorphological and biological applications. *Hydrological Processes*, **5**, 3–30.
Muller, J.-P. (1989) Real-time stereo matching and its role in future mapping systems. *Proc. 3rd UK Natl Land Surveying and Mapping Conf.*, Warwick, 17–21 April.
O'Callaghan, J.F. and Mark, D.M. (1984) The extraction of drainage networks from digital elevation data. *Comput. Vision, Graphics and Image Processing*, **28**, 323–344.
Palacios-Velez, O.L. and Cuevas-Renaud, B. (1986) Automated river-course, ridge and basin delineation from digital elevation models. *J. Hydrol.*, **86**, 299–314.
Palacios-Velez, O.L. and Cuevas-Renaud, B. (1989) Transformations of TIN data into a kinematic cascade. *Trans. Amer. Geophys. Un.*, **70**, 1091–1102.
Palmer, B. (1984) Symbolic feature analysis and expert systems. *Proc. 1st Int. Symp. Spatial Data Handling*, Zürich, **2**, 465-478.
Parry, J.T. and Turner, H. (1971) Infrared photos for drainage analysis. *Photogrammetric Engng Remote Sensing*, **37**(10), 1031-1038.
Petrie, G. (1990) The impact of analytical photogrammetric instrumentation on DTM data acquisition and processing. In Petrie, G. and Kennie, T.J.M. (eds.), *Terrain Modelling in Surveying and Civil Engineering*. Whittles, Caithness, Chapter 5.
Petrie, G. and Kennie, T.J.M. (eds.) (1990) *Terrain Modelling in Surveying and Civil Engineering*. Whittles, Caithness.
Peucker T.K. and Douglas, D.H. (1975) Detection of surface specific points by local parallel processing of discrete terrain elevation data. *Computer Vision, Graphics and Image Processing*, **4**, 375–387.
Peucker, T.K., Fowler, R.J., Little, J.J. and Mark, D.M. (1978) The triangulated irregular network. *Proc. Auto-Carto 3*. American Congress on Surveying and Mapping, Falls Church, Virginia, 516–540.
Quinn, P., Beven, K., Chevallier, P. and Planchon, O. (1991) The prediction of hillslope flow paths for distributed hydrological modelling using digital terrain models. *Hydrol. Processes*, **5**, 59–79.
Roy, A.G., Gravel, G. and Gauthier, C. (1987) Measuring the dimensions of surfaces: a review and appraisal of different methods. *Proc. 8th Int. Symp. Computer Assisted Cartography (Auto-Carto 8)*, Baltimore, Maryland, 68–77.
Scheidegger, A.E. (1965) *The Algebra of Stream-Order Numbers*. US Geological Survey Professional Paper no. 525B, 187–189.
Scheidegger, A.E. (1991) *Theoretical Geomorphology*. Springer-Verlag, London.
Schumm, S.A. (1956) The evolution of drainage systems and slopes in badlands at Perth Amboy, New Jersey. *Bull. Geol. Soc. Amer.*, **67**, 597–646.
Serra, J. (1982) *Image Analysis and Mathematical Morphology*, vol. 1. Academic Press, London.
Serra, J. (1988) *Image Analysis and Mathematical Morphology*, vol. 2. Academic Press, London.
Shreve, R.L. (1966) Statistical law of stream numbers. *J. Geol.*, **74**, 7–37
Skidmore, A. K. (1989) A comparison of techniques for calculating gradient and aspect from a gridded digital elevation model. *Int. J. Geogr. Inf. Sys.*, **3**(4), 323–334.
Smart, J.S. (1972) Channel networks. *Adv. Hydrosci.*, **8**, 305–346.
Smart, J.S. (1978) The analysis of drainage network composition. *Earth Surf. Processes and Landforms*, **3**, 129–170.

Smith, T.R. and Bretherton, F.P. (1972) Stability and conservation of mass in drainage basin evolution. *Water Resources Res.*, **8**, 1506–1529.

Soille, P. and Ansoult, M. (1990) Automated basin delineation from digital elevation models using mathematical morphology. *Signal Processing*, **20**, 171–182.

Speight, J.G. (1968) Parametric description of land form. In Stewart, G.A. (ed.), *Land Evaluation*. Macmillan, Melbourne, 239–250.

Stewart, G.A. (ed.) (1968) *Land Evaluation*. Macmillan, Melbourne.

Strahler, A.N. (1964) Quantitative geomorphology of drainage basins and channel networks. In Chow, V.T. (ed.), *Handbook of Applied Hydrology*. McGraw-Hill, New York, Section 4, Part II, 4–40.

Tang, L. (1992) Automatic extraction of specific geomorphological elements from contours. *Proc. 5th Int. Symp. Spatial Data Handling*, Charleston, South Carolina, 3–7 August, 554–566.

Tarboton, D.G., Bras, R.L. and Rodriguez-Iturbe, I. (1991) On the extraction of channel networks from digital elevation data. *Hydrological Processes*, **5**, 81–100.

Theodossiou, E.I. and Dowman, I.J. (1990) Heighting accuracy of SPOT. *Photogrammetric Engng Remote Sensing*, **56**(12), 1643–1649.

Tribe, A. (1991) Automated recognition of valley heads from digital elevation models. *Earth Surf. Processes and Landforms*, **16**, 33–49.

Vincent, L. and Soille, P. (1991) Watersheds in digital spaces: an efficient algorithm based on immersion simulations. *IEEE Trans. Pattern Analysis and Machine Intelligence*, *13*(6), 583–598.

Walling, D.E. (1987) Physical hydrology. *Prog. Phys. Geogr.*, **11**, 112–120.

Weibel R. (ed.) (1989) *Contributions to Digital Terrain Modelling and Display*. Geo-Processing Series no. 12, Department of Geography, University of Zürich.

Welch, R. (1990) 3-D terrain modelling for GIS applications. *GIS World*, **3**(5), 26–30.

Wolf, G.W. (1992) Hydrologic applications of weighted surface networks. *Proc. 5th Int. Symp. Spatial Data Handling*, Charleston, South Carolina, 3–7 August, 567–579.

Yoeli, P. (1984) Computer-assisted determination of the valley and ridge lines of digital terrain models. *Int. Yb. Cartogr.*, **24**, 197–205.

Section III
ECONOMIC AND PLANNING APPLICATIONS OF GIS

Chapter Seventeen

Mapping natural hazards with spatial modelling systems

G. WADGE, A. WISLOCKI AND E. J. PEARSON
NERC Unit for Thematic Information Systems (NUTIS), University of Reading

AND

J. B. WHITTOW
Department of Geography, University of Reading

Much of the effort in assessing the probability of natural hazard recurrence is ineffectual because it is not presented to the users of the information – planners, developers and the like – in an understandable way. We have developed a computer system that addresses this problem. This system is a combination of the ARC/INFO geographical information system (GIS) and the Nexpert Object expert system. Spatial data and associated data are held in the GIS; problem- or application-specific knowledge is held within the expert system. The interface between the two subsystems works at the transport level of UNIX communications. The operation of this approach is tested on two data sets concerned with landsliding in Cyprus and subsidence in Ripon. The former involves re-interpretation of a combined map/report presented originally as an engineering geology work. The Ripon subsidence information system models the spatial pattern of surface collapse as a series of alignments and places the results in the context of planning policy at the town level and development plans at the site investigation level.

1. INTRODUCTION

A hazard is a phenomenon that puts humans in a potentially dangerous or disadvantageous position. A natural hazard is one that is not created by human agency, such as an earthquake or a volcanic eruption. With imagination, a very long list of such hazards could be written. Many of these would be of such extremely low probability, for example, being struck by a meteorite, that they are of no practical concern. Others are certain to happen, and, again, need not concern us, such as hypothermia suffered by nudist sunbathers in Antarctica. Those natural hazards that are the subject of study are characterized by (i) having a probability of occurrence between these two extremes, and (ii) potentially affecting a large number of people. One of the key tasks of hazard studies is to evaluate this probability of occurrence. It is important to determine whether the hazardous phenomenon is a discrete event or a process. If it is an event it can be treated as a random variable. The probability is then given by the limiting value of relative frequency of occurrence derived from an infinitely long series of observations of such hazard events. This is usually done by studying the temporal frequency and magnitude of past hazard events.

Although humans are increasingly mobile we mainly live in permanent settlements with a complex but slowly changing spatial pattern. Many natural phenomena that can pose hazards to these settlements also have complex, variable spatial patterns (e.g. earthquakes, tornadoes). Another key task in hazard studies, therefore, is predicting the nature of the spatial interrelationship between human settlement and the hazard process. If we are to use our understanding of hazard to practical effect this has to be communicated to the people concerned, often through government authorities. To do this effectively, the hazard must be conveyed in terms that are understandable to non-technical experts and relevant to the planning or mitigation context.

GIS in general have the capabilities to analyse spatial patterns and, less so, to calculate statistical probabilities. Hence, they should be appropriate tools for the analysis of hazard (Wadge, 1988). We have evaluated this capability by developing a system with the following specific characteristics:

- Hazards from surface instabilities such as landslides are modelled on a commercial GIS.
- The system is useful to both earth scientists and planners.
- Expert system techniques are used to handle some application-specific knowledge.

This chapter illustrates our evaluation with reference to two case studies: landsliding in western Cyprus and subsidence due to gypsum dissolution in Ripon. Both of these rely heavily on data collected by British Geological Survey (BGS) geologists. Firstly, however, we examine how hazards are traditionally assessed; what data are required, which analytical methods are employed and what products result. Then we show how spatial modelling in GIS can be applied and how a particular combination of GIS and expert system techniques provides a powerful way of assessing hazard.

2. NATURAL HAZARD ASSESSMENT

The purpose of natural hazard assessment is to convey information on the likelihood of future hazardous events to the people concerned. Methods of assessment vary considerably, depending on the nature of the hazard and the data available (Hunter and Mann, 1992). They often involve one or both of the following components: (i) a temporal analysis, usually from a catalogue of past hazard events, from which a statistical measure of the probability of recurrence can be calculated, and (ii) a spatial analysis, usually based on a model, from which a map showing boundaries and relative categories of hazard can be derived.

2.1. Estimating probabilities of hazard

Hazard events can be quantified using: (i) the frequency–magnitude spectrum of past hazard events, and (ii) the spatial location relative to the individual hazard events. Fig. 1 illustrates how these data are related. High frequency–low magnitude events pose no hazard at a distance, though low frequency–high magnitude events do (Fig. 1a). Often the historical

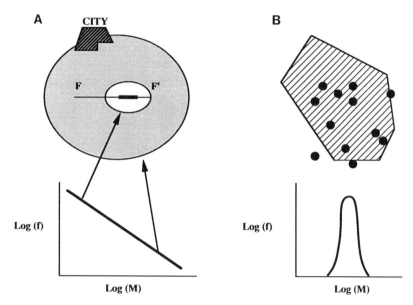

Fig. 1 Different hazards produce different spatial/temporal patterns. (a) An active fault (FF') produces large numbers (f) of small magnitude (M) earthquakes and few large magnitude earthquakes. The small magnitude earthquakes are no hazard to the city (small ellipse) but large earthquakes (large ellipse) are. (b) Point hazard events (circles) such as subsidences are essentially of uniform magnitude but spatial pattern may be apparently random

catalogue will not contain any information on these larger events and statistical extrapolation to the extreme must be made (Bardsley, 1988). Other hazards may display little or no such variation in magnitude but might have a complex spatial point pattern (Fig. 1b). In this latter case the probability can be expressed as a single value for each place. However, we might need to know the probabilities of a range of events at different magnitudes. If the event catalogue is sufficiently good this can be done directly from the empirical data, perhaps after curve-fitting. Alternatively, a theoretical distribution function may be justifiable, such as the Poisson distribution for volcanic events (Wickmann, 1966) of the form

$$P = 1 - \exp(-\lambda t) \qquad (1)$$

where P is the probability, t is time and λ is the average annual rate of events. Other, more subjective, techniques, such as Bayesian methods, may be used (Mann, 1988).

The probability of hazard caused by a continuous process, such as radon gas release from granites, obviously cannot be derived from an event catalogue. A model of the process with parameters estimated from field data is the usual way to produce the spatial probability of such a hazard.

2.2. Spatial modelling

Ideally, we should be able to identify the process responsible for the hazard, devise an appropriate mathematical representation of the process and use this to simulate the hazard over the area of interest using computer techniques. For example, Wadge, Young and

McKendrick (1993) used a simulation program to model the flow of lava over a digital elevation model. A statistical analysis of the eruption parameters of historical lava flows on the volcano constrained their probabilistic, Monte-Carlo series of lava flow simulations. The cumulative result of these simulations gave a map of the probability of inundation by lava.

Such a model simulation approach may not always be applicable because:

1. The hazard may be created by more than one process.

2. The process may be too poorly understood to be modelled.

3. Information on physical variables may not be available.

4. Simulation may be too time-consuming.

The alternative approach in spatial modelling is to use inductive rather than deductive reasoning. Provided there are some known examples of the locations of past hazard events then the ambient states of a number of environmental variables can be inferred and combined to calculate spatial weightings for future occurrence of those hazards. Several statistical methods have been used to derive global functions for inductive hazard mapping in GIS. They include calculation of point density of hazard events (e.g. Gupta and Joshi, 1990); linear regression (e.g. Jibson and Keefer, 1989); logistic regression (e.g. Bernknopf *et al.*, 1988) and discriminant analysis (e.g. Carrara *et al.*, 1991). Often the choice of environmental variables is *ad hoc* and there is uncertainty as to whether the measured variable values are the same as at the time of the hazard event. Also, cross-tabulation techniques ignore potentially useful information in the spatial autocorrelation of the data. Wadge, Wislocki and Pearson (1993) discuss the implementation of these spatial analysis techniques in GIS.

2.3. Cartographic representation

Traditionally, natural hazard assessments have been summarized by maps (e.g. Varnes *et al.*, 1984; Crandell *et al.*, 1984). Presenting the results of a GIS-based analysis of hazard as a digital version of an ordinary map should help to make the results understandable to a wide audience. Three main types of map for representing hazards can be distinguished:

1. Binary: there are two classes of map area, hazardous and non-hazardous. More than one hazard may be represented on the same map in this way.

2. Ordinal: several classes of hazard (e.g. high, medium and low). The choice of category boundary may be based on criteria relating to evacuation, mitigation or planning.

3. Ratio: continuous values of hazard, usually expressed as a recurrence probability (e.g. per year) or as a ratio of an exceedance threshold for each unit cell of the computer model.

One of the potential benefits of GIS for the creation of such maps is that they can be rapidly recreated to show different models or the effects of uncertainties. However, this has not been fully exploited, probably because of the desire from users for a single 'answer' on which to act.

2.4. Users' perspectives

Every hazard assessment should be addressed to a specific audience, usually authorities with responsibility for acting on this information. These users may be from very different backgrounds and have different concerns. Civil defence bodies are interested in evacuation and short-term mitigation priorities. Planners are concerned with land-use zoning. Developers and insurers will want to know the financial risk to properties. Because the backgrounds of the assessors (usually earth scientists) and the users (often non-scientific) may be different, there is the potential for poor communication and misunderstanding (Marker and McCall, 1990). The assessor may need to present the same results to separate user groups, which may entail separate processing and map drawing for each group. GIS has great potential to address this problem if it is used as part of a decision support system designed for these different user groups.

3. EXPERT-GIS

3.1. The need for knowledge bases

The distinction between knowledge (how to do things and why things happen) and data (types of things and their values) is fundamental to the design of computer systems to perform specific tasks. For some tasks it will be appropriate to embed the knowledge within ordinary procedural computer code. For example, an algorithm to perform an overlay operation on two sets of spatial data contains the knowledge of how to perform that operation. However, for other tasks it is better to keep this knowledge separate, within a knowledge base. This is appropriate when:

- The task is logically complex and there is more than one way to perform it, depending on data and circumstances.
- Several different domains of expertise are to be called upon.

A GIS for hazard assessment could use separate knowledge bases relating to different areas of expertise (earth scientist, spatial modeller, planner, developer) to fill its role as a decision support system for users from different backgrounds. We have used such knowledge bases, created using expert system software.

3.2. System configuration

Fig. 2 shows the configuration of our Expert-GIS for hazard assessment. It consists of the ARC/INFO GIS (version 6.0.1) and the Nexpert Object expert system development environment (version 2.0) running on a Sun Sparcstation under UNIX and OpenWindows. This version of ARC/INFO is a very large software package (containing about one million lines of code and two thousand functions). Important elements of this GIS that are generally useful for our hazard assessment tasks are: vector (ARC) and raster (GRID) spatial modelling functions, triangulation of surfaces (TIN) and a macro language (AML) that supports the construction of a windows-based user interface. The database, INFO, is a relational database, but of rather limited capabilities. One of the design criteria for the system was that the user would need no knowledge of how to use the GIS. The user interface must be entirely focused on the application, and we use ARC/INFO's AML to achieve this.

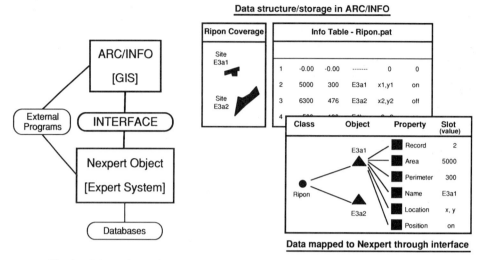

Fig. 2 Schematic configuration of the Expert-GIS for natural hazard assessment

Nexpert Object allows expert systems to be designed and run. These all consist of two main components: a knowledge base of rules, and a structural description of the data model that these rules operate upon. This data model is object-orientated, unlike ARC/INFO. However, the objects usually have properties whose values can be mapped to the equivalent fields of records in the INFO database. Complex hierarchical relationships of class and inheritance are supported.

ARC/INFO and Nexpert Object communicate through an interface. This potential was first demonstrated by Maidment and Djokic (1990). Our interface uses interprocess communications through the Transport Library Interface used by UNIX (Pan, Ferrier and Wadge, 1993). It uses the client–server model of communication. The calls to the transport level are managed by two sets of C programs and UNIX shell scripts, one for each mode of communication: Nexpert Object as client and ARC/INFO as server, and vice versa. This allows a network of workstations to access the system.

4. CASE STUDY 1: LANDSLIDING IN CYPRUS

4.1. Engineering geology map to GIS coverage

West of the Troodos Mountains of Cyprus is an area of chronic surface instability. This area was studied by engineering geologists from the BGS and the Cyprus Geological Survey, who produced complex engineering geology maps at a scale of 1:10 000 and an accompanying technical report (Northmore *et al.*, 1986). These maps show deep and shallow rotational landslides, translational landslides, debris flows and rockfalls classified as old, dormant and active. They are effectively unreadable by planners wishing to assess the hazard implications. The goal of the system in this case is to take this map information, together with data in the accompanying report and topographic information, and produce simpler hazard maps.

The engineering geology maps were drawn using conventional cartographic procedures. A considerable amount of interpretation on the part of the field geologist and cartographer has gone into their creation. The first task of the system is to convert this map into an ARC/INFO coverage. The formality of the vector topology data model imposes a greater

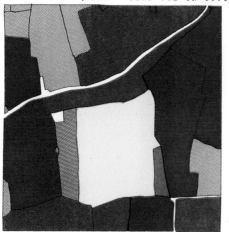
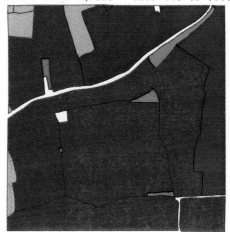
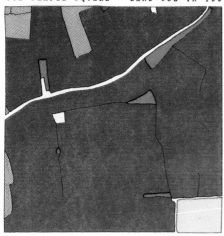
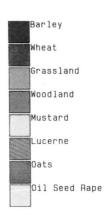

Plate I   Land use distribution in a sample 1-km square area in 1978, 1984 and 1990, demonstrating change in usage

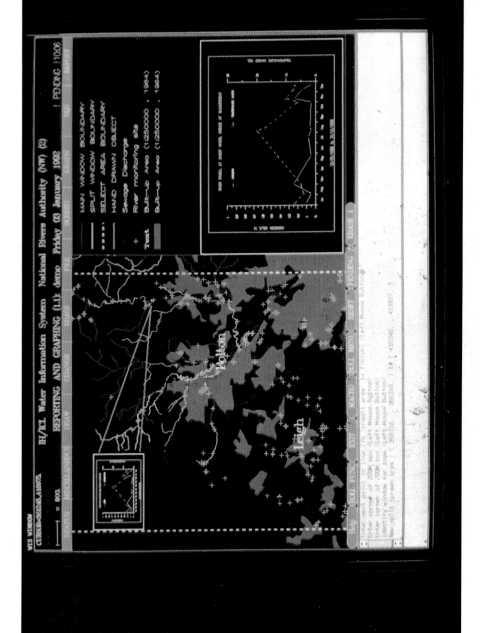

Plate II Sample display from Water Information System for catchments in northern England

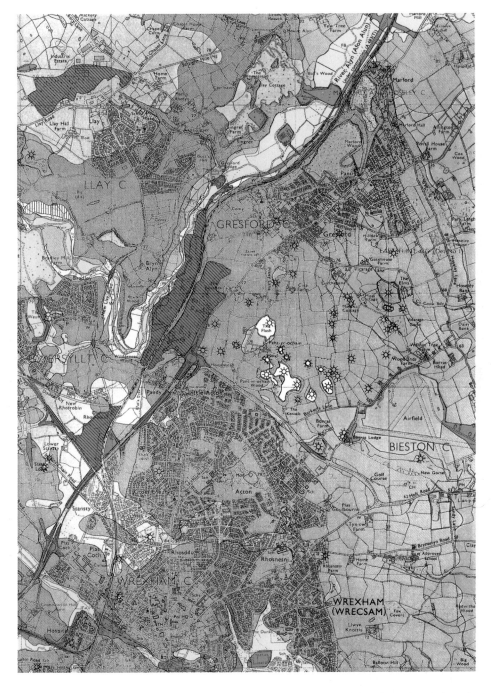

Plate III Example of thematic applied geological map, showing drift (glacial overburden) geology overlying area exploited for mineral deposits

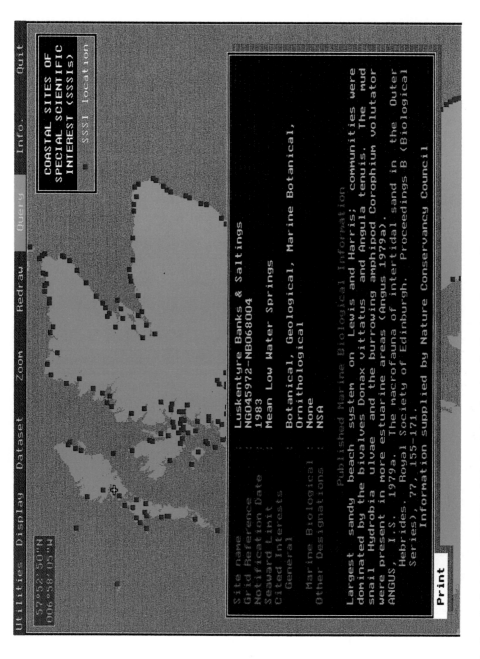

Plate IV Sample display from UK Digital Marine Atlas, showing distribution of coastal Sites of Special Scientific Interest in northern Scotland with pop-up data for one site

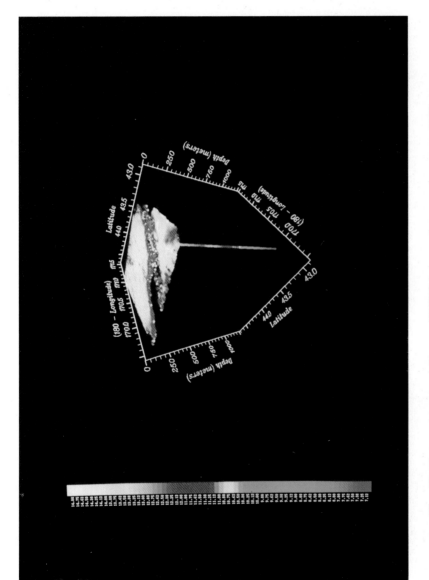

Plate V Ocean temperatures from an AVHRR image and a contemporaneous CTD cast

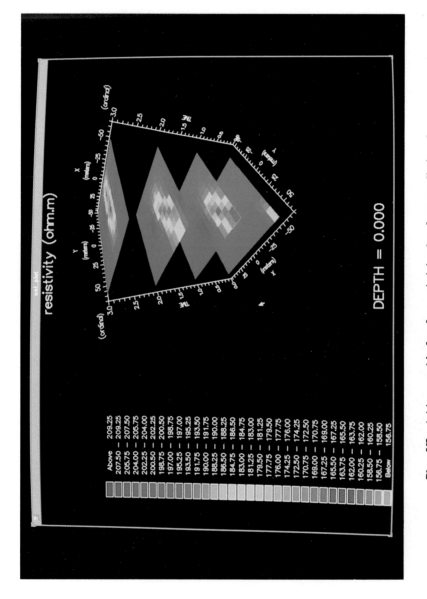

Plate VI A 'time-stack' of surface resistivity data from a pollution plume

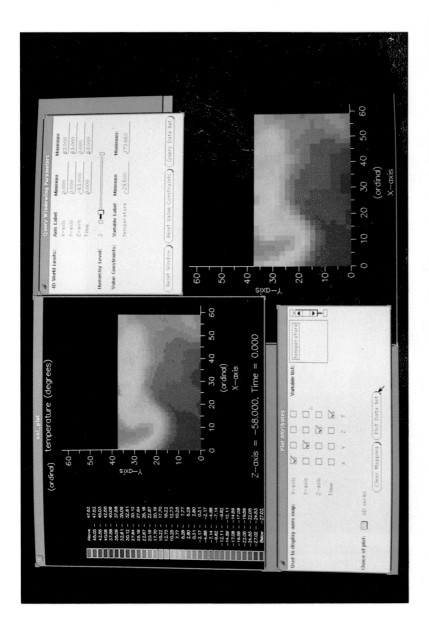

Plate VII Atmospheric temperature at fixed time and z values at different levels of generalization. (left) Full resolution; (right) half resolution

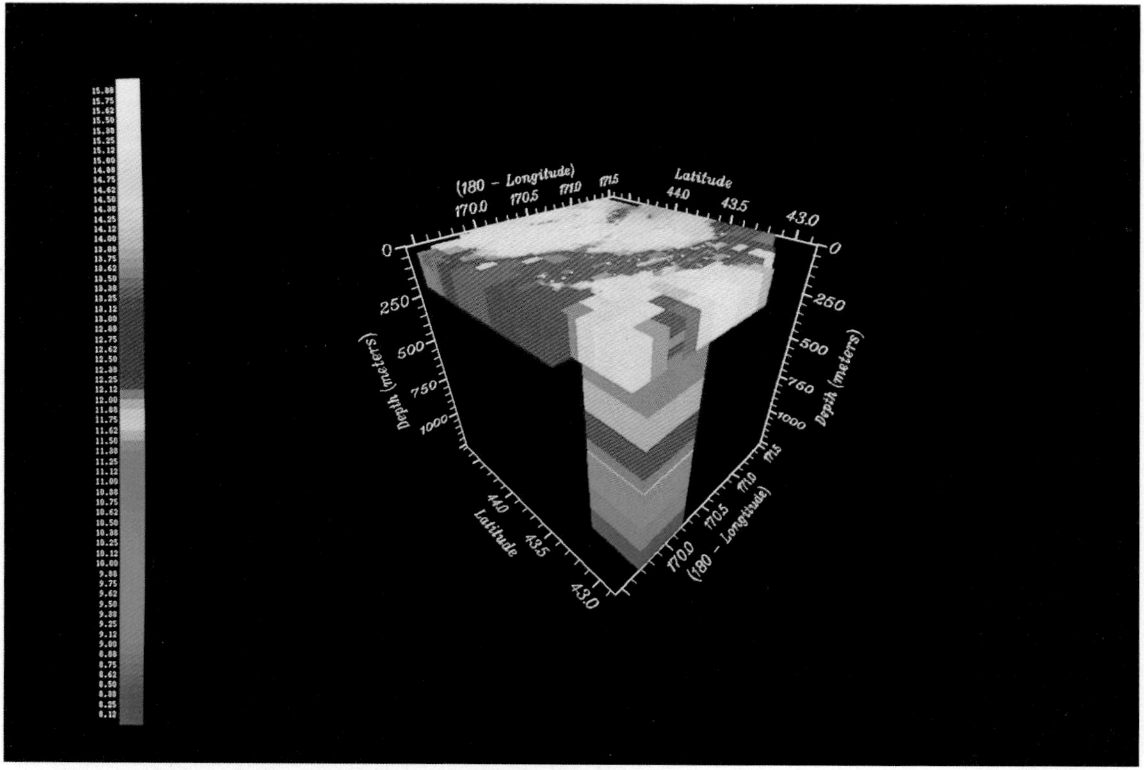

Plate VIII Interpolated ocean temperatures

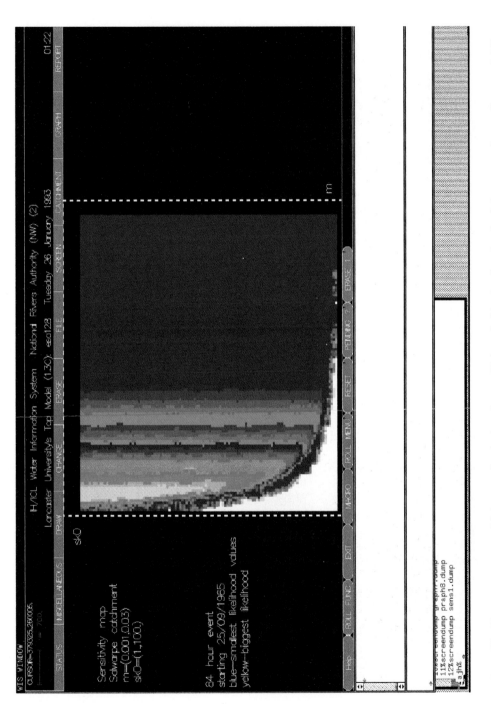

Plate IX Two-parameter sensitivity map for the 'maximum efficiency' objective function for Salwarpe catchment (for m and sk0)

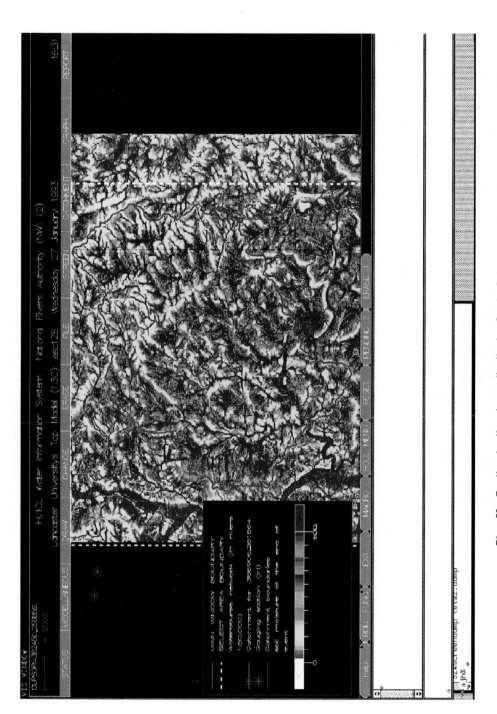

Plate X Predicted soil moisture distribution for the Salwarpe catchment

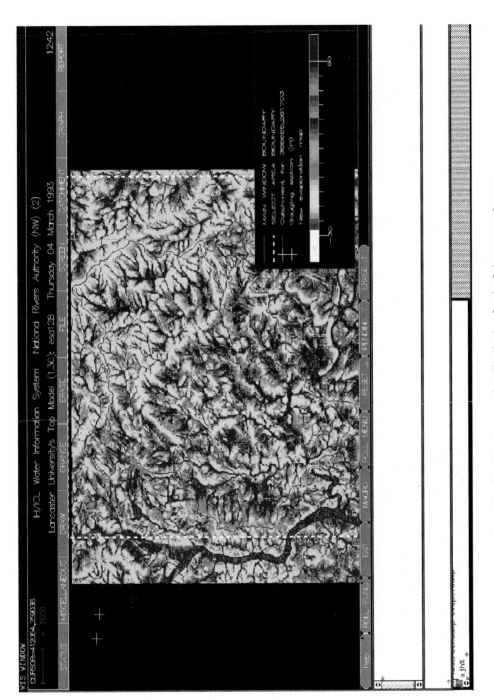

Plate XI Predicted evaporation distribution for the Salwarpe catchment

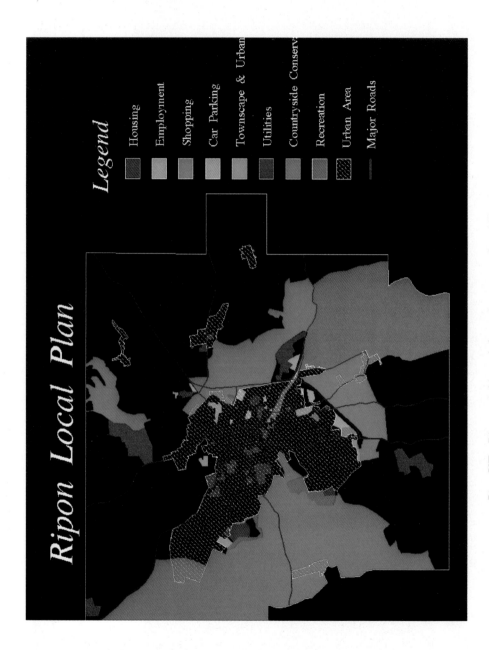

Plate XII Ripon subsidence information system: Local Plan browser

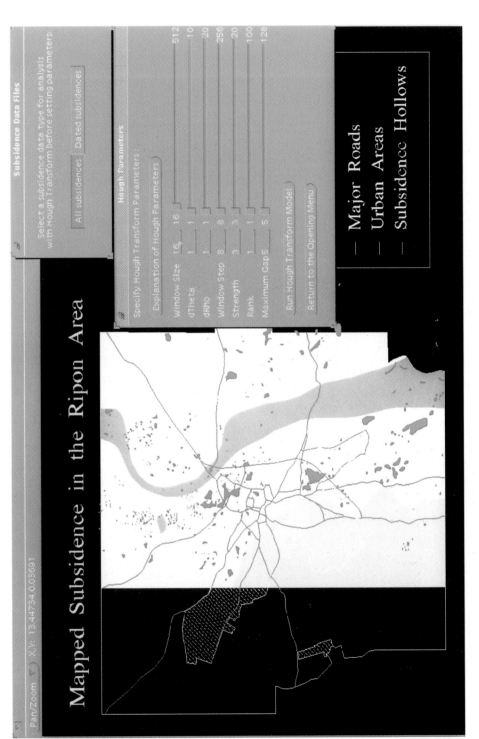

Plate XIII Ripon subsidence information system: subsidence database and alignment analysis. The user can freely choose the values of seven parameters for a Hough transform analysis of collapse alignment

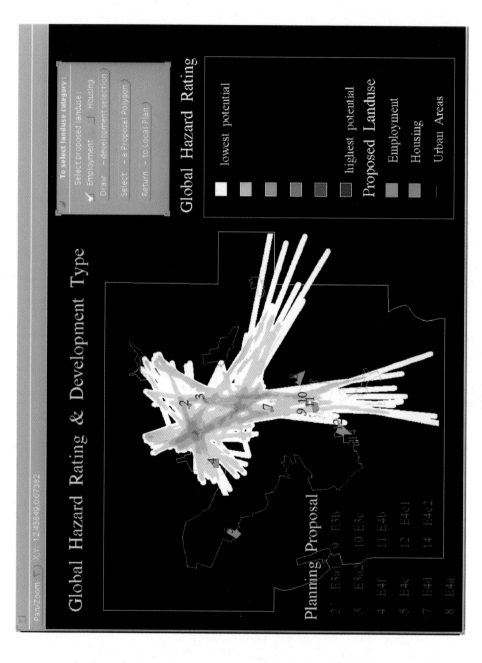

Plate XIV Ripon subsidence information system. A single model output of alignment analysis is converted to a global mapping of alignment intersection density for Ripon. Employment planning proposals are superimposed

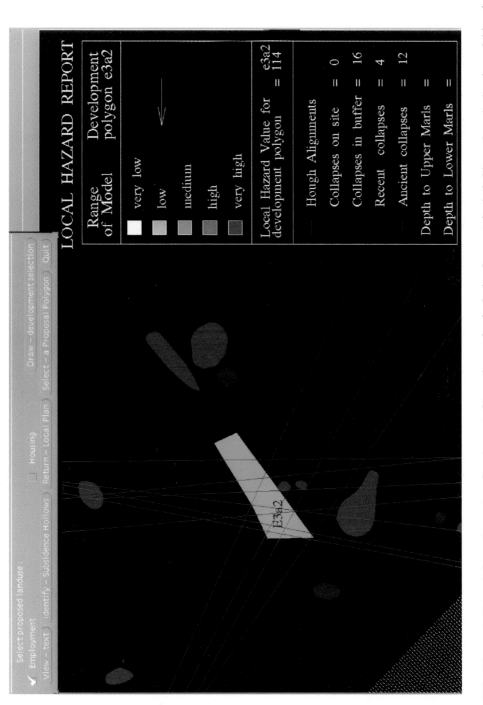

Plate XV Ripon subsidence information system. Local report of hazard around a single development site (pale blue polygon), showing neighbouring collapses and alignments

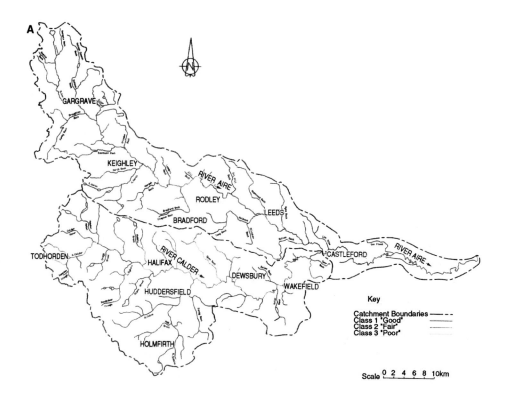
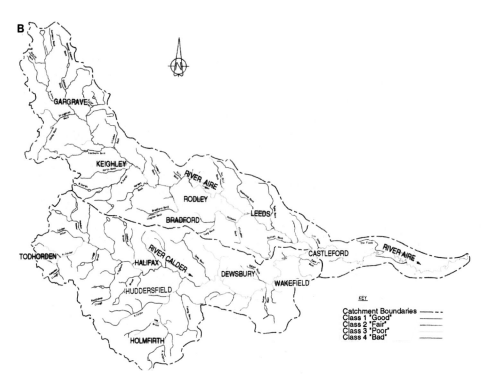

Plate XVI National Rivers Authority, Yorkshire region, Aire and Calder catchments. A: River quality objective, B: river quality 1989

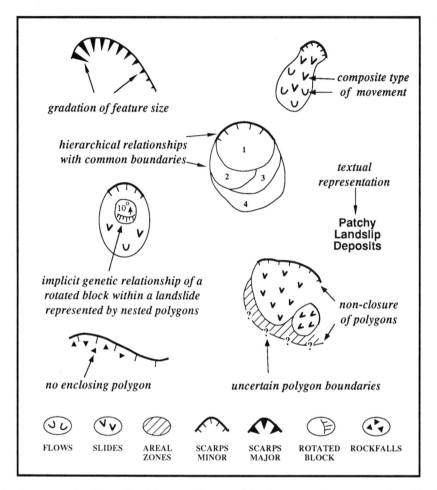

Fig. 3 Summary of problems faced in converting a traditional cartographic representation of landslide hazards into a GIS vector representation. The data were taken from the maps of Northmore *et al.* (1986)

rigour on the logic of how a landslide is represented than was required on the original maps. 'Open' polygons, graduations in symbolic representation and implicit spatial relationships are some of the problems encountered (Wadge, Wislocki and Pearson, 1993) (Fig. 3). Also, in this case, it is clear that a hierarchical representation of features derived from the same movement event, such as a rotated block polygon within a landslide polygon whose boundary is partly coincident with the source scarp, would have been a valuable data model to use for direct GIS input.

4.2. Hypothesis testing

The report of Northmore *et al.* (1986), like many similar technical reports, proposes a number of hypotheses to explain the spatial and temporal occurrence of landslides. Some of these hypotheses can be tested by inductive methods within the system. For example:

- Failure planes of major, rotational landslides occur at the junction of the clay-rich rocks and the limestones.

- The commonest (most recent) landslide mechanism is the reactivation of old slides.
- Translational landslides in clay-rich rocks are close to ambient limiting equilibrium.

Such testing helps to justify the choice of variables in a complex mapping of hazard because, in general, the tests involve only one or two variables at a time and the effects of spatial autocorrelation can be better appreciated. This process of hypothesis testing is well suited to the backward-chaining inference process that Nexpert Object supports, and Pearson, Wadge and Wislocki (1992) describe how this is implemented for the third of the above hypotheses.

5. CASE STUDY 2: SUBSIDENCE IN RIPON

5.1. Patterns and effects of gypsum dissolution

Gypsiferous strata are highly prone to dissolution by groundwater in and around Ripon. The gypsum occurs in two beds within limestones that outcrop in the town and dip gently eastwards (Cooper, 1986, 1989). The dissolution process is more rapid along vertical joint planes, forming caves. In places the roofs of the caves are unstable, and break up, forming cylindrical cavities that produce circular collapse pits at the surface. These pits are typically 20 m–50 m in diameter and can be over 10 m deep. In places they amalgamate into larger, complex depressions. Because they form above caves, which are essentially linear features, the collapses tend to be aligned (Fig. 4).

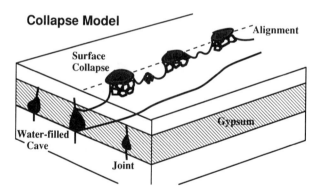

Fig. 4 Collapse model to explain subsidence hazard from gypsum dissolution in Ripon. Caves form preferentially along joints within the gypsum (shaded zone). Cylindrical roof breakdown intersects the surface as aligned circular collapses

The formation of the surface collapse pits causes subsidence of overlying buildings. This may be sufficient to necessitate demolition or engineering remedial measures to span the zone of subsidence with a rigid foundation. Measures used to mitigate possible future collapse have included deep piling.

5.2. Local authority planning and developers' needs

Planning to alleviate the effects of subsidence in Ripon is the responsibility of Harrogate Borough Council. In particular, the Council has the general responsibility (Department of the Environment, 1990) to:

1. Indicate major areas of unstable ground and any specific policies that may apply.

2. Evaluate development applications to ensure that any potential subsidence has been taken into account.

3. Decide on any further site investigations that must be made to satisfy (2) or other conditions to ensure development site stability.

The responsibilities of (1) are currently carried out in the Ripon Local Plan (Harrogate Borough Council, 1992), which contains general statements about subsidence potential and specific statements on each proposed development site. Two points should be emphasized: firstly, there is no specific body of legislation governing safety requirements for natural subsidence, only guidance (Department of the Environment, 1990), and secondly, the onus to investigate the likelihood of subsidence at a particular site is on the developer. From the planners' perspective the need is to make clear the general pattern of possible subsidence. From the developers' point of view, the need is to be able to evaluate the local level of hazard and the amount and type of site investigation required to satisfy the local authority (Wislocki et al., 1992).

5.3. Hazard assessment

The BGS database on the Ripon collapses contains about 400 items, about 44 of which are historical collapses (taking place during the past 200 years). The number reported in the past 20 years is 14. For the total area underlain by gypsum this gives a recurrence rate of about 0.1 collapses per year per km^2. This rate must be higher on alignments. One alignment east of Ripon contains 3 collapses in 20 years in 0.24 km^2: a rate of about 6 collapses per year per km^2.

In order to map the locations of these alignments the Hough transform technique of detecting alignments of point features (e.g. Wadge and Cross, 1989) is used. This works by transforming the (x,y) locations of the centroids of each collapse polygon into circular coordinates and searching this space with user-determined tolerances of alignment. A circular search window restricts the maximum length of an alignment (window diameter), and the minimum number of collapses and the maximum distance between any two collapses in an alignment must be set. The results of such an analysis can be mapped as a series of lines. No one model provides a definitive answer because the alignments are not perfect, uniform and of known size. A model with a low tolerance of alignment, long total length and separation threshold and low number of non-historical collapses required will produce a large number of alignments, though of lower hazard significance, than a model produced by short alignments of large numbers of highly aligned, historical collapses.

There are two known effects that reduce the completeness of the collapse database, and hence affect the alignment analysis:

1. The floodplain of the River Ure masks collapses.

2. The incidence of collapses within the urban areas is under-reported.

5.4. Ripon subsidence information system

The structure of the system is shown in Fig. 5. A digital version of relevant parts of the Local Plan allows the user to browse the various combinations of planning themes in map

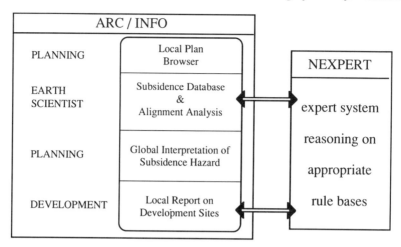

Fig. 5 Schematic functionality of the Ripon subsidence information system. The focus of user interest is shown on the left

form, employment proposal sites, conservation areas and the like, together with parts of the text itself, scanned and displayed as a raster image (Plate XII). This allows the subsidence hazard to be seen in the context of the other planning considerations.

The subsidence database can be interrogated and edited interactively by use of a cursor on a map of subsidences. Alignment analysis uses an external Fortran program called via an AML macro. The user can select freely the seven parameters required (Plate XIII). However, this selection is passed to a rule base in Nexpert Object acting in server mode. Nexpert Object then decides if this combination of parameters is reasonable. If not, some of the parameters are modified to produce a set that will give non-spurious results.

The third main function of the system is to provide global interpretations of subsidence hazard in map form. These are: (i) the area underlain by gypsum, (ii) areal density of collapses, and (iii) intersection densities of collapse alignments. These maps provide a means of zoning of subsidence hazard and a relative scale for local investigations (Plate XIV).

The Local Report functionality enables the hazard at specific development sites to be assessed (Plate XV). The following factors are taken into account:

- Number of collapses on site and within a 200 m buffer zone
- Number of collapse alignments passing through site
- Global hazard rating
- Ure flood plain or urban area
- Depth to base of the lower gypsum beds

A Nexpert Object knowledge base assesses the factors to provide a local hazard assessment that developers can use to plan their requirements for site investigation work.

6. DISCUSSION AND CONCLUSIONS

The concept of an information system (decision support system) designed for several distinct user groups raises the interesting issue of the breakdown of confidence in the traditional 'expert advice' relationship. If a developer has access to such a system does he need to consult the geologist at all? In the case of the Ripon Subsidence Information System, it is unlikely that a developer would have a version of this system. The more natural homes for it are, firstly, the BGS and Harrogate Borough Council. It could then form the basis of a closer understanding and dialogue on the problems of subsidence.

The principal conclusions of our work are:

- A single computer system for hazard assessment that, at least partly, satisfies the needs of earth scientists and planners/developers is desirable and possible.
- Expert-GIS is a demonstrably useful approach for hazard assessment.
- It is feasible to reinterpret previous technical maps and reports of hazard into a more useful form using GIS.
- Natural hazard GIS must be focused on the needs of the users of the hazard assessments, and a single generic system to cope with all eventualities is impractical.

ACKNOWLEDGEMENTS

Many people gave generously of their advice and time during the course of this project, particularly in the collection of data. In particular, we would like to thank Drs Tony Cooper and Paul Gostelow of BGS. This work was supported by NERC grant GST/02/488.

REFERENCES

Bardsley, W.E. (1988) Toward a general procedure for analysis of extreme random events in the earth sciences. *Mathl Geol.*, **20**(5), 513–528.

Bernknopf, R.L., Cambell, R.H., Brookshire, D.S. and Shapiro, C.D. (1988) A probabilistic approach to landslide hazard mapping in Cincinnati, Ohio, with applications for economic evaluation. *Bull. Ass. Engng Geol.*, **25**(1), 39–56.

Carrara, A., Cardinali, M., Detti, R., Guzzetti, F., Pasqui, V. and Reichenbach, P. (1991) GIS techniques and statistical models in evaluating hazard. *Earth Surf. Processes and Landforms*, **16**, 427–445.

Cooper, A.H. (1986) Subsidence and foundering of strata caused by the dissolution of Permian gypsum in the Ripon and Bedale areas, North Yorkshire. In Harwood, G.M. and Smith, D.B. (eds), *The English Zechstein and Related Topics*. Geological Society Special Publication, no. 22, 127–139.

Cooper, A.H. (1989) Airborne multispectral scanning of subsidence caused by Permian gypsum dissolution at Ripon, North Yorkshire. *Q. J. Engng Geol.*, **22**, 219–229.

Crandell, D.R., Booth, B., Kazumadinata, K., Shimozuru, K., Walker, G.P.L. and Westercamp, D. (1984) *Source-book for Volcanic-Hazards Zonation*. UNESCO Natural Hazards no. 4, Paris.

Department of the Environment (1990) *Development on Unstable Land*. Planning Policy Guidance no. 14. HMSO, London.

Gupta, R.P. and Joshi, B.C. (1990) Landslide hazard zoning using the GIS approach – A case study from the Ramganga catchment, Himalayas. *Engng Geol.*, **28**, 119–131.

Harrogate Borough Council (1992) *Ripon Local Plan*.

Hunter, R.L. and Mann, C.J. (eds) (1992) *Techniques for Determining Probabilities of Geologic Events and Processes*. Oxford University Press, Oxford.

Jibson, R. and Keefer, D. (1989) Statistical analysis of factors affecting landslide distribution in the New Madrid seismic zone, Tennessee and Kentucky. *Engng Geol.*, **27**, 509–542.

Maidment, D.R. and Djokic, D. (1990) Creating an expert geographic information system: The ARC–Nexpert interface. Report, Department of Civil Engineering, University of Texas, Austin.

Mann, J.C. (1988) Methods for probabilistic assessments of geologic hazards. *Mathl Geol.*, **20**(5), 589–601.

Marker, B.R. and McCall, G.J.H. (1990) Applied earth-science mapping: the planners' requirement. *Engng Geol.*, **29**, 403–411.

Northmore, K., Charalambous, M., Hobbs, P.R.N. and Petrides, G. (1986) *Engineering Geology of the Kannaviou, 'Melange' and Mamonia Complex Formations – Phiti/Statos Area, S.W. Cyprus*. EG and RPRG Report, EGARP-KW/86/4, British Geological Survey, Keyworth.

Pan, P.S.Y., Ferrier, G. and Wadge, G. (1993) Knowledge-based GIS: interfacing ARC/INFO with Nexpert Object, in preparation.

Pearson, E.J., Wadge, G. and Wislocki, A.P. (1992) An integrated expert system/GIS approach to modelling and mapping natural hazards. *Proc. 3rd European GIS Conf. (EGIS'92)*, Munich, 23–26 March, 762–771.

Varnes, D.J. and The International Association of Engineering Geology Commission on Landslides and other Mass Movements on Slopes (1984) *Landslide Hazard Zonation – A Review of Principles and Practice*. UNESCO Natural Hazards no. 3, Paris.

Wadge, G. (1988) The potential of GIS modelling of gravity flows and slope instabilities. *Int. J. Geogr. Inf. Sys.*, **2**(2), 143–152.

Wadge, G. and Cross, A.M. (1989) Identification and analysis of the alignments of point-like features in remotely sensed imagery – volcanic cones in the Pinacate Volcano Field, Mexico. *Int. J. Remote Sensing*, **10**(3), 455–474.

Wadge, G., Wislocki, A.P. and Pearson, E.J. (1993) *Spatial analysis in GIS for natural hazard assessment*. In Goodchild, M.F., Parks, B.O. and Skyaert, L.T. (eds.), *Geographic Information Systems and Environmental Modelling*. Oxford University Press, Oxford.

Wadge, G., Young, P.A.V. and McKendrick, I. (1993) Mapping lava flow hazards using computer simulation. *J. Geophys. Res.*, in press.

Wickmann, F.E. (1966) Repose period patterns of volcanoes. *Ark. Miner. Geol.*, **4**, 291–367.

Wislocki, A.P., Wadge, G., Pearson, E.J. and Whittow, J.B. (1992) Expert system/geographic information system approaches to planning in areas of instability. In Rideout, T (ed.), *Proc. Planning and Environment Special Group*. Institution of British Geographers, Swansea, January.

Chapter Eighteen

An evaluation of GIS as an aid to the planning of proposed developments in rural areas

D. A. DAVIDSON AND A. I. WATSON
University of Stirling

AND

P. H. SELMAN
Cheltenham and Gloucester College of Higher Education, Cheltenham

This chapter reports the results of a project that assesses the potential contribution of GIS to rural environmental planning through the investigation of four case studies (a skiing development, an overhead powerline development, an afforestation scheme and an opencast coal mine). The project thus seeks to establish whether the adoption of a GIS can yield cost-effective, helpful and accessible information for policy formulation and decision-taking. Documentation was already available for the case studies that were selected; thus, the project was concerned with evaluating GIS processing in retrospect. The final part of the project involved the presentation of the GIS results from the case studies to a workshop of senior planners. They gave favourable responses to GIS capabilities of spatial data management, landscape simulation and visibility analysis; the main criticisms focused on costs and ease of availability.

1. INTRODUCTION

Many planning authorities are either in the process of acquiring or have recently acquired a GIS. Typical applications of GIS in land use planning include policy production (e.g. Lindhult *et al.*, 1988), property enquiries (e.g. Cryan and Gentile, 1990), monitoring landscape change (e.g. Hooper, 1988; Bird and Taylor, 1990), transportation and retail planning, development control and environmental impact assessment (for example, Clark, Gurnell and Edwards, 1990; McAulay, 1991). One application of considerable importance, prompted by statutory requirements for environmental assessment, has been evaluation of the visual impacts of new development proposals. Hitherto, this has involved laborious manual techniques or protracted and expensive photographic editing. Increasing use is now

Geographical Information Handling – Research and Applications. Edited by P. M. Mather
© 1993 John Wiley & Sons Ltd

being made of digital terrain models (DTMs) and associated overlays (Davidson, 1992). Applications have included mine workings (Kent, 1985), transmission lines (Hadrian, Bishop and Micheltree, 1988; Hull and Bishop, 1988) and afforestation (Kellomaki and Pukkula, 1989; Davidson, 1991). There is considerable potential for much wider use of GIS to allow the rapid and routine analysis of development proposals. However, such routine usage of GIS technology by planning authorities immediately raises key questions of cost-effectiveness, the appropriateness of the output to the planning issues, and technical problems associated with assembling diverse data sources (Selman, Davidson and Watson, 1991).

Rural environmental planning in Britain takes two main forms: (i) strategic planning at the county or regional level, and (ii) development control, whereby planners react to specific proposals. For the former, there is considerable potential for GIS as an aid to data management, monitoring and policy formulation (Aspinall, Miller and Birnie, 1993). As an example, regional planning authorities in Scotland have been devising forestry indicative strategies as a means of providing a framework for their responses to forestry grant applications. GIS can be used as an efficient means of overlaying different data sets in order to map areas according to indicative strategy map classes. A different example is given by Eweg (1992), who explains how a GIS can assist with exploring and evaluating possible approaches to rural restructuring in an area in the Netherlands.

The helpfulness and cost-effectiveness of GIS to the process of development control in the rural environment is far less clear and merits research. This chapter reports the results of a research project designed to evaluate the extent to which GIS processing and output can produce timely, cost-effective and appropriate output as an aid to the planning of proposed developments in the rural environment.

2. PROJECT METHODOLOGY

The basis to the project was the retrospective evaluation of GIS processing and output produced for four contrasting case studies in Scotland. The selected case studies are a skiing development, an overhead powerline development, an afforestation scheme and an opencast coal mine. Such case studies are considered typical of proposed major developments in rural areas and are often controversial. The selection of these case studies for the project was done in close collaboration with a steering group consisting of planners from regional authorities, a conservation agency and a commercial consultancy. The project methodology can be summarized in the following six stages:

1. Selection of suitable case studies for which a range of information was readily available on file.

2. Identification by members of the steering group of the key planning issues associated with each of the case studies.

3. Digitizing map data, DTM creation, overlays, production of plots.

4. Logging of time required to perform the various GIS processes.

5. Obtaining regular comment and suggestions from the steering group regarding data selection, processing and output.

6. Evaluation of the results by means of a workshop attended by 35 invited planners and other environmental professionals.

Particular emphasis was given to disseminating the results by means of the concluding workshop, as well as to evaluating GIS output. The latter was achieved at the workshop by means of discussion of the output for the four case studies and by participants completing questionnaires. For GIS processing, use was made of the LaserScan system, which offers a suite of programs to perform automated cartography, polygon processing, vector–raster conversions, DTM with overlay capabilities and visibility analysis. The software was run on a MicroVAX II computer with display on a Sigmex 6000 terminal. In this project, particular priority had to be given to good quality output. Use was made of a high quality A0 plotter as well as an A3 colour electrostatic printer. Unfortunately, it is impossible in this volume to reproduce the large plots produced for evaluation by the planners who attended the workshop.

3. CASE STUDIES

3.1. Aonach Mor skiing development

The Aonach Mor skiing development is east of Fort William in the Ben Nevis range. The planning application for this development was submitted to Highland Regional Council in 1986; permission was granted and the development opened in December 1990. It consists of a car park and a cafe at the bottom, and a gondola which goes to the top station, where there is a restaurant, bar and shop. The skiing facilities include a chairlift and six ski tows. The planning issues identified by the steering group are listed in Table 1.

Table 1   Planning issues for the four case studies

Case Study 1: Aonach Mor skiing development
Visual impact of the base station, infrastructure, the new access road and removal of trees by the Forestry Commission
Location of gondola towers
Impact on vegetation
Impact on water courses
Effect on public water supply
Siting of snow fences
Effect on summer recreation
Increased risk of forest fires

Case Study 2: Skye overhead powerline
Visual impact on existing settlements, the National Scenic Area, and views from roads
The powerline to follow existing routes as far as possible
Siting of powerline as near as possible to field or forest boundaries to minimize disruption
Siting of terminal substation in an unobtrusive position
Powerline route to take into account access during construction and for maintenance
Powerline route not to be located beneath the high tide level

Case Study 3: Tinto Hills afforestation scheme
Recreational importance of the area and public access
Visual impact
Nature conservation
Archaeological sites
Impact on water courses
Land ownership
Sites of Special Scientific Interest
Agriculture

(continues)

Table 1 (*continued*)

Case Study 4: Lochwood farm opencast mine
Location within the green belt and urban fringe
Effect on groundwater levels
Effect on recreation within the local area
Visual impact of excavation and spoil heaps
Siting of powerlines
Ground stability
Effect on local agriculture and forestry

A core area of 20 km^2 was initially digitized, although this was subsequently extended to 56 km^2 so that visibility studies could be carried out for a wider area. Data on contours, roads, tracks, footpaths, all the features of the skiing development, snow fencing, the Site of Special Scientific Interest (SSSI), erosion risk assessment, vegetation and forestry were digitized. It took an experienced operator a total of 70 hours to input the data and a further 24 hours for processing. Particular emphasis was given to determining whether GIS processing could offer assistance in terms of assessing visual and environmental impact.

3.1.1. Visual impact

DTMs with a pixel size of 20 m were created from the contour data based on Delaunay triangulation. These models were then draped with a variety of vector files to provide realistic images of the development. The LaserScan software includes a module for calculating visibility. In the first instance, visibility was determined from the 20 m DTM, but each calculation took 10 hours; a change in pixel size to 50 m reduced this time to 1 hour. One key GIS operation was to calculate the visibility of the skiing development according to different forestry patterns. In summary, the output that was produced to assess the capability of GIS to evaluate impact consisted of the following: (i) visibility map for the top station, (ii) visibility map for an alternative top station location, (iii) visibility map for the chairlift, showing areas with different degrees of visibility, (iv) visibility map from the summit of Ben Nevis, (v) visibility maps of the skiing development from a local beauty spot (Commando Memorial, Spean Bridge), taking into account different forestry patterns.

3.1.2. Environmental impact

Polygon processing of the soil, vegetation and slope data was undertaken to provide prediction of areas most likely to be eroded. The vector files on soils and vegetation were converted into raster format and were assigned values. The map for the angle of slope was calculated using the DTM, which output the information to a raster file. Raster files were then combined so that any combination of soil and vegetation could be selected for different slope classes. The results were then vectorized and converted into polygons to yield maps showing variation in erosion risk and thus environmental impact.

3.2. Isle of Skye: overhead powerline

A 132 kV powerline across the Isle of Skye from Broadford to Ardmore was part of a major project by the North of Scotland Hydro-Electric Board to link the Western Isles to the mainland electricity supply. Planning permission was sought from Highland Regional

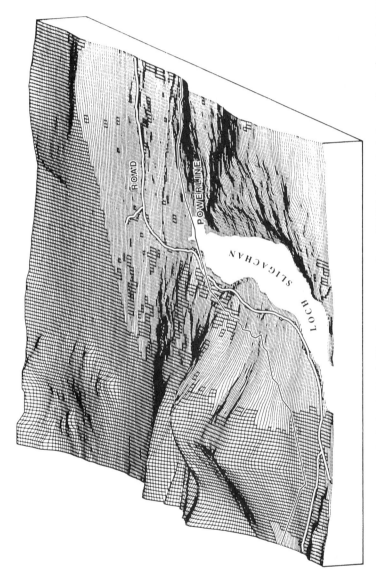

Fig. 1 Portrayal of the visibility analysis for the Loch Sligachan area as a draped DTM; in the shaded area, five pylons at most can be seen

Council in 1986 and was subsequently approved. The overhead powerline consists of high amenity woodpole pylons, which were designed to replace the conventional steel towers. A substation at Ardmore was also part of the application and has since been completed. The planning issues associated with this development are given in Table 1.

Time constraints meant that two sample areas of this powerline route were investigated using GIS techniques. Areas of 45 km^2 around Loch Sligachan and 50 km^2 at the northern end were digitized for contours, powerline route, including location of individual pylons, national scenic areas, SSSIs, roads, field boundaries and forestry. A total of 39 hours were spent digitizing data and a further 66 on processing. Digital data were not available from the Ordnance Survey at the time of the project.

It was considered that the main planning issue that would benefit from GIS analysis was visual impact – the substation and the pylons. Thus, a visibility map was produced for the substation and, to aid presentation, was draped over a DTM of the area. For the powerline in the sample areas a visibility plot was made for each pylon. As each was created, it was added together with the raster maps from the previous pylons. This was achieved by giving the pixels visible from each pylon a score of 1 and then accumulating these scores to produce a composite visibility map. A command program was written to perform such calculations. Using this approach, visibility maps were produced for alternative powerline routes as well as for pylons of different heights. The resultant maps tended to appear rather complicated, with a range of shadings for pylon visibility. To aid presentation the DTMs were draped with a shading pattern corresponding to those areas within which at most five pylons were visible, as illustrated in Fig. 1.

3.3. Tinto Hills afforestation scheme

The Tinto Hills lie to the south-east of Glasgow and rise to a height of 710 m. An application by a commercial forestry company to plant 740 hectares of forestry under the forestry grant scheme was put forward in 1986. Permission was not granted and a renewed application was submitted in 1988 to plant a reduced area (279 hectares). This application was approved, but the company entered into a management agreement with the Nature Conservancy Council so that the area was not planted, but compensation was paid to the applicants.

For an area of 100 km^2, the following were digitized: contours, rivers, roads, tracks, footpaths, villages, existing forestry, archaeological sites and ancient monuments, water catchments, SSSIs, land of good agricultural quality, the first forestry proposals and their subsequent revisions. Data input took 54 hours, with a further 24 hours spent on processing and output.

As can be seen from Table 1, visual and environmental impact were key planning issues with this forestry proposal. Emphasis was given to simulating views of the Tinto Hills from tourist stopping points along the main road. Realism was aided by giving the draped forestry areas a height above the land surface. Competing land uses such as archaeological sites, good agricultural land and footpaths were also draped on to the DTMs.

3.4. Lochwood farm opencast mine

Lochwood farm lies near the eastern boundary of Glasgow and in 1986 a company submitted a planning application to extract coal by opencast methods from an area of 1.8 hectares.

Permission was not granted, mainly because of the location of the site within a green belt and because the area had been targeted for recreational development as part of the Easterhouse project.

As this was mainly a site-based case study, only a small area of 6 km^2 was digitized including pre- and post-development contours, site development plans, streams, roads and villages. Such digitizing took 17 hours, with a further 5 hours spent on preparing plots and generating DTMs.

The main planning issue was whether it was desirable to have such a development within the green belt and near to an urban area. The application of GIS seemed to be limited to automated cartography, for example the simplification and better presentation of site plans; DTMs were also of potential value as an effective way of illustrating land form before, during and after development of the mine.

4. DISCUSSION

The responses of the participants in the workshop are summarized in Table 2. In broad terms the findings accord with the comments of McAulay (1991), who states that a small number of basic GIS functions are likely to be useful or essential in the preparation of every environmental statement. These are the overlaying and sieving functions, illustrated by the erosion risk maps for the skiing development. He further stresses the use of a GIS to model or simulate environmental effects. At the workshop very positive comments were made on the DTMs portraying the skiing development, the overhead powerline and the afforestation scheme. Such models with drapes of alternative development proposals were judged to be of particular value in providing effective landscape images, though some participants commented that realism could be improved by using better shading patterns for trees and other landscape features. The participants felt that such models were far more useful than conventional maps in giving a feel for areas, and furthermore would provide useful reminders of the development proposals after site visits. The DTMs were well received for the skiing, overhead powerline and forestry proposals, given the importance of visualizing the setting of these developments in the landscape. In addition, high confidence could be given to the precise location of the development features on the models – a very difficult task to do manually on landscape drawings by artists. DTMs of the opencast proposal were far less successful, primarily because they portrayed a site rather than a landscape.

Table 2 A summary of the advantages and disadvantages of using GIS to aid rural environmental planning, compiled from the responses of planners who attended the workshop

Advantages noted by planning authorities
Ability to link several data sets
Ability to yield information in a variety of formats
Presentation of clear visual information to a variety of audiences – especially officers, members and general public
Rapid analysis on a relatively routine basis – no need for manual solutions or complex 'one-off' programs
Ability to perform polygon processing operations
DTMs can help determine visibility and intrusiveness of objects placed into the terrain
Overcomes inconvenience of map boundaries – especially important for extensive rural sites
Excludes extraneous information – maps include irrelevant and distracting material

(continues)

Table 2 (*continued*)

GIS plots produced precisely for the desired areas
Visual displays can show pre- and post-development situation – 'before and after' simulation
Can explore 'what if' questions
Ability to focus attention on key planning issues raised by individual developments, e.g. susceptibility to soil erosion, avoidance of 'skylining', encroachment into designated areas
May reduce (though not eliminate) need for site visits, and will make them more productive
Can simulate screening effect of landscaping measures
Ability to store and update vast amounts of data
Not necessarily more expensive than technical alternatives, such as photomontages and model-building
Can give more rapid interpretations of policy and assist in determination of complex developments when working against tight time limits
Aids determination of development control policies in rural areas, so may help to reduce 'blanket restrictive' reputation of rural development control
Ability to pinpoint precise location and boundaries of the development within the landscape
Ability to produce high quality maps and plots
Can produce plots highlighting different features, in different colours, and at different scales, i.e. can customize output so that it is relevant to a particular task
DTMs can aid recollection after site visits
Ability to zoom in to critical areas

Possible problems noted by planning authorities
Cost and time
Incompatibility of systems
Problems of acquiring own system – need to be assured of steady throughput and in-house technical expertise
Easy to be seduced by graphics – worth needs to be proven in terms of accessibility, efficiency, helpfulness and cost-effectiveness
Did not appear to assist detailed policy issues very greatly – more valuable with respect to 'broad-brush' planning considerations
Some technical problems with 'intelligibility' of plots and ornamentation with realistic symbols

As indicated in Table 1, visual impact was an issue with all four case studies. The ability to map the visibility of development proposals, taking into account changing surface features such as forestry, was identified as an extremely useful capability. It was thus possible to compare visual impacts for different skiing development designs or overhead powerline routes.

The cost-effectiveness of GIS processing and output was judged at the workshop on the reported hours for the different operations. One factor that emerged as significant was the considerable time spent on data input. All source data had to be manually digitized since at the time of the project, Ordnance Survey digital data were not available. The dominant view of the planners was that, however useful the output in terms of evaluating development proposals, cost and time constraints would exclude routine usage of GIS for such planning purposes. Rather than having planning departments undertake such work, the view was that the applicants for proposed developments ought to carry out such GIS analysis integral to the preparation of their environmental statements. If met by developers, the costs associated with such processing would constitute a very small part of overall projects. Furthermore, the purchase of topographic data would reduce time and costs. The ideal would be for planning authorities to be able to receive and input GIS files as part of environmental statements and to be able to compare immediately development options using their own systems. Such a situation is not likely to become common in the near future, but with such interaction between developers and planners, GIS has considerable potential for playing a key role in the planning of specific developments in rural areas.

ACKNOWLEDGEMENTS

Thanks are expressed to members of the steering group for their help, to the Countryside Commission for Scotland for access to files on the case studies and to Sandra Winterbottom, research assistant for the project. This article is based on research funded by the Natural Environment Research Council under grant GST/02/492. Longman Group (UK) kindly gave permission for the reproduction of Fig. 1 (fig. 7.8 from D.A. Davidson, *The Evaluation of Land Resources*), and the European Conference on Geographic Information Systems (EGIS) allowed the use of material previously presented at EGIS'92.

REFERENCES

Aspinall, R.J., Miller, D.R. and Birnie, R.V. (1993) Geographical information systems for rural land use planning. *Appl. Geog.*, **13**, 54–66.

Bird, A.C. and Taylor, J.C. (1990) GIS for landscape monitoring: an operational system for national parks in England and Wales. In Harts, J. *et al.* (eds.), *Proc. 1st European GIS Conf. (EGIS'90)*, Amsterdam, 10–13 April, 68–76.

Clark, M.J., Gurnell, A.M. and Edwards, P.J. (1990) A GFIS approach to management decision making for the coastal environment. In Harts, J. *et al.* (eds.), *Proc. 1st European GIS Conf. (EGIS'90)*, Amsterdam, 10–13 April, 189–198.

Cryan, S.M. and Gentile, A.R. (1990) The utilisation of a GIS for applications in an urban environment: an instrument for local authorities. In Harts, J. *et al.* (eds.), *Proc. 1st European GIS Conf. (EGIS'90)*, Amsterdam, 10–13 April, 27–35.

Davidson, D.A. (1991) Forestry and GIS. *Mapping Awareness*, **5**, 43–45.

Davidson, D.A. (1992) *The Evaluation of Land Resources*. Longman, Harlow.

Eweg, R. (1992) Exploring possible directions of development in a rural area with GIS. In Harts, J. *et al.* (eds.), *Proc. 3rd European GIS Conf. (EGIS'92)*, Munich, 23–26 March, 19–28.

Hadrian, D.R., Bishop, I.D. and Micheltree, R. (1988) Automated mapping of visual impacts in utility corridors. *Landscape and Urban Plann.*, **16**, 261–282.

Hooper, A. (1988) Monitoring of landscape and wildlife habitats. In Park, J.R. (ed.), *Environmental Management in Agriculture*. Belhaven Press, London, 21–34.

Hull, R.B. and Bishop, I.D. (1988) Scenic impacts of electricity transmission towers: the influence of landscape type and observer distance. *J. Environ. Mgmt*, **27**, 99–108.

Kellomaki, S. and Pukkula, T. (1989) Forest landscape: a method of amenity evaluation based on computer simulation. *Landscape and Urban Plann.*, **18**, 117–125.

Kent, M. (1985) Visibility analysis of mining, and waste tipping sites – a review. *Landscape and Urban Plann.*, **13**, 101–110.

Lindhult, M.S., Fabos, J., Brown, P. and Price, N. (1988) Using geographic information systems to assess conflicts between agriculture and development. *Landscape and Urban Plann.*, **16**, 333–343.

McAulay. I. (1991) Environmental impact analysis: a cost-effective GIS application? *Mapping Awareness*, **5**, 36–40.

Selman, P., Davidson, D. and Watson, A. (1991), An application of geographic information systems in rural environmental planning. *Town Plann. Rev.*, **62**, 215–223.

Chapter Nineteen

Property-based GIS: data supply and conflict

D. J. FAIRBAIRN
Department of Surveying, University of Newcastle upon Tyne

1. BACKGROUND

The well-being and development of the property profession in Britain is based on the existence of millions of land parcels, each possessing location, extent and associated characteristics, such as ownership and use. The potential task involved in recording and utilizing the information related to such property is important, enormous and problematic: this is recognized by those currently proposing the establishment of some form of National Land Information System (Dale, 1992), which would have important implications for property professionals. Data availability and data conversion are the key aspects in improving the take-up of GIS in general (Department of the Environment, 1987) and in the fields of property management and development in particular (Courtney, 1990). These problems are examined here in the context of two case studies examining property data within a GIS from an English perspective.

2. PROPERTY-BASED DATA

2.1. Ordnance Survey map data

Digital topographic data is an essential element in a property GIS and, although there are some alternative sources such as the Goad shopping centre plans, the vast majority of such data is obtained from the Ordnance Survey. The history of Ordnance Survey's evolution has had a significant influence on the handling of property information in Great Britain. Unlike many nations, Britain has no cadastral mapping agency specifically tasked to map property boundaries and produce specialist documents defining land parcels. The systematic large-scale mapping of Britain presents the locations of tangible topographic features in the landscape. It so happens that land parcel boundaries here invariably follow such features, such as hedges, fences and roads, and despite some argument to the contrary (Barrett, 1986) and evidence of difficulty in a small minority of land transfers, it is felt that there is little need to move from this 'general' boundaries system to one that relies on precisely coordinated boundary markers, such as those used in continental Europe. The large-scale

---

Geographical Information Handling – Research and Applications. Edited by P. M. Mather
© 1993 John Wiley & Sons Ltd

Ordnance Survey maps, therefore, form the basis of the graphical information presented by the Land Registry (see Section 2.2) when defining land parcel boundaries.

The Ordnance Survey is considerably far advanced in its programme of digitizing these large-scale maps. The urban 1:1250 series is complete, and the 1:2500 agricultural and rural map series is on target for a 1995 completion date. It is clear, therefore, that Ordnance Survey can readily supply a vital data source. There are, however, a number of limitations. In addition to the irregular progress in the county-by-county digitizing of the 1:2500 series, the potential user must consider the costs (for purchase or lease), the imposition of a copyright licence to cover repeated use within a database, the digital media on which the data is supplied, and the problems relating to the data structure, which can make it difficult to construct the very land parcels that would be at the heart of a property GIS. All of these factors can affect the incorporation of Ordnance Survey data into a GIS.

Ordnance Survey has made strenuous efforts recently, within the context of its commercial goals of cost recovery, to consider the large range of customers for its digital data. As a result, there is considerable flexibility for the user in data supply, in obtaining updated information at intervals governed by change on the ground, in copyright licensing and in leasing plans.

2.2. Land Registry data

Her Majesty's Land Registry (HMLR) provides a national map-based system in England and Wales for recording the title to land holdings. It is the most comprehensive source of data about land parcels, and its recent 'opening up' (see Section 2.2.2) provides an important opportunity for those creating and maintaining property databases. The basic idea of registration of title, the English form of land registration, is to reflect on a register those matters concerning land that will be of importance to a potential purchaser and that otherwise would only be discoverable by inspection of the land, enquiries of the occupants and perusal of a miscellany of documents.

2.2.1. Content

The Register is divided into three parts:

1. The Property Register, which describes the land parcel and refers to a map or plan; it also contains notes of interests held for the benefit of the land, such as easements for which the land is dominant tenement. The boundaries shown on the map or plan are general, except in a small number of cases (estimated at 40), where they can be stated as being fixed (Land Registration Rules 1925, r.278; SI 1925 No. 1093).

2. The Proprietorship Register states the nature of the title, the name, address and description of the registered proprietor, and any cautions, inhibitions or restrictions affecting his right to deal with the land.

3. The Charges Register lists those rights that adversely affect the land, such as mortgages and easements for which the registered land is servient tenement.

2.2.2. Access

Since 3 November 1990, and the passing of the Land Register (Open Register) Rules 1990 (SI 1990 No. 1362), which brought the Land Registration Act 1988 into force, it is open to anyone to search the Register and obtain a copy of an entry. The information available is limited to that described above; the following further restrictions, many of which could be regarded as drawbacks to the use of HMLR data as the foundation for a property database, should be noted.

2.2.3. Limitations

2.2.3.1. Geographical restrictions The entire country (England and Wales) has been subject to compulsory registration only since December 1990 (although the vast majority of areas were so designated long before then). However, titles are only registered when the land is conveyed on sale or when registrable leases are created or assigned on sale. Thus, it was estimated in 1990 that only about 13 million out of a possible total of 22 million registrable titles in England and Wales were actually registered. This figure is increasing by approximately 500 000 each year (Pryer, 1990).

2.2.3.2. Souvenir plots The Land Registration (Souvenir Plots) Rules 1972 (SI 1972 No. 985) indicate that land which is sold in small plots, say 1 square foot, often in order to make compulsory purchase of the whole area impractical, need not be registered.

2.2.3.3. Land-related burdens 'Burdens, whether imposed by legislation or the common law, affecting land owners or occupiers generally, such as the liability to pay rates, the consequences of planning laws, criminal or tortious liability in relation to land' do not appear on the Register (Law Commission Report No. 158, para. 2.1, 1987). Thus, penalties for building on land without planning permission (Town and Country Planning Act 1990, s.57) or for adversely affecting a neighbour's enjoyment of his land in circumstances that would constitute a private nuisance are not spelled out in the Register (these are, in effect, restrictions of land use).

2.2.3.4. Overriding interests Overriding interests are those that bind a purchaser of registered land even though they do not appear on the Register (Land Registration Act 1925, s.3 (16)). These include (Land Registration Act 1925, s.70; Land Registration Rules 1925, r.258; Coal Industry Act 1987):

- Rights of common, drainage rights, customary rights (for example, public rights of way, defined in the Wildlife and Countryside Act 1981 s.53), public rights, profits à prendre (for example, grazing or shooting rights) and certain easements and other analogous rights, including ancient manorial rights.
- Various rents and charges having their origin in tenure, most, if not all, of which were repealed by the Law of Property Act 1922.
- Liability to repair the chancel of a church, which rarely arises, but can have severe repercussions when it does (Law Commission Report No. 152, para. 3.4).
- Liability in respect of embankments and sea and river walls, which is generally treated as a local land charge and therefore appears on a separate register.
- Further rights under local land charges, such as a restriction on land use by a Minister or local authority, which will be entered on the Local Land Charges Register (a separate register).

- Certain charges and annuities relating to tithes, most of which have been abolished and some of which should appear on the Register anyway (Law Commission Report No. 158, para. 2.93).
- Rights acquired, or being acquired, by adverse possession (otherwise known as 'squatter's rights').
- Further occupiers' rights, which can endure through different ownerships of the land.
- Defects in the title which prevented registration with an absolute title and necessitated registration with a possessory, qualified or good leasehold title (which are less watertight forms of title).
- Leases with 21 years or less to run (not included because of the vast amount of short-term letting that goes on).
- Certain minerals and mining rights.

2.2.3.5. Financial burdens Although the names of mortgagees and other proprietors may be revealed, the extent of their financial interest is not held on the Register.

2.2.3.6. Fees A personal inspection of the Register is subject to the payment of a fee of £12.00 per title (The Land Registration Fees (No.2) Order 1990; SI 1990 No. 2029).

There are further minor circumstances where an entry in the Register may be incomplete or changeable. These may take place as a result of the inclusion of nominee titles, previous errors, court cases and agreements of relevant parties.

The limitations described would obviously lessen the effectiveness of HMLR data within a property GIS, although compared to the situation before 1990, the availability of such data has an enormous potential impact.

2.3. Other external data

Although the combination of Ordnance Survey digital mapping and HMLR computerized data may prove to be the cornerstone of any property-based GIS (and indeed is the eventual goal of the Land Registry; Fenn, 1991), there is, potentially an enormous amount of further information that the property manager or developer would find of value and that could be included in a GIS. To exemplify this, a study was made of the management of land parcels and features by English Heritage on Hadrian's Wall in northern England.

This study was designed to include as much spatial data as possible relating to the management and use of land parcels along the Wall, including the Ordnance Survey digital data and HMLR information already described. Further data, from a potentially wide range relating to legal restrictions and easements, was also to be included. One of the difficulties in setting up a working land management GIS is to gain access to such data. Public rights of way are legally defined on a copy of an Ordnance Survey map held by local authorities, the 'definitive map', which is at a scale of 1:25 000. This scale has not been digitized by Ordnance Survey and, therefore, all such data must be digitized 'in-house' for subsequent incorporation. Tree Preservation Orders and Listed Building records are also held by local authorities: these are significant items. Areas of Outstanding Natural Beauty, National Parks, Conservation Areas and other more straightforward administrative area boundaries define zones of regulation for land and building management and development. Sites of Special Scientific Interest (SSSI) also fall into this category, as will future defined areas of agricultural 'set-aside'. Such data is more likely to be available in digital form (SSSI data sets can be purchased readily from a commercial supplier) but it is considered that all the data sets mentioned have a role to play in the development of property GIS. A fuller description of the data integration used in this study can be found in Fairbairn (1993).

Further examples of handling and disseminating spatial data in the public sector can be found in the wide variety of initiatives being undertaken by local authorities and central Department of the Environment activities. This includes the national Register of Contaminated Land*, the National Street Gazetteer (for highway authorities and utility companies) and field trials of the Land and Property Gazetteer and a Local Land Charges enquiries system.

2.4. Internal data

2.4.1. Paper records

A further case study (Fairbairn and Peedell, 1991) considered an organization that has also used significant amounts of internal, paper-based records, the Tyne and Wear Development Corporation (TWDC), an autonomous private body set up by the government in 1987. TWDC has the responsibility for creating the conditions in which the private sector will make investment – in industry, commerce and housing – in a traditionally low investment area of the UK. Independent of local authorities, it performs the function of a planning authority for a small but complex urban area of 2400 hectares along the banks of the lower reaches of the rivers Tyne and Wear. In this organization, the acquisition and disposal of land is recorded as part of its property management remit.

The TWDC handles property data using the Surveyors' Land Information Management PACkage (SLIMPAC), a GIS package developed in the Department of Surveying at the University of Newcastle upon Tyne. A combination of AutoCAD with DBase IV, full graphics and relational database management functionality are present in SLIMPAC. Further manipulative capability has been added to introduce analytical techniques, such as automated topological structuring, restriction of database search by area, and network analysis. Interrogation of both graphical screen and database menus gives considerable flexibility in use, and SLIMPAC has been installed in a number of commercial utility/property management agencies throughout Europe.

The property terrier of TWDC has four distinct application areas, each using internally produced and held data. Firstly, there is the TWDC Ownership file, comprising all freehold land owned by the Corporation; secondly, there is the list of properties rented by TWDC from external landlords (TWDC Leasehold/Licence Interests Held); thirdly, there is a file of properties leased by TWDC (Rent Roll (Lettings)/Tenancies Granted); and, finally, a file denotes the properties sold (Freehold Disposals). For each of these, a series of forms has been encoded within SLIMPAC to allow for the map-based querying and display of management information about land parcels within the TWDC area. Detailed information on description of property, dates of acquisition/disposal, area of property, name and address of tenant/landlord, term of lease, maintenance responsibilities and agreements (rights and easements, planning permissions and so on) is readily available. This information was already in DBase form before SLIMPAC was installed and the GIS merely replicated the data and attached it to 'intelligent features' (essentially SLIMPAC-created, topologically-structured points, lines and polygons) on the digital map.

---

*This proposal has now been shelved.

2.4.2. Further information

When confronted with the task of encoding data other than that held on paper records, more difficulties may be envisaged. The TWDC has duties in managing and developing property, the latter involving planning the overall development of a large area; but also, similar to a commercial property developer, speculatively acquiring, reclaiming (in the case of derelict land), improving (both land and buildings) and disposing of individual parcels and units. This may well involve activities which, by their very nature, engender considerable risks, particular financial risks.

Variations in willingness to take risks are an important element in the theory of decision making (Byrne and Cadman, 1984) and it might be useful to determine how to incorporate a spatial element into models of assessing risk, uncertainty and undertaking decision making in the property speculation field. The key element in such decisions made in property development as a whole is information, and GIS should be able to prove a major help to the property developer. Requirements may be for items that summarize land and property acquisition costs, performance of rents, building costs, interest rates, investment yields and the performance of property portfolios. Some of this information may be directly measurable from past or present trends, but there are, in addition, numerous more intangible factors affecting the basic intuitive 'feel' for market conditions that appears to be the driving force for the experience of many property professionals (Byrne and Cadman, 1984).

Each site and building, existing or potential, is different and although this may consequently mean that an objective computer-based modelling of property development is bound to be generalist in approach, it also indicates the potentially important role of location in determining some of the primary factors mentioned above.

2.4.3. Rent, costs and yields

The three major factors of rent, costs and yields may require some form of spatial component for their prediction. Internal data, both recorded and in the form of 'unwritten feelings', must be incorporated into the property GIS to allow such predictions to be made accurately. For example, initial costs of land parcels and buildings are dependent on a number of factors – use, potential use and planning restrictions, location, physical nature of the site (size, slope), neighbours, present owner, potential for development, and so on. The market costs will obviously take these into account: spatial information may well provide bargaining counters. Rents and associated yields are also, potentially, subject to a similar long list of variables, again many with a spatial character.

3. MERGING OF DATA

The breadth of data source, exemplified above, available for incorporation into property GIS is growing and, as other applications of GIS technology become considerably more widespread and analytical procedures become more sophisticated, a potentially huge array of data from a multitude of different sources may be integrated and used.

The compilation, filtering and presentation of multiple data sets is the everyday work of spatial data handlers as wide ranging as cartographers and foresters, planners and geologists. Each may be required to resolve differences and inconsistencies in data accuracy, scale, currency, level of resolution and reliability. Such problems are also confronted in the digital environment.

The example of TWDC was chosen to show potential conflicts and inconsistencies when merging spatial data sources: in this case merely the results of combining one graphical data set with one alphanumeric file. The problems shown are typical of the those to be expected when data is being integrated within a GIS: obviously, they will be multiplied considerably when a larger number of data sets are to be merged.

The 1:1250 scale Ordnance Survey digital data covering part of the TWDC area on the banks of the River Tyne in Newcastle was examined, along with data extracted from the property gazetteer held by the City Council of Newcastle upon Tyne. The latter contains a variety of information associated with properties and land parcels, and is based on the Joint Information System (JIS), a pilot urban gazetteer system, initiated in 1975, which was developed and maintained by local authorities in the now-defunct county of Tyne and Wear.

3.1. Graphical data conflicts

3.1.1. Sheet edge mismatching

Examples of this conflict do appear in the study area, which is covered by a number of paper map sheets, each digitized as a separate file. Such inconsistencies are of great consequence when further topological manipulation (e.g. construction of polygons representing land parcels) is attempted across the false divide which is the sheet edge. Current Ordnance Survey digital mapping is edge-matched and does not, therefore, suffer from this problem, but in general inconsistencies resulting when data of differing qualities are merged at sheet edges should not be overlooked. The solution to this particular problem may be time consuming and complicated, but it is often undertaken using proprietary software solutions. At best, a 'mean fit' or 'shift' of mismatching lines can be invoked, a common generic function within a GIS (e.g. ADJUST, MAPJOIN and DISSOLVE in ARC/INFO). At worst, it involves total field re-survey of the area around the sheet edge.

3.1.2. Variation in age of data

The preceding conflict may be a result of variation in the date of survey/map publication of adjacent map sheets. However, age variations may apply in other situations, particularly those connected with revision of digital data. Ordnance Survey is able to ensure that customers have digital data for all topographic features at 1:1250 and 1:2500 that is as up to date as possible. Depending on the option chosen, updates can be obtained at monthly intervals with varying threshold levels of change (20, 30, 40 and 50 units). However, for data digitized from some other series, or in other organizational settings, it may be that revision is only undertaken for certain elements of the map, and that, as a result, boundary information, for example, is not of the same vintage as road information.

3.2. Alphanumeric data conflicts

3.2.1. National Grid coordinate determination

This can often be encountered in the context of postcode-to-National Grid conversion. The Post Office postcode information gives a non-specific location for a number of land parcels and properties, which is primarily used for the efficient operation of the mail delivery system in the UK. However, it also provides a widely used and memorable form of spatial referencing for land parcels. Programs exist that, given the postcode, will return the National

Grid coordinate, which is a more precise and specific locational indicator. This conversion is inevitably approximate, since the average postcode covers a widely varying, and occasionally extensive, area although the resultant National Grid coordinate is given to a nominal resolution of 10 m (more practically, 100 m). Investigation into the fidelity of the transformation has revealed that it may well show up gross errors, leading to some parcels being located a considerable distance from their true location (Raper, Rhind and Shepherd, 1992). Changing postcode geography (18 000 alterations per year) may well create further problems, and, despite its initial apparent utility as an urban locational identifier, the programs should not be recommended for property information systems at a high level of land parcel discrimination and resolution. In fact, in the case of the Newcastle City gazetteer, National Grid coordinates were not converted from postcodes but were read directly from large-scale mapping, which, as a manual operation, may lead to further, different errors being introduced.

3.3. Conflicts resulting from combining the data sets

It is the conflicts that arise when data sets are combined or merged that potentially pose the most significant problems. These are illustrated in Fig. 1.

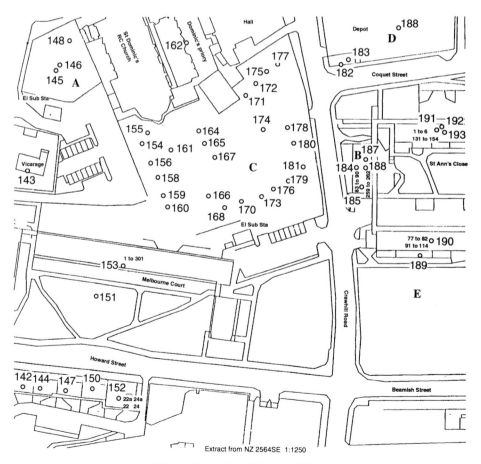

Fig. 1   Examples of data conflict

Property-based GIS: Data Supply and Conflict 269

3.3.1. Multiple floors of one parcel with differing uses

This is a twofold problem, since it relates to both database entry and graphical display of information. There are numerous occasions in this study where a property having a number of floors has multiple usage but only one database entry. For example, parcel **A** has a retail outlet on the ground floor, with a light engineering workshop above, and residential accommodation on the top floor. Neither of the two packages being used to construct the demonstration property information system for this area, SLIMPAC and ARC/INFO, efficiently allows multiple entries into the database for one parcel. The problems involved in portraying such multiple uses in map form have occupied cartographers engaged in traditional mapping for many years. Particular areas of concern include land use mapping in urban areas and topographic mapping of overhangs and buildings straddling roads.

3.3.2. Multiple coordinated entries within the same parcel

In this case there are a number of database entries plotted as being within the bounds of a single land parcel. This could be regarded as being a potential solution to the previous problem, but once again there are difficulties in constructing database entries which duplicate forms within one graphical feature. The situation in parcel **B** is different, in that it portrays a land parcel that has internal walls not indicated in the Ordnance Survey digital data. Each database entry refers to a discrete parcel, but these parcels are not adequately identified on the map and in the graphical digital data.

3.3.3. Coordinates clearly within no parcel

Area **C** is an example showing points clearly outside a parcel – there is no property within a considerable distance. There are a number of potential reasons for this, the obvious one being that the land parcel or property is new and the digital map data is outdated in this area.

3.3.4. Coordinates in the wrong parcel

Point **D** exemplifies a linkage between database and graphic that reflects incorrect attribute coding. The combination of a large building that has the outward appearance on the map of a warehouse with a database entry stating 'residential' exemplifies possible inconsistency in combining the two elements.

3.3.5. Coordinates just outside a parcel

This particular conflict (exemplified by entry 189 at Point **E**) could potentially result from the situation described above, where the postcode-to-National Grid conversion gives a 'near-miss' to the attempt to merge the database entry with the graphic. Strategies for considering points on or near the boundary lines of administrative regions may involve the use of probability and the epsilon zone and, although this exercise is at considerably larger map scale than that considered by most of the GIS error handling literature and the boundary lines (in this case, of properties) are more concrete, such approaches could be valid.

In many cases, a GIS manager may have no option other than to rely on 'second-hand' data, supplied by external organizations or interpreted by individuals. Increasingly, users are becoming aware of the potential shortcomings of such data and are requesting some form of error or accuracy assessment of the data. This 'accuracy tagging' could be as simple as a date-stamp, or as complex as a root mean standard error of locational accuracy. With multiple data sets being integrated in the common contemporary practical GIS, such attached indicators may prove difficult to interpret and handle.

Data quality concerns apply not only to positional data, but also to attribute data, which may have been interpreted using a qualitative assessment. Even where there is little chance of error in categorizing attribute data (such as the property uses coded in this exercise, which are fairly clear cut), the actual classification scheme that governs possible categories may be unsuitable for a variety of applications. Thus, a land use classification with ten classes may be suitable for a small-scale enquiry by structure planners or a property speculator, but may be too coarse a division for development corporation planning.

Having examined the data sets available, and the application area in question, the conclusion is that when working with spatial data which attempts to model the real world for recording or analytical purposes, there is no substitute for ground checking to ensure data correctness, completeness and currency. Fishwick, Hadley and McVeigh (1990) reach a similar conclusion after checking a data file for land use in Nottingham.

4. APPLICATIONS

With the data access, collection and integration problems so far mentioned and the propensity for the property management and development sector to rely on traditional 'feel for the market' approaches, it may not be surprising that the impact of GIS on this sector is limited. The depressed nature of the property market (commercial and industrial) during the late 1980s and early 1990s may be a further reason for the reluctance to embrace new technology. However, the data used in this industry is inherently spatial, the competitive nature of the sector rewards those willing to take risks, both in speculative dealing and in introducing new working methods, and the expanding international scope of the property business suggests that the efficient management of the data used will become increasingly important.

4.1. Commercial systems

Thus, a number of commercial systems have been developed which specifically address the needs of the property sector. Examples of these are given in Table 1.

Table 1   Categories of property information system

Category	Application	Graphic output	Example system
Recording system	Property management/terrier	No	SKYLINE (Fraser Williams)
'Value-added' system	Property development	No	Focus (Property Intelligence)
Land information system	Property management/terrier	Yes	Trace Mapping (Trace Solutions)
Geographical information system	Property development/speculation	Yes	Goad GIS (Chas. Goad)

Future developments will rely on the increasing sophistication of the 'value-added' packages and information systems, plus the customization of more general-purpose GIS packages, such as ARC/INFO, SmallWorld and SLIMPAC, for property applications. At the core, however, is the problem of data availability and conversion. The property profession will only embrace GIS where significant financial benefits can be proven, and this depends on complete, accurate and integrable data sets.

5. CONCLUSIONS

The research described in this chapter has indicated that there is a role for a recording or procedural GIS in the property development and management sector and that this role will become ever more apparent as the growing number of property data supply firms find their services and data being used by more and more speculators and property portfolio managers. It is also clear that 'low cost' systems, such as SLIMPAC, have a role to play in meeting such needs: there are also many other systems on the market – some specific, some more general-purpose – capable of being applied in property management. The more analytical GIS packages, which are more extensive and expensive, appear to have a more limited utility for the property sector: as it seems difficult to encode the 'feelings' that govern decision making in this area, there will always be the need for human input. Thus, the marginal benefits of going beyond a purely 'operational' GIS (Paschoud and Rix, 1991) may well be very small.

The issues raised by data availability may appear to be more difficult. Although the opening up of the Land Registry is a potentially important new data source of great benefit to property firms and property data suppliers (particularly at the low prices that are being charged), there is an enormous range of further data sources that need to be accessed and converted. The search for spatial data, their purchase or lease, the copyright implications, and the associated conversion costs are the major stumbling blocks in the efficient and widespread adoption of GIS in this important sector of the national economy.

REFERENCES

Barrett, M. (1986) The Open Register and boundaries. *Land and Miner. Surveying*, **4**, 7.
Byrne, P. and Cadman, D. (1984) *Risk, Uncertainty and Decision-making in Property Development*. E&FN Spon, London.
Courtney, S. (1990) GIS in commercial property. *Mapping Awareness*, **4**(9), 21–23.
Dale, P. (1992) What is an NLIS? *Mapping Awareness*, **6**(9), 25–28.
Department of the Environment (1987) *Handling Geographic Information. Report of the Committee of Enquiry chaired by Lord Chorley*. HMSO, London.
Fairbairn, D.J. (1993) The development of a land information system for Hadrian's Wall. *GIS Europe*, **2**(1), 23–27.
Fairbairn, D.J. and Peedell, S. (1991) Analytical GIS in property management and development. *Technical Papers, ACSM 51st Annual Convention*, vol. 2, Baltimore, Maryland, 104–110.
Fenn, R.J. (1991) HM Land Registry information systems strategy. Paper presented at the AGI Meeting on Land and Property Assets Recording, Durham.
Fishwick, M., Hadley, C. and McVeigh, W. (1990) GIS and city planning. *BURISA Newsl.*, 94, 7–8.
Paschoud, J. and Rix, D. (1991) GIS requirements for property based systems. *Mapping Awareness*, **5**(2), 15–19.
Pryer, E.J. (1990) Land registration: ending unwanted 'Victorian values'. *Land and Miner. Surveying*, **8**(11), 9–11.
Raper, J., Rhind, D.W. and Shepherd, J. (1992) *Postcodes: The New Geography*. Longman, Harlow.

Chapter Twenty

The impact of GIS on local government in Great Britain

I. MASSER AND H. CAMPBELL
Department of Town and Regional Planning, University of Sheffield

The research described in this chapter examines the impact of geographical information systems (GIS) on local government in Britain. It makes use of a combination of methods to assess the extent of the take-up of GIS facilities in British local government and explore its impacts on administrative practices. The extent of take-up was evaluated by a comprehensive telephone survey of all 514 local authorities, while the impacts of the introduction of GIS are explored in greater depth in the context of twelve case studies of selected authorities with at least two years' experience of GIS.

The results of the telephone survey indicate that one in six local authorities in Great Britain had acquired GIS facilities by April 1991. The level of take-up varies considerably with local authority function, size and location. It is greatest at the county and Scottish regional level, where three out of five authorities have already acquired facilities. The findings of the twelve case studies suggest that most GIS applications in British local government are concerned with operational rather than strategic activities. A striking feature of the case studies is the limited impact that GIS has had on most of the authorities, even after at least two years' experience. Only three out of the twelve systems have reached the stage where at least one application is being used by end users.

Taken as a whole, the findings of the research suggest that GIS in British local government is at a crossroads, because the rate of take-up has slowed considerably during the last 18 months. They also highlight the need for further work on a number of detailed theoretical and practical questions relating to the impact of organizational issues on the successful utilization of GIS.

1. BACKGROUND

Technological progress in the last few years has removed many of the barriers that inhibited the development of GIS. It is generally agreed that the potential of this technology to store, manipulate and display spatial data is considerable. However, the introduction of GIS technology involves the complex process of managing change within environments that are typified by uncertainty, entrenched institutional procedures and individual staff members with

---

Geographical Information Handling – Research and Applications. Edited by P. M. Mather
© 1993 John Wiley & Sons Ltd

conflicting personal motivations. Given these circumstances, personal, organizational and institutional factors are likely to have a profound influence on the extent to which the opportunities offered by GIS will be realized in practice (Audit Commission, 1990; Campbell, 1990a, 1991; Department of the Environment, 1987; Willis and Nutter, 1990).

One of the most important groups of users of GIS is local government. The range of potential applications in this field is considerable, extending from property registers and highways management to emergency and land use planning. It is often assumed that the cost of equipment and data preparation, when combined with the capacity of GIS to integrate data sets from a wide variety of sources, makes the development of departmental systems inappropriate (Bromley and Selman, 1992; Gault and Peutherer, 1989; Grimshaw, 1988). Strategic and efficiency benefits associated with increased levels of information sharing between departments suggest that the implementation of GIS will be accompanied by an extension of corporate activities, which in turn has significant implications for the development of administrative practices in local government.

Given the theoretical advantages of adopting a corporate approach, there are an increasing number of accounts indicating that the realization of these benefits in practice may be more difficult than generally envisaged (Cane, 1990; John and Lopez, 1992; Openshaw *et al.*, 1990; Van Buren, 1991; Winter, 1991). It is clear from these descriptions that users are encountering mixed results even in similar types of organizations. Furthermore, research that has examined the implementation of other forms of information technology suggests that realizing the potential of such systems is often a highly problematic process (Eason, 1993; Moore, 1993; Rogers, 1993). There is no reason for assuming that successful implementation of GIS technology will prove more straightforward than that of other computer-based systems. However, despite the interest GIS has provoked, particularly in local government, little systematic analysis has been undertaken that provides information on the extent of the diffusion of this technology or on the impact that GIS applications are having on the organizations in which they are being implemented. As a result, there is a need to evaluate the underlying assumptions concerning the processes of adopting and implementing GIS, to provide a benchmark against which the isolated accounts of GIS development can be assessed.

2. OBJECTIVES

The research examines the impacts of GIS on local government in Great Britain. Particular attention has been paid to the extent to which institutional, organizational and personal factors are impeding the take-up and implementation of GIS technology. The research therefore provides factual information on one of the key issues cited in the Chorley Report (Department of the Environment, 1987), and points to possible strategies for resolving the difficulties that are being encountered in practice. The objectives of the research were:

1. To assess the problems and benefits associated with the implementation of GIS in local government.

2. To evaluate the impact of GIS on existing organizational practices in local government with particular reference to the development of corporate activities.

These objectives were operationalized by splitting the study into two complementary parts. The first provided an overview of the take-up of GIS technology throughout British local government and examined the key benefits and problems associated with the implementation of this technology. Given this general context, the second part investigated the detailed impact of GIS technology on the activities and procedures of local authorities and the extent to which organizational issues affect the outcome of the implementation process. The research findings discussed later indicate that these objectives were achieved, although in doing so, many further questions were raised.

3. METHODS

A combination of methods were used, reflecting the differing emphasis of the two objectives. These included a comprehensive telephone survey of all 514 local authorities in Great Britain and twelve in-depth case studies. A telephone-based survey approach was adopted because of the large number of postal questionnaires that were being circulated to local authorities and the resulting concern about the level of response. The 100% response rate supports the adoption of this method, in addition to respondents giving valuable subsidiary information during the interviews. This method also removes the ambiguity that exists in some surveys over the respondents' perceptions of the definition of GIS, because the capabilities of the software and the precise nature of its use in the host authority can be related to the operational definition. The research adopted a broad interpretation of GIS, which included automated mapping and facilities management type systems but excluded thematic mapping and computer-aided design (CAD) packages. One respondent, generally the project manager, was interviewed with respect to each separate system present within a particular authority. The survey took place over the period from February to June 1991. The database developed as a result of the survey was updated in September 1992 by conducting a further round of telephone interviews with the 44 authorities that had been identified by the initial survey as intending to purchase GIS technology within the coming year.

The selection of the twelve case studies was based on a preliminary analysis of the findings of the telephone survey. The choice of authorities was based on three criteria. Firstly, each authority must have had GIS technology present within the organization for at least two years (i.e. before June 1989), to avoid the research findings being limited to the initial teething problems associated with the introduction of computer technology. Secondly, a range of organizational contexts were chosen based on a typology of styles of approach to implementation that were developed from the survey findings. These ranged from systems in which all departments in the authority were participating, to situations where GIS software was being developed by a single department. This criterion also ensured that the full range of types and styles of local authority were included. Thirdly, it was essential that the authority was willing to cooperate.

The twelve case studies were undertaken between May 1991 and February 1992, and built on the methodology developed by Campbell in a study examining the use of geographic information in local authority planning departments (Campbell, 1990b). To prepare further for the case studies, exploratory fieldwork was undertaken in New England to examine the extent to which time affects the experiences of organizations in implementing GIS. Given the longer experience of local authorities in handling GIS in the United States, these investigations proved instructive in highlighting the crucial role of organizational factors in achieving the successful utilization of such systems (Campbell, 1992). As a result of this experience interviews were undertaken with all the staff connected with the introduction of the GIS, including senior management, potential users, computer specialists, project managers and technical support staff. This was a crucial aspect of the methodology because much of the existing research in this area is based solely on a single interview with either the project manager or computer specialists. The extent of the fieldwork required for the case studies varied considerably, ranging from one day to two weeks, depending on the scale and complexity of the project. In several cases a preliminary discussion with the project manager was supplemented by a second period of fieldwork. The interviews were supported by direct observation of the various organizational contexts and a review of the existing documentation.

4. RESULTS

The main results of the research will be presented in two parts, reflecting the organization of the study. The first part provides an overview of GIS adoption and implementation in British local government based on the main findings of the comprehensive survey, while the second highlights some of the key issues raised by the case studies with regard to the impact of GIS on the activities of British local government.

4.1. Overview of GIS adoption and implementation

The extent of current interest in GIS in local government in Great Britain can be seen in Table 1. This shows that 356 out of the 514 local authorities (69.3%) surveyed were considering introducing GIS in one form or another in April 1991. However, the introduction of GIS was at an early stage in most authorities. Only 85 authorities (16.5%) had already acquired a GIS, while a further 44 authorities (8.6%) had firm plans to purchase one in the twelve months following the survey.

Table 1 Plans for GIS in local authorities in Great Britain

Plans for GIS	Number	%
Already have GIS facilities	85	16.5
Firm plans to acquire GIS within one year	44	8.6
Considering the acquisition of a GIS	227	44.2
No plans to introduce a GIS	158	30.7
Total	514	100.0

The findings indicated that there was a high level of awareness of GIS in local government in Britain, even among authorities that had no plans to invest in such systems. However, there were marked variations between local authorities of different types, and also between different regions within Great Britain, with respect to the extent of both current and anticipated GIS availability.

The first two columns of Table 2 show the number of authorities that had already acquired a GIS, by local authority type. Around 59.3% of all counties and regional level authorities in Great Britain already had GIS facilities. In contrast, only 7.2% of shire districts and 7.5% of Scottish districts had GIS facilities. The proportion of metropolitan authorities with GIS fell mid-way between the two extremes, at 31.9%. These findings reflect differences in size, with larger authorities much more likely to adopt GIS than the small shire and Scottish districts.

Table 2 Plans for GIS by type of local authority

Local authorities	Already have GIS		Total
	Number	%	
Shire Districts	24	7.2	333
Metropolitan Districts	22	31.9	69
Shire Counties	32	68.1	47
Scottish Districts	4	7.5	53
Scottish Regions	3	25.0	12
Total	85	16.5	514

In addition to the variation in take-up between different types of local authorities, the survey also indicates a distinct north/south divide in the level of GIS acquisition. Nearly three-quarters of all authorities that already have GIS are located in the southern half of Great Britain, and the overall proportion of authorities with GIS in the south is nearly double that of the north (see Table 3). The difference is most apparent with respect to metropolitan authorities, where 47.5% of all authorities have GIS facilities compared with only 10.3% in the north. Although some allowance must be made for those authorities that make use of the facilities provided for them by agencies such as the Merseyside Information Service, the contrast between north and south is striking.

Table 3 Percentage of authorities who already had GIS facilities, by local authority type and region

Local authorities	South	North	Great Britain
Shire Districts/Scottish Districts	8.6	5.2	7.3
Metropolitan Districts	47.5	10.3	31.9
Shire Counties/Scottish Regions	71.0	46.4	59.3
All authorities	20.1	11.4	16.5

With these considerations in mind, the survey went on the investigate the types of GIS systems being implemented, including the time of adoption, software acquired and the main benefits and problems being experienced during implementation. This analysis is therefore based on systems rather than authorities. A system is regarded as a distinct piece or combination of software, which one or more departments within a local authority are implementing. For instance, a situation where several departments are developing separate applications based on the same software is considered as one system. Table 4 indicates that a total of 98 systems had been purchased by 85 local authorities. Shire counties were the most likely to have more than one system within an authority, with 44 systems in 32 counties. Evidence from subsidiary information provided by respondents suggested that the presence of several systems within one authority was more likely to reflect the desire of certain departments to have complete control of the form and speed of system development than the perceived inability of GIS software to perform a variety of functions.

The findings indicated that there has been a gradual growth since the early 1980s in the number of GIS being purchased, with nearly 40% of all systems acquired in 1990. Because the survey was conducted over the period from February to June 1991, the figures for 1991 are incomplete, but there is a slight indication of a declining rate of take-up. It is probable

Table 4 Level of GIS adoption by authorities and systems

Local authorities	Number of authorities possessing GIS	Number of GIS
Shire Districts	24	24
Metropolitan Districts	22	22
Shire Counties	32	44
Scottish Districts	4	5
Scottish Regions	3	3
Total	85	98

The survey identified four further systems, one adopted by a joint board and three by metropolitan research and information units.

that the recession, uncertainty over the future structure of local government and changes in the internal management of authorities have contributed to this situation. Monitoring will be necessary to ascertain whether 1990 represents the peak period of system take-up. Analysis of the length of time authorities have been implementing GIS suggests growing experience amongst authorities, with the average length of time since system take-up being slightly over two years. This increasing experience is important as some authorities are beginning of replace their existing facilities. Among these are a small number that decided to evaluate a microcomputer-based system prior to committing more substantial resources, and a number where the original software proved unable to satisfy their requirements. This discussion focuses on the configurations of equipment currently in use.

Table 5 provides a breakdown of the software packages introduced by local authorities. It should be noted that systems with GIS capabilities that are essentially being used to perform other activities such as CAD have been omitted from this exercise. In addition, in instances where packages were purchased to provide specialist GIS facilities to supplement an existing system such as SPANS, it is the main system that is recorded in the analysis. A striking feature of the survey findings is the market share of around 22% held by ARC/INFO (ESRI), with Alper Records (a system developed by a small British company based at Cambridge), and GFIS (IBM) holding a further 12% and 10%, respectively. These three packages therefore account for around 45% of the systems adopted by local authorities, with the remaining 55% accounted for by seventeen packages and four home grown systems. The range of products purchased by authorities tends to indicate the overall immaturity of the market, with as yet only tentative indications of specialization on a few core products.

Analysis of these findings in relation to time highlights a number of issues with respect to the diffusion of GIS software. These results indicate that the market share held by ARC/INFO has been growing from around 17% of all systems in 1987 to 30% in 1990. At the same time the adoption of GFIS has declined from around 35% in 1988 to 2% in 1990. It is also clear that there has been a steady increase in the variety of systems being acquired by local authorities. For instance, in 1986 four different systems were purchased, in 1988, eight and in 1990, sixteen. A number of more detailed issues are raised by examining the take-up of GIS software by local authority type. It is clear from this that while ARC/INFO is dominant within the counties and regions, accounting for 41% of all systems, this pattern is not repeated throughout local government. In the metropolitan districts, Hoskyns G-GP (a system developed by a small British company) represents nearly 23% of the total number of systems introduced. This reflects the original development of the software by the Research and Intelligence section of the now abolished Greater London Council and the

Table 5 GIS software adopted by British local authorities

Software	Shire and Scottish Districts (%)	Metropolitan Districts (%)	Counties and Regions (%)	Total (%)
ARC/INFO	3.4	9.1	40.4	22.4
Alper Records	24.1	9.1	6.4	12.2
GFIS	3.4	9.1	14.9	10.2
Hoskyns G-GP	3.4	22.7	2.1	7.1
Axis Amis	13.8	9.1	2.1	7.1
McDonnell Douglas GDS Maps	6.9	9.1	2.1	5.1
Coordinate	10.3		2.1	4.1
Wings			8.5	4.1
Other[1]	34.5	31.8	21.3	27.5
	(n = 29)	(n = 22)	(n = 47)	(n = 98)

[1]Other systems purchased include: SIA Datamap, PC ARC/INFO, Planes Spatial, Viewmap, Tydac Spans, LaserScan Metropolis, GDMS, Siemens Sicad, Pafec Dogs, Maps in Action, Atlas GIS, Hoskyns PIMMS and four home-grown systems.

subsequent interest this provoked on the part of a number of London boroughs. As a result, five of the seven systems purchased are in London. A similar presence in Scotland is enjoyed by Alper Records but the reasons behind this are less clear. The striking feature of the take-up of software in the shire districts is the broad range of systems and the absence of a clear market leader. In more general terms these findings indicate a greater demand on the part of shire counties for software that provides a variety of GIS facilities, while the other authorities tend to favour systems with more limited capabilities, such as automated mapping and facilities management.

Given the characteristics of the systems being implemented, respondents were asked to identify the main benefits and problems associated with GIS implementation. Table 6 shows that the main group of benefits was perceived to be improved information processing facilities. This includes such advances as improved data integration, increased speed of data provision, better access to information and an increased range of analytical and display facilities. It is striking that these benefits were regarded as most important by 60% of the respondents, with 31% stressing better quality decisions, including managerial, operational and strategic considerations, and only around 6% linking the main benefits of GIS to achieving savings. Given the work on cost–benefit analysis that has been undertaken, the very low level of importance attached to the role of GIS in achieving time, staff and financial reductions is particularly interesting. Respondents suggested that while financial justifications are important in gaining agreement from elected members and senior management to purchase a GIS, they perceived that in practice these savings were likely to be limited. Furthermore, there is no evidence to suggest a strong link between GIS and improved decision making. In instances where better decisions were regarded as the prime benefit to be gained from GIS implementation, just over 38% linked these advances to routine operational decisions, while the figures for strategic and managerial activities were 28.8% and 25%, respectively.

The findings shown in Table 7 demonstrate that a range of technical, data-related and organizational difficulties is being experienced by those implementing GIS in local government. Most respondents emphasized that the introduction of GIS into their organization

Table 6 The most important group of benefits associated with the implementation of GIS in local government

Benefits	Shire and Scottish Districts (%)	Metropolitan Districts (%)	Counties and Regions (%)	Total (%)
Improved information processing facilities	68.4	70.4	51.1	60.5
Better quality decisions	27.0	13.6	42.5	31.5
Savings	4.6	11.4	4.2	5.9
Other		4.5	2.1	2.0
	(n = 29)	(n = 22)	(n = 47)	(n = 98)

did not prove a straightforward process. Only in four cases was it stated that no problems had been encountered. Overall, there was a slightly greater emphasis on data-related issues, such as the lack of compatibility between existing data sets, problems of maintaining up-to-date data sets, the cost of data capture and the cost and availability of Ordnance Survey digital data, than on the other sets of problems.

Table 7 The most serious group of problems associated with the implementation of GIS in British local government

Problems	Shire and Scottish Districts (%)	Metropolitan Districts (%)	Counties and Regions (%)	Total (%)
Data-related	21.8	31.8	42.5	34.0
Technical	47.7	27.3	17.0	28.4
Organizational	27.0	27.3	29.8	28.4
Other		4.5	6.4	4.1
No problems	3.4	9.1	2.1	4.1
System not sufficiently developed			2.1	1.0
	(n = 29)	(n = 22)	(n = 47)	(n = 98)

Given these general trends, particular difficulties were raised by each category of local authority. For instance, technical problems were most pronounced in the shire and Scottish districts, whereas the counties and regions had fewer technical difficulties, but data-related matters were proving more serious. It is probable that the decreasing level of technical problems encountered by the counties and regions through the metropolitan districts to the shire and Scottish districts is a reflection of the greater resources and experience available within the larger authorities. It is noticeable that organizational difficulties are relatively constant regardless of local authority type.

The additional comments of respondents gave some more detailed insights into the difficulties being experienced by local authorities implementing GIS. Around 20% mentioned resource issues such as cost and a shortage of staff time as inhibiting the introduction of GIS, while a significant number were highly critical of the activities of vendors. In particular, there was concern about the extravagant claims of vendors prior to software purchase, their lack of appreciation of the local government context, the limited scope for linkage between systems and inadequate post-sales support. The importance, as well as the difficulty, of establishing the financial viability of vendors was also stressed. Criticism was also levelled at the Ordnance Survey. In particular, dissatisfaction was focused on the incomplete digital coverage for Britain, which leads to partial data sets for most local authority areas, and the quality of some of the data produced, with regard to the speed of update and edge matching of maps. However, in addition to these external influences, many of the comments of respondents related to the specific organizational circumstances in which GIS were being implemented. These included lack of previous technical experience in GIS, limited awareness among the technical specialists of user needs and lack of appreciation by senior staff that the implementation of GIS involves the complex process of managing change. A number of respondents were also concerned about the tension between developing a corporate system, which tends to involve a long lead time and the imposition of standards, and the immediate service needs of departments. More specifically, given the computer storage space required for handling geographic information, some respondents were concerned about the introduction of direct charging for mainframe time. Such circumstances were often associated with an internal reorganization into business units, leading to a number of less wealthy departments such as planning feeling that they would have to pull out of some of the mainframe-based GIS projects.

The survey provides a useful overview of the diffusion of GIS in British local government. Overall, the results indicate considerable general awareness about the potential of such systems, with GIS software already present within 16% of authorities. The introduction of GIS has been most striking in the counties, regions and the larger district authorities, especially in the southern part of Britain (see Masser, 1993). It has also been found that within some of the counties there may be more than one GIS. Furthermore, in terms of the systems being adopted there has been an increasing move towards the ARC/INFO software and workstations as the supporting hardware. Important issues were also raised in terms of the benefits and problems associated with GIS implementation. Given the emphasis that has been placed on the contribution of GIS to improve strategic decision making and cost savings through advances in efficiency, it is striking that these respondents associate the main benefits with the more straightforward activities of improved information processing facilities. The findings also indicate that a range of technical, data and organizational problems are experienced during the implementation of GIS technology. These findings, however, provide only a broad overview of GIS development in local government; the detailed impact of these systems was examined in the case studies.

4.2. The impact of GIS on British local government

The twelve case studies included a range of local government environments, with four of the systems examined in shire districts, two in metropolitan districts, a further two in a large Scottish district and the remaining four in shire counties. Of these systems, three were being implemented departmentally, while the development of the other nine involved several departments. The systems had also been present within the authorities for at least two years. The following discussion provides a brief summary of the main case study findings. The overview concentrates, firstly, on the types of application being developed and the level of utilization that has been achieved and, secondly, given these findings, on the processes that appeared to influence whether GIS are successfully implemented in local government.

The case study findings indicate that the vast majority of GIS applications being developed aim to assist with operational activities, regardless of the more sophisticated facilities that may be available within the software. Furthermore, despite the common use of the same GIS software by several departments in nine of the case studies, most of the applications concentrate on the needs of no more than one department. The main departments involved with the development of GIS applications are concerned with technical service type activities such as planning, highways, estates, architecture and surveying, assisted in some cases by the central computing department of the authority. The key application area for most authorities has been to exploit the digital data available from the Ordnance Survey to develop automated mapping facilities. The emphasis given to this application probably explains the significant role played by planning departments, as they have traditionally had responsibility for meeting the cartographic needs of authorities. In addition to automated mapping, the other main areas of application include grounds and highways maintenance and estate management. There is very little current or planned use of complex spatial analysis techniques, with most local authorities only perceiving a need for basic display and query facilities. It should also be noted that active interest in GIS has not yet permeated through to the often large community service type departments of housing, education and social services. This is probably a reflection of the considerable administrative pressures that these departments have faced, partly as a result of frequent legislative changes by central government, as well as the limited resources for additional activities such as new data handling initiatives.

The case study findings indicate the tendency for GIS technology to be employed differently even in apparently similar environments developing similar applications. No two case studies, or in many ways users, were utilizing these facilities in exactly the same manner. In general, staff in local government regard GIS simply as a mapping system or a query facility. These general perceptions of the technology are at variance with the accepted understanding of other groups, such as researchers and vendors. These results highlight an important theoretical issue concerning the tendency for each of those involved with a technology to 'reinvent' the exact nature of that technology (Rogers 1993, 1986). As a result, there is a need for further work to examine the extent to which users reinvent GIS technology and also to what extent there is a shared understanding of its nature and purpose, because misconceptions are likely to lead to the development of technology and facilities that nobody wants.

A striking finding of the research is the limited impact GIS has had on most of the authorities, even after two years' experience. Only three case studies had reached the stage where at least one application was being employed by end users. The remaining seven were either still developing the system or had achieved an operational application but it was not being utilized by users. These results indicate that the lead time to the development of a working application is often considerable because of the time-consuming nature of data capture. Very few of the authorities already had complete data sets or their existing information in a suitable format for input. This means that the implementation of GIS must be sustained over several budgetary cycles. Two of the case studies had abandoned the development of the software they had originally purchased and were reconsidering whether GIS facilities could provide them with any real benefits. The research also demonstrates that a technically operational system will not necessarily be employed by users simply because it is there. Overall, these findings raise an important theoretical issue about the nature of success and failures in relation to the implementation of computer-based systems. In terms of the research, use was regarded as the basic indicator of success.

With these considerations in mind, factors that appear to account for the successful implementation of GIS were explored in greater depth. In terms of the basic features of the systems, there were no similarities between the more or less successful in terms of technical characteristics, such as software and hardware, or the organizational structures adopted, for example the involvement of several departments or just one. In addition, the technological problems faced by the less successful authorities seemed to be no greater than the others. Previous work in this area suggests that there are three necessary and generally sufficient conditions for the effective implementation of computer-based systems (see Campbell, 1990b; Masser, 1992). These are:

1. An information management strategy that identifies the needs of users and takes account of the resources at the disposal of the organization.
2. Commitment to and participation in the implementation of any form of information technology by individuals at all levels of the organization.
3. A high degree of organizational and environmental stability.

This framework was used as the basis for exploring the experiences of the case studies. The key question, therefore, was 'What factors appear to increase the probability of successful GIS implementation?'

The findings of the research indicate that the implementation of GIS in local government is as much social and political as technical in nature. Technical problems tend to reinforce existing organizational difficulties rather than be responsible for the failure of the process of implementation. The findings of the analysis suggest that there are some organizational cultures that are inherently receptive to the development of innovations, such as GIS. Out of the twelve case studies, two appear to have the capacity to take on the organizational changes implied by the introduction of GIS and sustain the process in the presently highly dynamic context of British local government. It was evident that the skill levels and expertise among the staff in both these authorities were higher than in other similar environments. However, while the all-round expertise of these individuals was undoubtedly important, the existing culture was in many ways responsible for gathering these individuals into its service. As a result, it appears that innovative environments attract innovative individuals, and presumably the reverse is also the case. In both these authorities it was evident that there was a long tradition of being at the forefront of new innovations in the local government sector, including a wide range of policy areas as well as information processing. The outward characteristics of the projects in the two authorities were very different, but both demonstrated a fundamental capacity to treat change as an opportunity. In both cases, mistakes were made, but rather than this outcome tending to thwart initiative, it seemed to be regarded as a process of education from which the authority would move forward. It therefore appears probable that the long-term utilization of GIS seems most likely to be sustained in such an environment.

However, while very few organizations are inherently innovative, it was possible to identify four factors that appeared to enhance the chance of success. These are:

1. Simple applications, producing information that is fundamental to the work of potential users.
2. User-directed implementation, which involves the participation and commitment of all the stakeholders in the project.
3. An awareness of the limitations of the organization in terms of the range of available resources.

4. A large measure of stability with respect to the general organizational context and personnel, or, alternatively, an ability to cope with change.

The features contributing to successful implementation overlap to a considerable degree. For instance, it is unlikely that the key needs of users will be identified without the users themselves taking a leading role in system implementation. It was also evident from the case studies that short-term success can be achieved for a relatively small project, based on the expertise and political skills of an enterprising individual, sometimes referred to as a 'champion'. The critical issue in this case is whether the organization can take the innovation and sustain it. Experience suggests that this is often doubtful, as even a most expert individual cannot ensure successful implementation in a vacuum. It is also possible that organizations will achieve technical success without fulfilling the four factors. In other words, operational GIS applications are developed but they do not become a routine part of the work of users. The analysis of the case studies suggests that this is most likely to occur in situations where system development is controlled by the computer specialists. A further factor that may thwart development is instability, as changing circumstances alter priorities, with organizations that have set goals that are marginal to the needs of users likely to see the system rejected. It is also often difficult for organizations to alter priorities once a course of action has been established. Overall, the results suggest that departmental systems can fail as easily as any of the various forms of corporate development. However, by involving several departments it is likely that the number of variables will increase.

The findings of the case studies generally confirm the usefulness of the three conditions that provided the framework for the analysis. It is important, for instance, that an information management strategy is devised that identifies the information needs of the users and the types of service they require, as well as considering the resources at the disposal of the organization. The form of this initiative, that is to say, whether it has been formally ratified or not, appears to be largely irrelevant. Of far more importance is that the process has been undertaken. Commitment and participation by staff throughout the organization are crucial and must be set within a user-centred framework. The relationship between stability and utilization is perhaps the most complex, with the impact of such forces appearing to depend on the capacity of the organization to withstand change.

This research provides the first systematic attempt to examine the impact of GIS on a group of organizations. Given these results, there is clearly a need for far more theoretical and practical research if many GIS are not to become redundant and the technology devalued. The findings of this work highlight similar issues to those noted by researchers exploring the implementation of other forms of computer technology (see, for example, Barrett, 1992; Danziger et al., 1982; Danziger and Kraemer, 1986; Eason, 1988, 1993; Hirschheim, 1985; Kanter, 1983; Mumford and Pettigrew, 1975; Pfeffer, 1981). In particular, the crucial impact of organizational cultures on the effective implementation of computer-based systems has been identified. Furthermore, these findings indicate the need for greater understanding of the process of diffusion of an innovation, including the importance of the concept of reinvention in relation to GIS (Rogers, 1986, 1993). In Section 5, specific areas requiring further research are examined.

5. FUTURE RESEARCH PRIORITIES

The findings of this research suggest that GIS is at a crossroads in British local government. The recent update of the initial survey indicates that while one in four authorities now have a GIS, the pace of take-up has slowed considerably by comparison with 1990 (Campbell, Craglia and Masser, 1993). This poses the question as to whether this simply reflects the current recession and uncertainty about the future of local government, or is a more profound comment on the value of GIS. It will therefore be important to monitor the future diffusion of this innovation and explore the reasons behind the actions of the various organizations involved with GIS technology.

There is also a need for further research that examines a number of detailed theoretical and practical questions concerning the impact of organizational issues on the successful utilization of GIS. These are:

1. Investigation of the concept of data sharing in relation to the geographic information needs of users. With the exception of Ordnance Survey maps, this research has found it difficult to identify a latent demand for data sharing within local government. What are the experiences of other contexts?

2. Improving understanding of the role of information in the decision making processes of organizations. This area of concern must underpin any work seeking to identify user requirements, both in terms of the form of information needed and the type of service best equipped to meet these requirements.

3. The nature of the organizational contexts in which GIS are being implemented needs to be explored. This research has indicated that not all environments are the same. Greater understanding of the social and political processes that influence the particular characteristics of individual organizations would facilitate system implementation.

4. The roles of individuals as against the organizational culture needs detailed consideration. This research suggests environments that are inherently innovative can sustain change. Is there more general evidence for this? What is the role of 'champions' in the process?

5. Investigations into the impact of instability and uncertainty on system implementation need to be undertaken, particularly in the context of the current changes in local government.

Research into many of these issues will have implications for the successful implementation of a wide variety of forms of information technology. However, given the current prominence of GIS, it is vital that further research is undertaken in these areas if resources are not to be wasted and if GIS is to avoid becoming yet another failed technology.

REFERENCES

Audit Commission for Local Authorities in England and Wales (1990) *Management Papers: Preparing an Information Technology Strategy: Making IT Happen.* HMSO, London.

Barrett, S. (1992) Information technology and organisational culture: Implementing change. *EGPA Conf.*, Maastricht, August.

Bromley, R. and Selman, J. (1992) Assessing readiness for GIS. *Mapping Awareness and GIS in Europe*, **6**(8), 9–12.

Campbell, H. (1990a) The use of geographic information in local authority planning departments. Doctoral Thesis, Department of Town and Regional Planning, University of Sheffield.

Campbell, H. (1990b) The organisational implications of geographic information systems in British local government. *Proc. 1st European GIS Conf. (EGIS'90)*, Amsterdam, 10–13 April, 145–157.

Campbell, H. (1991) Organisational issues in managing geographic information. In Masser, I. and Blakemore, H. (eds), *Handling Geographic Information*. Longman, Harlow, 259–282.

Campbell, H. (1992) Organisational issues and the implementation of GIS in Massachusetts and Vermont: Some lessons for the UK. *Environ. Plann. B*, **19**(1), 85–95.

Campbell, H., Craglia, M. and Masser, I. (1993) GIS in British local government: an up-date. *BURISA Newsl.*, **107**, 2–5.

Cane, S. (1990) Implementation of a corporate GIS in a large authority. *Proc. 1st European GIS Conf. (EGIS'90)*, Amsterdam, 10–13 April, 158–166.

Danziger, J.N. and Kraemer, K.L. (1986) *People and Computers: The Impacts of Computing on End Users in Organisations.* Columbia University Press, New York.

Danziger, J.N., Dutton, W.H., Kling, R. and Kraemer, K.L. (1982) *Computers and Politics: High Technology in American Local Government.* Columbia University Press, New York.

Department of the Environment (1987) *Handling Geographic Information. Report of the Committee of Enquiry chaired by Lord Chorley.* HMSO, London.

Eason, K.D. (1988) *Information Technology and Organisational Change.* Taylor and Francis, London.

Eason, K.D. (1993) Gaining user and organisational acceptance for advanced information systems. In Masser, I. and Onsrud, H.J. (eds), *Diffusion and Use of Geographic Information Technologies.* Kluwer, Dordrecht, 27–44.

Gault, I and Peutherer, D. (1989) Developing GIS for local government in the UK. Paper presented at the European Regional Science Association Congress, Cambridge.

Grimshaw, D.J. (1988) The use of land and property information systems. *Int. J. Geogr. Inf. Sys.*, **2**(1), 57–65.

Hirschheim, R.A. (1985) *Office Automation: A Social and Organisational Perspective.* John Wiley, Chichester.

John, S.A. and Lopez, X.R. (1992) Integrating data derived from within the British local government institutional framework. *Proc. Ass. Geogr. Inf. Conf.*, Birmingham, 27–29 November. Westrade Fairs Ltd., Rickmansworth, 1.18.1–1.18.8.

Kanter, R.M. (1983) *The Change Masters.* Simon and Schuster, New York.

Masser, I. (1992) Organisational factors in implementing urban information systems. *Proc. Urban Data Mgmt Symp.*, Lyons, 16–20 November. UDMS, Delft, pp. 17–25.

Masser, I. (1993) The diffusion of GIS in British local government. In Masser, I. and Onsrud, H.J. (eds.), *Diffusion and Use of Geographic Information Technologies.* Kluwer, Dordrecht, 99–115.

Moore, G.C. (1993) Implications from MIS research for the study of GIS diffusion: Some initial evidence. In Masser, I. and Onsrud, H.J. (eds), *Diffusion and Use of Geographic Information Technologies.* Kluwer, Dordrecht, 77–94.

Mumford, E. and Pettigrew, A. (1975) *Implementing Strategic Decisions.* Longman, Harlow.

Openshaw, S., Cross, A., Charlton, M., Brunsdon, C. and Lillie, J. (1990) Lessons learnt from a post-mortem of a failed GIS. *Proc. Ass. Geogr. Inf. Conf.*, Brighton, 22–24 October. Westrade Fairs Ltd., Rickmansworth, 2.3.1–2.3.5

Pfeffer, J. (1981) *Power in Organisations.* Pitman, Boston.

Rogers, E. (1986) *Communication Technology: The New Media in Society.* Free Press, New York.

Rogers, E. (1993) The diffusion of innovations model. In Masser, I. and Onsrud, H.J. (eds.), *Diffusion and Use of Geographic Information Technologies.* Kluwer, Dordrecht, 9–24.

Van Buren, T.S. (1991) Rural town geographic information systems: issues in integration. *Urban and Regl Inf. Sys. Ass. (URISA) Proc.*, **3**, 136–151.

Willis, J. and Nutter, R.D. (1990) A survey of skills needs for GIS. In Foster, M.S. and Shand, P.S. (eds), *AGI Yearbook 1990*, 295–303.

Winter, P. (1991) Selling a corporate GIS. *Proc. Ass. Geogr. Inf. Conf.*, Birmingham, 20–22 November. Westrade Fairs Ltd, Rickmansworth, 2.6.1–2.6.5.

Chapter Twenty-One

An information system for the Eastern Thames Corridor

A. CHURCH, S. JOHN AND J. W. SHEPHERD
South-East Regional Research Laboratory, Birkbeck College, University of London

M. FROST
Department of Geography, King's College, University of London

AND

A. MACMILLAN
The London and South East Regional Planning Conference (SERPLAN), London

1. THE BACKGROUND

The ESRC/NERC Joint Programme on Geographical Information Handling was established to meet five main objectives: to explore the potential of high-profile GIS applications; to integrate socio-economic and environmental data; to develop links between academic organizations and government agencies; to generate collateral due to inputs from other government and private sector organizations, and to disseminate results in a direct and effective manner (ESRC/NERC, 1989). At the South-East Regional Research Laboratory (SERRL) in the Department of Geography, Birkbeck College, a specialism in GIS applied to urban and regional planning suggested that the development of a geographically referenced database for the Eastern Thames Corridor (ETC) would be likely to meet all these criteria:

- The subject matter had a high policy profile, not least after the ETC area was 'championed' by the then Secretary of State for the Environment Mr Michael Heseltine.
- At the heart of the subregional planning problem was the perception by central and local government of the complex interrelationships of socio-economic and environmental processes operating at a range of geographic scales.
- Because of its topicality and the need for large amounts of information to cover the range of planning issues involved, close collaboration with public agencies in the project would be essential.
- These projected links with public sector organizations would ensure early and effective dissemination of the results of using GIS technology in a planning context.

The project was conceived from the start as a piece of 'action research' in which the building of the GIS database and the work carried out upon it would be part of the ongoing process of learning from and informing the planning agencies and other organizations in the ETC area. Given its pivotal position in the process of devising a strategy for the ETC, the collaboration of SERPLAN in the project was essential. This was secured to the extent that SERPLAN became a joint grant holder in the project and SERRL was represented at all meetings of the ETC Working Group of SERPLAN for the duration of the project. Through this link, many valuable contacts within local government and other agencies (e.g. English Nature and the Countryside Commission) were made and insights gained into the nature and role of information in the planning process. The benefits to be derived from collaboration were confirmed when the results of a study commissioned by SERPLAN into the problems of the ETC from a property development point of view were made known. This suggested that the development potential of the ETC was seriously hampered by two factors: the lack of a clear *identity* for the area and the lack of up-to-date and well-presented *information* on the area as a whole (Deloitte, Haskins and Sells, 1989). Although focused on the concerns of the property industry, the report added further to an awareness that an ETC GIS would make a practical contribution to research for policy.

1.1. The Eastern Thames Corridor

The regional context and character of the ETC were essential to the rationale of the research. Regional planning guidelines for the South-East state the need to redress the balance of attraction for development between the more affluent areas to the west of London and the less affluent areas to the east (SERPLAN, 1989). The main causes of the slower pace of development and the economic problems in the ETC have been identified in successive SERPLAN reports as being the nature of the industrial structure, the transport infrastructure, a degraded environment, a poor image and constraints (environmental and economic) on the development of vacant and derelict land sites.

An important reason for using GIS technology in the project lay in the indeterminate and fragmented nature of the ETC, giving rise to the need for flexible and efficient access to data. As identified by SERPLAN the ETC does not conform to any existing administrative boundaries, but is roughly defined by the A12/A127 on the north side of the Thames and the A2/M2 to the south (SERPLAN, 1987). It is an area of 378 square miles (roughly half the size of the GLC), covering the estuarine parts of the counties of Kent and Essex, ten London boroughs, including the London Docklands Development Corporation (LDDC), and the whole or part of eleven district councils (Fig. 1). Only the LDDC, one London borough and three district councils are fully contained within the boundary of the ETC. At the functional level, the ETC forms part of several travel to work areas, none of which is wholly subsumed within the subregion.

The geography of the ETC is also extremely varied, leading to its representation in a range of different data types. Settlement in the ETC ranges from very high density (inner city) areas and lower density (inter-war) suburbs within Greater London, to many free-standing small and medium-sized towns (including a former New Town at Basildon) and ribbon developments situated along the main roads. These centres of population are interspersed with agricultural land of varying types and degrees of viability, as well as large areas of degraded farmland, abandoned mineral workings and older industrial sites. In contrast, the Thames floodplain in both Kent and Essex is characterized by extensive areas of

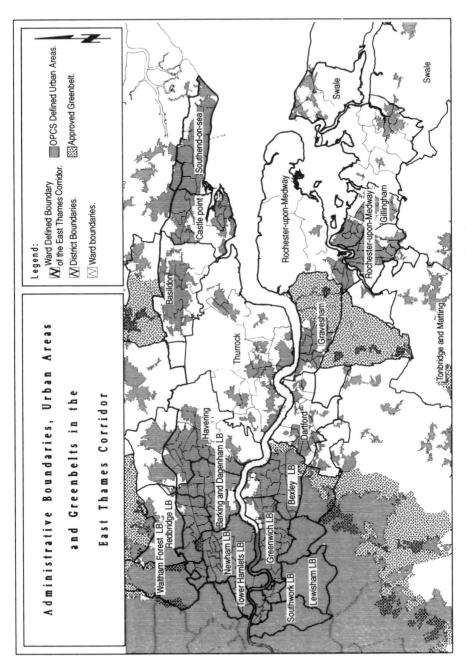

Fig. 1 The Eastern Thames Corridor administrative areas

virtually unpopulated mudflats and saltmarsh. There are environmental sites of international importance and no fewer than 29 Sites of Special Scientific Interest (SSSI) within the ETC, some of which, such as Rainham Marshes, are of considerable size. A large part of the ETC is also designated as Metropolitan Green Belt, some of it of poor environmental quality which, it has been suggested, might be better used for development purposes. In contrast to 'classic' subregional planning approaches (Cowling and Steeley, 1973) it was the very disparateness and lack of apparent 'structure' that made the ETC such an attractive proposition for a GIS study.

To summarize, both the administrative structure of regional and local government and the characteristics of the ETC area were therefore seen as providing an ideal setting for a realistic and challenging test of the role and value of geographic data management and analysis in subregional planning. The potential demands on a GIS thus lay in providing solutions to the following:

- A difficulty in creating a rigid territorial definition of the area, which suggested the need for a *flexible approach* to subregional definition depending on data sources and types.
- A wide range of 'natural' and built environments within a comparatively small area indicated a correspondingly *wide range of descriptive data* in different forms and to different quality standards.
- The administrative fragmentation of the area (both territorially and hierarchically) indicated the need for the *data integration* capabilities of GIS.
- A continuously evolving policy framework offered a setting in which the rapid data handling and *presentational facilities* of GIS would be readily accepted and appreciated.
- A direct *involvement with regional and local planners* provided the opportunity to develop a substantively relevant GIS interface for non-expert users.

1.2. The ETC GIS project

In brief, the overall aim of the ETC GIS project was to develop a coherent, accessible, up to date, presentationally flexible and, above all, relevant data source for the Corridor. As far as possible this was to be done using available data sources. In the course of establishing the GIS four interrelated research aims were to be met:

1. To assess the data availability, data management and data quality problems of a GIS in a multi-subject and multi-agency environment.

2. To assess the impact of the ETC GIS on decision making in regional, county and local planning authorities.

3. To use the ETC GIS and other data sources in the substantive analysis of the process of local economic and land use change in the ETC.

4. To contribute to training opportunities and programmes, and develop links between SERRL and appropriate government agencies through which to disseminate results and promote the ETC and the ETC GIS.

This chapter presents an overview of some of the work carried out on the ETC GIS project in the period 1989–1992. It is both a generalized and a selective account of this research in so far as some of the detail of particular topics was contained in the project reports to ESRC/NERC (SERRL, 1992, 1993). For reasons of space, much detail of the research has been culled and some aspects of the work undertaken (on land use change, for example) have been omitted altogether. Fuller accounts of the main research topics have already appeared elsewhere and these are indicated in the text. The structure of the chapter is as follows. Section 2 is a description of the planning background to the ETC, which points out the nature of the interaction between strategic goals and local issues/initiatives as these have evolved since the importance of the Corridor was first established in 1985. These issues are indicative of the sorts of information required to develop an ETC GIS and the structure such data might take to address the issues. This is followed in Section 3 by a description of the approaches taken in putting together the ETC GIS database and the attitudes towards GIS in local government in the Corridor. Section 4 describes an example of the use of the ETC GIS in action, represented by a study of employment structure and travel to work in the ETC. Finally, Section 5 attempts to draw some general conclusions from the overall research in relation to the role of GIS in planning.

2. THE BACKGROUND TO THE ETC INITIATIVES

The ETC was first identified in SERPLAN's 1985 regional review as the part of the South-East most in need of investment, but there was little more than acknowledgement of this in the government's Regional Planning Guidance (Department of the Environment, 1988) that followed. Examination of the problems of the area and the action needed to realize its development opportunities was left to a SERPLAN officer-level working party, established in 1986 and drawn from its member authorities in the Corridor, which produced or commissioned a number of reports on the area. These efforts made significant headway. On 10 October 1991, the government wrote to SERPLAN announcing two strategic decisions shaping the future of the Corridor. Firstly, the Minister for Planning said that the government saw the development of the Corridor as making an important contribution to the development of London, the South-East and the country as a whole. He therefore invited cooperation in a study that the government was about to commission on the development potential and infrastructure needs of the Corridor.

At the end of March 1993, the consultant's report was published, as a result of which the Secretary of State formally recognized the great potential of the area, announced how he proposed to set about the preparation of a planning framework for the area and produced first thoughts on implementing the scheme. The second announcement, made in October 1991, was that a decision had been taken to route the fast rail link between the Channel Tunnel and London through the ETC. This recognition of the part that the rail link could play in realizing the potential of the Corridor was something that SERPLAN had pressed on government, pointing out that, in addition to being a vital transport facility, the link should seek to meet three equally important planning objectives:

> It should form an integral part of the national and international rail network, facilitating travel within and through the region. The route and stations should assist in promoting regional economic objectives, especially development of ... the East Thames Corridor. The link should be routed and designed so as to minimise adverse environmental impact.
> (SERPLAN, 1991).

There were further developments during 1992. British Rail announced that the rail link would be designed to take freight, and made available papers on environmental standards for the route and on the terms of reference for its development. Formal consultation arrangements with local authorities have been set up at ministerial level, both for the rail link and for the development of the ETC. Finally, at the end of March, the Secretary of State for Transport announced that he had received a report from Union Railways (a subsidiary of British Rail) setting out options for the route of the rail link and possible locations for new stations and junctions, some of which will almost certainly be within the ETC. The 'preferred route' is now the subject of public consultation.

2.1. The SERPLAN 1986 study

The first SERPLAN report on the ETC was a detailed study of the needs and opportunities represented by the Corridor (SERPLAN, 1987). This showed that the Corridor had an overall unemployment rate of nearly 15%, half as high again as the regional average, with local peaks of 25%, and an employment structure dependent on vulnerable manufacturing industry, where the number of jobs had fallen by 12% between 1971 and 1981. The study also found over 1600 hectares of land in the Corridor outside Docklands already allocated in local plans for employment-generating uses. Well over half that land was unused, or virtually unused, as against a quarter that was old industrial land. Much of the land potentially available for development had been discarded by private and public industry in the wake of economic decline and technological change, and was now ripe for recycling. The effect of de-industrialization and the decline of port activities had left the area with a legacy of environmental degradation, poor accessibility and a shortage of skills for growing economic activities.

The question was what could be done to encourage investment, which was desirable both to revitalize an area of urban decay and to provide jobs for the residents, and which, if successful, would tend to ease pressures on other 'overheated' parts of the region? An important task, therefore, was to raise the profile of the Corridor by giving it a name and an identity, and by providing information about the opportunities it contained. This was important in drawing the attention of the private sector to the opportunities presented by the Corridor and to heighten awareness of local planning and of the policies needed to encourage uptake of development sites. The 1987 study also showed the location of available sites, how they could be made more accessible by some priority road and rail investment, and where poor environment was likely to prove a constraint (Fig. 2).

2.2. The study review

A review of the 1987 study carried out in mid-1988 showed that nearly half the land previously identified for employment-generating uses was being developed or was firmly programmed to be developed (SERPLAN, 1988). This was partly because of the development boom (then at its height) and partly because of proximity to Docklands activity. Some of the progress was, however, undoubtedly attributable to the influence of the earlier publicity and to the fact that by then almost half the government's main road investment in London was concentrated in the ETC boroughs. Despite the significant progress, though, the study noted some continuing and some new problems. In particular, despite the level of investment in roads, there was seen to be a general lack of coherence in transport planning, and housing demands were being met at the expense of employment-generating uses.

2.3. The consultants' study

SERPLAN and the local authorities in the Corridor were still not satisfied that either transport or land use strategies were sufficiently advanced to make things happen once the most

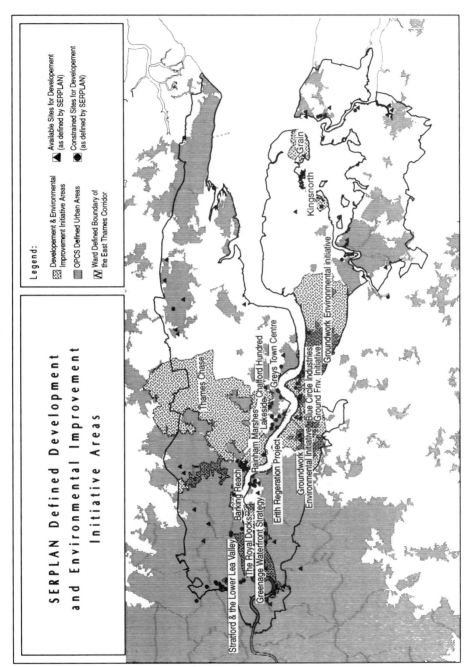

Fig. 2 Eastern Thames Corridor, constrained sites (SERPLAN, 1987)

obvious opportunities had been taken, or that the private sector was adequately involved. With financial help from the government, SERPLAN commissioned consultants to 'propose the actions needed to create the demand for the sites available for development in the Eastern Thames Corridor to be taken up and developed'. Their report concluded that there was indeed a 'window of opportunity' to secure an enhanced programme of development in the ETC (Deloitte, Haskins and Sells, 1989) but to take advantage of this opportunity the immediate needs were:

- To forge a common sense of purpose between the public and private sectors.
- To reduce the 'planning hurdles' in the way of would-be developers and employers.
- To set out easily understood objectives for the long term while ensuring that implementable actions and projects are pursued in the short term.

The tasks necessary to meet these objectives were:

- Strategically, to promote growth in general by identifying and encouraging 'flagship' projects – major projects that would themselves raise the profile of the area and might be expected to attract other schemes and activity in their wake.
- In the medium term, to develop a strategic framework acceptable to all the parties to coordinate all their activities.
- To market the concept of the ETC; identify infrastructure projects that could be brought forward through private sector involvement; identify projects to make use of the Thames waterfront.
- To formulate a training plan to improve the existing skill base in the light of future labour requirements in the Corridor, and to begin to attract appropriate growth industries into the Corridor.

2.4. The future of the Eastern Thames Corridor

This considerable activity on the part of SERPLAN and local government secured the importance of the ETC to the overall planning strategy for the South-East (SERPLAN, 1990a) and in October 1990 SERPLAN published a further document, *Action in the East Thames Corridor* (SERPLAN, 1990b). This consisted of two parts: an action plan and a series of case studies of proposals for actual sites. The action plan developed the policy stances that had evolved through the earlier work. The main elements of the plan were concerned with economic activity, the environment, transport, housing, and marketing and promotion. It was at this point that the first use of the ETC GIS, then in the early stages of development at SERRL, was made. Albeit at a fairly elementary level (i.e. mainly drawing maps), this development was of considerable significance in demonstrating to sometimes sceptical planners the value of integrated, geographically referenced information systems in the flexible handling and attractive presentation of information about the ETC.

Much has been accomplished in the past six years. The Corridor is now seen in the context of a potential restructuring of London as east–west links (CrossRail and the Jubilee lines) are put in place, and with the fast rail link emphasizing the increasing importance of relations with neighbouring parts of the Continent. The primary responsibility at this stage is intended to pass to the individual boroughs to make sure that readily available and serviced sites continue to come forward and to be marketed on a site-by-site, rather than a broad Corridor, basis. Meanwhile, the nature of the feasible development of the area and hence of the necessary implementation mechanisms depend both on the national economic future and more immediately on the outcome of the government's consideration of the ETC study. The announcement in March 1993 that the 'eastern approach' to London for the Channel Tunnel rail link was confirmed and that the government would support its development with finance pointed the way forward in this respect. The main endeavour now is to maintain the impetus that has been achieved for and within the ETC and to ensure that the future of the Corridor is determined in a regional context. In this, the ability to deliver relevant, accessible, appropriately structured and geographically referenced information to the planners is clearly a key requirement.

3. THE ETC DATABASE

3.1. The starting point

The starting point, and continuing raison d'être, of the ETC research project was the development of a land and economic development information system (ETC GIS) which would both improve our knowledge of the nature and availability of information on the ETC and in itself act as a resource for a number of substantive studies. In doing so the aim was not to follow the 'classic' prescription of the GIS user needs study (Aronoff, 1989) – there was neither the time nor the resources available for such an approach – but to establish to what extent it was possible to set up a working GIS for the ETC using relatively readily available data resources.

Our approach to the definition of the database was based upon two considerations: firstly, an acceptance that very large quantities of data already exist whose categories and spatial units are predetermined and inflexible (e.g. the population census), and which must, to a considerable extent, define the key facets of the database structure; and secondly–and this is to some extent a countervailing tendency – an awareness of both the research issues to be explored regardless of data availability and the statutory concerns of the local government and other agencies operating within the Corridor. Rather than developing a database that would furnish the planners with all or even the majority of the information they required to develop planning strategies for the ETC, our aim was to develop a structure that would best articulate available data for planning purposes. Such a database would (i) facilitate an investigation of how well available data could serve the subregional planning process in a GIS context and (ii) draw attention to the 'added value' aspects of GIS in bringing the data habitually used by planners together in a structured and integrated manner.

There are, therefore, two groups of data issues requiring further consideration. One group relates to the substance or meaning behind the data and is treated below. The other refers to fitting the data to a particular GIS model and to related technical issues. The latter is discussed in John and Lewis (1992) and Lopez and John (1993).

In substantive terms the main criteria for search and inclusion of data sets were:

- To reflect the key subregional planning issues in the ETC as set out in the SERPLAN and other documents, which implied the need for a *wide range* of data.
- To aid the process of monitoring the progress of revitalization in the Corridor, which implied *time series* data.
- To support understanding of the processes of local economic change and development in the Corridor, which implied exploring *relationships* between the variables representing the data.

- To establish the affects of environmental and infrastructural constraints on key development sites identified by SERPLAN within the ETC, which implied the need for data measured at *local and subregional* scales.

3.2. The data search

The institutional scope of the search for relevant and usable data for the ETC is shown in Fig. 3. Given the emphasis on exploiting data that are available, it is important to grasp the nature of the complex web of organizations whose statutory tasks and policy perceptions define the scope of available information. In broad terms we can conceive of three types of data deriving from such a framework:

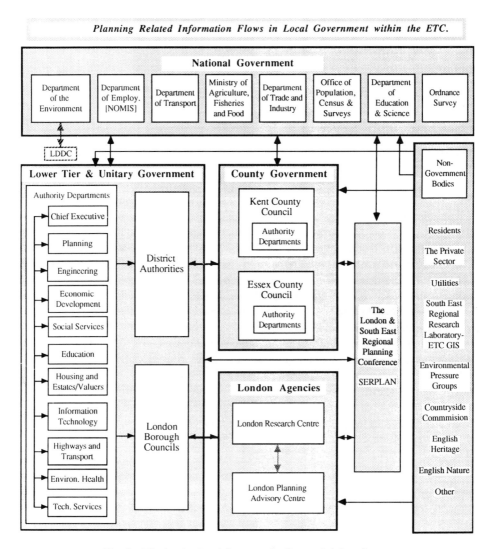

Fig. 3   The institutional framework of potential data flows

- Data that are nationally consistent and comprehensive, which provide enough local detail to be relevant to all types of provider, such as census, small area statistics or postcodes (Clark and Thomas, 1990; Raper, Rhind and Shepherd, 1992; Shepherd and Brewer, 1992).
- Data derived from the carrying out of administrative functions, which may or may not be consistent across authorities, depending on the nature of the function under consideration (Charlton and Ellis, 1991; Huxhold, 1991).
- Data derived from *ad hoc* censuses or surveys, which may cover only part of the area of interest, for example, the North Kent marshes area, and which may suffer from various forms of discontinuity in time, space and physical representation (Townshend, 1991).

An important implication of this is that certain topics may be addressed relatively easily, because comparatively rich reserves of pertinent information are in existence, while for other topics data resources remain sparse. The demographic structure of small areas is one obvious example of the first data type, whereas data on ecological diversity are an example of the last. These differences derive to a significant degree from the existence in certain fields of long established statutory responsibilities, the fulfilling of which has involved the development of substantial bureaucracies, whose activities have in turn generated large amounts of data (with accompanying boundary delimitations), albeit of a traditional kind. Much recent work in GIS has focused on drawing such familiar data into an IT configuration, embodying database and mapping functionality. Our experience, derived partly from the ETC GIS project, suggests that future work must strive to make maximal use of less familiar, though substantial, data reserves by adding value through spatial analytic capacity. Such data include those derived from 'indirect' processes – the Department of the Environment's Land Use Change statistics are one example (Bibby and Coppin, 1993) – and those that represent the recycling of digital data originally designed for other purposes, such as postcodes and business listings (Bibby, 1992; Shepherd and Brewer, 1992)

To such sources, however, must be added those data types that presently appear to lie in a conceptual and technological limbo. In terms of those sources that presented themselves in this project they appear to relate mainly to environmental issues. In the case of contacts with English Nature, for example, there were particular problems with species and habitat data that did not appear amenable to satisfactory inclusion in the ETC GIS, while the data on monuments and sites of historic interest in the Corridor presented problems of representation at the geographic scales typically represented in the database. There is scope for challenging and valuable research in geographic data handling in these contexts.

Given the particular aim of the ETC GIS project, that is to say its overriding concern with maximizing the utility of available data for purposes defined within the specific context of subregional planning, the prime objective of the following paragraphs is to focus on the interface between GIS and the needs of a particular set of users in local government and academia. Of necessity, certain important topics must receive less attention than they merit. This applies in particular to the technical structuring of the geometric data within the GIS, the characteristics of the 'data dictionary' and the approach adopted to compiling the database in a collaborative environment (Table 1).

3.3. The ETC–GIS interface

The ETC–GIS interface was originally conceived as a response to the need to set up a GIS in a multi-agency environment and to respond to user needs. It was developed to illustrate

Table 1 The stages of database construction

Stage	Description
Phase I: Search and design	
1	Database specification: identification of goals, objectives, user requirements and data requirements
2	Identification of sources: bibliographic and other searches
3	Initial contact with sources: assessment of data availability and quality
4	Reappraisal of data needs
5	Database structuring, based on initial knowledge of user requirements
Phase II: Data Collection	
6	Data collection: letter and telephone contacts, site visits and meetings
7	Data processing: conversion of raw data to ARC/INFO coverages
8	Data integration: assembly in relation to interface requirements
Phase III: Reassessment	
9	Data dictionary: recording codes, fields, categories, and so on
10	Meta-data: data processing type, contact names, conditions of access and so on
11	Identification of new data needs based on prototype uses
12	Return to Stage 3
13	Reassessment of database structure
Phase IV: Implementation	
14	Applications and user assessment

the functionality of the database to relevant organizations, to provide an information source for the ETC and to stimulate discussion on the issues of data integration, sharing, management and GIS implementation in local government. In this context there was a need to provide a means whereby users with a wide range of GIS skills and experience could perform queries and some analysis on the data. Identifying what constituted 'meaningful' functionality and providing a simple method of executing this formed part of the assessment of GIS among local authorities in the ETC.

3.4. The Eastern Thames Corridor: GIS in local government

The county, district and other public bodies comprising the ETC played a key role in the data assessment, collection and assembly stages of the project in two main ways. Firstly, they provided invaluable information on the level of *uptake* and *use* of GIS among public authorities in the Corridor, thus paving the way for an assessment of GIS needs among planners operating at the subregional level. Secondly, as potential *users* of the GIS database to be assembled, their views on the data to be included and the structuring of the data to permit appropriate access and analysis were obviously essential. This aspect of the research was invaluable in developing a planner-orientated 'front-end' to the ETC GIS.

In-depth (up to two-hour) interviews were conducted with the planning officers concerned with spatial data management and GIS issues in each of the local government bodies in the study area (eighteen local authorities, two county councils, the LDDC and the London Planning Advisory Committee (LPAC)). The questionnaire was designed to examine a wide range of data-related issues, including the level of utilization of GIS, spatial data user needs, data collection and maintenance procedures, user awareness of data quality requirements, the role of spatial information and GIS technology in local government, the potential influence of GIS on planning and decision making, and the degree to which each authority had made progress towards implementing GIS. Interviews were undertaken in early 1991.

3.5. The role of GIS in local authorities

Of the twenty-two ETC authorities interviewed, eighteen had made an assessment of GIS in relation to their planning work and seven of these had implemented some form of GIS. In percentage terms, as Table 2 shows, although the ETC represents a small sample, the results are remarkably similar to those reported by Campbell and Masser in their nationwide study (1991). The assessment studies undertaken were often detailed and in many cases involved contact with potential suppliers. GIS awareness was high, although a full understanding of its potential role in local government was less well developed. The six authorities that had not undertaken assessment studies had made conscious decisions not to do so, either because of budgetary constraints or because GIS was not seen to be an appropriate technology in the light of their resources and needs.

Table 2 GIS uptake nationally and in the ETC

Local authorities	Considering introduction of GIS (%)	Have acquired GIS (%)	Have plans for purchase of GIS (%)
National survey (Campbell and Masser, 1991)	69	16.5	8.6
ETC survey (SERRL)	68	23	5

Most respondents saw GIS as serving the more mundane tasks of planning rather than those that were more intellectually challenging. Thus, while 60% of interviewees envisaged GIS as assisting in such tasks as local land searches, accident monitoring or social service workload planning, only 35% saw it as having the potential to contribute to strategic decision making. Campbell and Masser also concluded that, given the emphasis that has been placed on the contribution of GIS to improved strategic decision making, it is striking that staff in local authorities continue to associate such systems with the more straightforward activities of improving information processing facilities. As a result, the main benefits of GIS were perceived to include increased speed of data provision, better access to information and an increased range of analytical and display facilities, rather than better analysis of data.

One reason for this emphasis seemed to be a lack of a sustained impetus for developing an awareness of the potential of GIS within an authority. Only twelve of the twenty-two local ETC authorities had a GIS strategy, and only eight of these twelve were satisfied with its progress. The main reasons for this lack of satisfaction were that such strategies lacked direction, suffered from lack of resources, there was little enthusiasm among officers and members and, despite acknowledged benefits, there were interdepartmental and interauthority differences of views on data sharing (John and Lopez 1992; Bromley and Coulson, 1989). No less than 75% of the interviewees recognized the need for more information technology awareness in chief officers, members, planning officers and technical staff although, paradoxically, only 43% felt there was a need for formal training to promote awareness.

In addition to lack of resources, other commonly cited disincentives to GIS development were the time required for implementation and, especially, the poor quality and availability of data. No fewer than nineteen of the twenty-two authorities interviewed considered that there was a basic need to improve spatial data collection methods within the authority in order to improve their spatially referenced databases and/or their GIS. The changes most commonly envisaged as being necessary included more regular updates, accuracy improvements, better resolution of survey data and the addition of new data sets. The extent to which data quality appeared to hold back planning analyses was indicated by the fact that over three-quarters of the respondents indicated that they were reluctant to use advanced spatial analysis for local planning given data inadequacies. A particular area of confusion and admitted ignorance was the crucial issue of data ownership and copyright. More positively, among those planning officers who *were* satisfied with the progress of their GIS strategy all cited the interest of officers and members as a key element of its success, not least because this was usually linked to availability of resources. Questions concerning incentives to the development of GIS drew similar responses: its potential to increase productivity, its ability to carry out new tasks, pressure for better information, the interest of officers and members, and the inherently geographical nature of local authority data were all seen as strong arguments for the introduction of GIS.

3.6. The ETC 'front-end'

The views of planners were also sought on the main types of data access, analysis and presentation required of the ETC GIS. These included discussions about such matters as the main thematic groupings of information required for convenience of access, the nature of the geographical comparisons to be made using the data and at what scales they were required, the level of complexity of exposition of the pull-down menus needed to 'drive' data selection and presentation, and the depth of exposition required in the 'meta-data' describing the individual data sets. In the course of such discussions a number of weaknesses of GIS – especially in relation to data transformation and spatial analysis – from the planners' point of view became apparent. Although it was not possible to meet all of these in the development of the ETC GIS itself, they are an obvious 'next step' and should, perhaps, be better recognized by the GIS software developers (Bibby and Shepherd, 1992).

As a result of these discussions, the data were assembled into distinct groups of 'coverages' numbering over 100 in all (for a selection see Table 3) and held in an appropriate form within the GIS software. In itself, this process of 'front-end' development was an extremely important educational process, both for researchers as database developers and for planners as GIS users. A basic 'toolbox' of functions was developed to allow the user to move between data themes and across spatial areas. The user is given the freedom to move spatially from the largest extent of data (the South-East region), through the counties, districts and finally the wards that constitute the ETC, and at each level manipulate data relating to population, employment, travel to work patterns, land use change, environmental sites, development sites and infrastructure. At all levels, it is possible for the user to identify single features and select groups, as methods of investigating the data behind the maps. These GIS facilities, developed using the ARC/INFO programming language AML, are accessed via pull-down menus, forms and on-screen 'buttons' (Fig. 4). In the course of the project over 300 people and 190 organizations were able to view the ETC database via the GIS interface.

Fig. 4　The ETC–GIS interface

Table 3 Data sets held in the ETC database, as of October 1992

The South-East	Five-county (London, Kent, Essex, Surrey, Hertfordshire)	Three-county (London, Kent, Essex)	The ETC	London
Administration				
SE outline				
Counties	Counties	Counties	Counties	
SE-clip	Counties-clip (to the ETC)	County-clip		
Districts	Districts	Districts	Districts	Districts
			Districts-clip (to the ETC)	
		Wards	Wards	Wards
	EDs	EDs	EDs	EDs
	Kent-EDs		ETC_clip	
Look-up tables		Wards.Lut	Road-defined polygon	
		Districts.Txt	Ward-defined lines	
			Ward-defined polygon	
Employment				
Counties			Districts	
Unemployment 1984–1989	County unemployment 1984–1989	County unemployment 1984–1989	Unemployment 1983–1991	District unemployment 1983–1989
			District-clip	
		Unemployment 1983–1991	Wards	
			Unemployment 1983–1991	Ward unemployment 1983–1989
			Unemployment by economic population	Unemployment by economic population
			Employment 1984, 1987, 1989	Employment 1984, 1987, 1989
			Change in unemployment and employment	
			Employment: cover	
			Totals 1984, 1987, 1989	
			Employees in employment	
			Self-employed	
Look-up tables				
county relate				
unemp.lut				
name.lut				

Table 3 (*continued*)

The South-East	Five-county (London, Kent, Essex, Surrey, Hertfordshire)	Three-county (London, Kent, Essex)	The ETC	London
Environment				
Generalized greenbelts	Generalized greenbelt	Generalized greenbelt	Generalized greenbelt	
SE-approved-greenbelt	Approved-greenbelt	Approved-greenbelt	Approved-greenbelt	
	AONB	AONB	AONB	
	Landfill: Friends of the Earth	Landfill: Friends of the Earth	Landfill: Friends of the Earth and SERPLAN	Landfill: Friends of the Earth and SERPLAN
		Landfill: SERPLAN	ETC_SSSIs	ETC_SSSIs
			ETC_Nature Reserves	ETC_Nature Reserves
			Vulnerability.dat	
			National Status.dat	
Infrastructure				
SE BR stations	BR stations	BR stations	BR stations	BR stations
	Rail	Rail	Rail-ETC	Underground and rail
SE urban areas	Urban areas	Urban areas	Urban-ETC	
SE urban-OPCS	Urbanbound	Urbanbound	Urbanbound	
	A-roads	A-roads	A-roads	A-roads
	B-roads	B-roads	B-roads	B-roads
	Motorways	Motorways	Motorways	Motorways
	Roads	Roads	New and proposed infrastructure	
			New roads	
	SAMS	SAMS	New rail	
Land				
Land use change 1985–1989_SE	Land use change 1985, 1986, 1987		Land use change 1985–1991	Land use change 1985–1991
Land use 1990_SE			Rural to urban change	
Districts: aggregated land use change statistics	Districts: aggregated land use change statistics		Net change in urban and housing land by population	
			Percentage of land changing from and to urban and rural uses	

(*continues*)

Table 3 (continued)

The South-East	Five-county (London, Kent, Essex, Surrey, Hertfordshire)	Three-county (London, Kent, Essex)	The ETC	London
Population				
Districts	Districts	Districts	Districts	Districts
Population 1961, 1971, 1981, 1991	Population 1961, 1971, 1981, 1991	Population 1961, 1971, 1981, 1991	Population 1961, 1971, 1981, 1991	Population 1961, 1971, 1981, 1991
Population change 1981–1991			Wards	Wards
Population density			Total population	Total population
			Population density 1981, 1986, 1987, 1988	Population density 1981, 1986, 1987, 1988
			Population change	
			Enumeration districts	
			EDs population 1981	
			EDs *de facto* population	
Sites				
			Available development sites	Available development sites
			Constrained development sites	Constrained development sites
			Development and environmental initiative areas:	Development and environmental initiative areas:
			Barking, BCI, Chase, Grain, Greenwich, Groundwork, Kings, Rainham, Stratford, Chafford, Grays, Lakeside, Royal Docks, Bexley	Barking, Chase, Greenwich, Rainham, Stratford and Lea, Royal Docks, Bexley
Travel				
			Travel-to-work—male and female	
			ETC wards to inner London	
			ETC wards to outer London	
			ETC wards to the City	
			ETC wards to Westminster	

ED, enumeration district.

3.7. The ETC multimedia project

A more experimental aspect of database development was represented in a small-scale project designed to explore the potential for multimedia GIS as a vehicle for presenting a wider range of information to users who were not in need of the sorts of quantified information normally associated with GIS or who had a need for archiving and accessing a much wider range of graphic data. Multimedia, a rather novel aspect of GIS technology at the time of the ETC project (though rapid strides have been made since; Raper, 1993), offered an approach to investigate the functionality and data requirements of a group of users who did not have access to full-blown GIS, or who wished to make a different use of the SERRL ETC database.

The system was developed using HyperCard, an inexpensive and widely available hypertext control and programming package written for Apple Macintosh machines. This was used in the design of the interface to the system, enabling the linkage of maps, photographs, other graphics such as histograms or three-dimensional views, sound, text and selected numerical data embedded within a '*geo*-graphical' frame of reference (Lewis, 1991; Lewis and Rhind, 1991). Among the data themes that could be addressed for the ETC as a whole or for selected sub-areas were satellite imagery, aerial and other photographs and colour maps at 1:250 000 and 1:50 000 scales. In principal, too, there would have been no technological barrier to incorporating, for example, the birdsong of species resident in ETC SSSI or an aural guide to a historic Thames-side area.

This system was presented to an interested subset of the more than 100 persons who viewed the main ETC GIS database, and their reactions were gauged. The benefits of this approach were seen as speed of access to information, the low cost and relative ease of development, maintenance and update, and ease of use. The graphic qualities of the system made it of particular relevance in an educational context and to those concerned with developing a public relations image for particular parts of the Corridor or real property within it (Deloitte, Haskins and Sells, 1989).

4. THE ETC DATABASE IN ACTION

As an example of the use of the ETC GIS in empirical research on the ETC, this section makes use of a number of diverse data sets relating to population, employment and travel to work. In addition, it employs several of the boundary data sets onto which the numerical and other attribute data can be projected using the ETC GIS interface. The research is reported in an abridged form, compared with the original papers from which it is taken, but is presented in a way that reflects the substantive research aims rather than the technical aspects of the GIS. In this respect, the research problems were approached in conjunction with SERPLAN, which, until the establishment of the ETC database, was not able easily to access and manipulate well-defined and detailed economic data for the area.

4.1. Population change in the ETC, 1981–1988

As this chapter was going to press, the small area statistics from the 1991 Census of Population were just becoming available. The population analyses reported here, therefore, are for the period 1981–1988, the latter date being that of the last available set of Registrar General's estimates for small area and projections based thereon. In this period the total population of the ETC had grown from 2 020 731 to 2 057 694, an increase of 1.8%

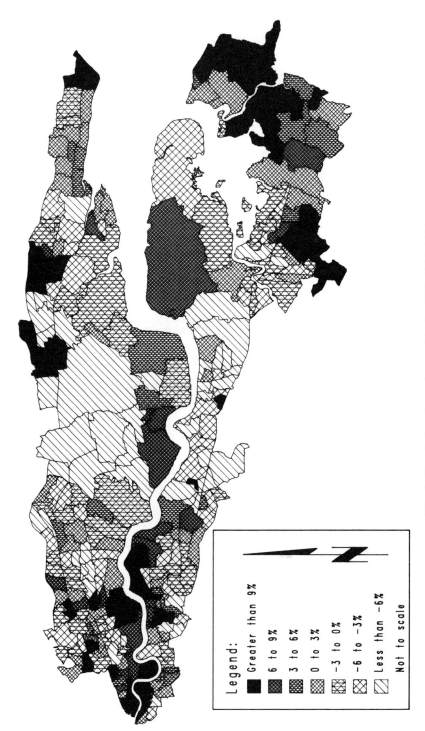

Fig. 5 Population change by ward in the ETC, 1981–1988

compared to a 1% decline for Inner London, a 0.3% increase in Outer London, and a 4.2% increase in the rest of the South-East.

The pattern of population change between 1981 and 1988 is shown in Fig. 5. Wards recording population growth in excess of 9% appear at a number of locations in the ETC. In Inner London the effect of the Docklands redevelopment and subsequent in-migration is apparent, with high growth rates in riverside wards at the western end of the ETC. Some of the high growth rates in riverside wards in Outer London on the south bank of the Thames are partly the result of low initial bases in low density areas. The non-riverside wards of Inner London to the north of the Thames that have registered high rates of growth are wards with high concentrations of ethnic minority groups who have recorded high rates of fertility, and the growth probably stems from natural change rather than in-migration. Outside London certain wards on the edge of established population concentrations exhibit high growth rates, including wards south of the Medway Towns, around Sittingbourne, on the outskirts of Basildon and Canvey, and east of Southend. Population increase is also marked in the Thurrock and Grays area, with the exception of the Tilbury ward. Patterns of land availability and housing construction on the fringes of established settlements are partly responsible for this map of change.

Almost half (45%) of all ETC wards recorded a decline in population between 1981 and 1988. In London these are spread evenly throughout inner and outer districts. In Essex many of the rural wards to the east of London experienced declines of over 6%. South of the Thames, the area between Dartford and the Medway Towns contains many wards where population declined, including rural wards and those in the towns of Northfleet and Gravesend. Similarly, the central areas of the Medway Towns recorded falls in population. The map of population change is thus the first indication of the varied patterns of social and economic change recorded in the ETC and leads us to guard against any treatment of the Corridor as an area of undifferentiated economic fortune. Examination of the pertinent economic data was designed to shed light on this level of variation.

4.2. Economic change and spatial data

For many parts of the South-East of England the mid- and late-1980s were a period of rapid development and economic growth. Until this project began no economic studies of the ETC had been undertaken, even though it was believed to be the South-East's least successful subregion. The research was therefore designed to analyse economic change in the ETC in the context of the South-East region as a whole and for broadly functional areas within it.

An initial problem facing the economic analysis, therefore, was the issue of the definition of the ETC as a whole in economic terms and of areas within it. The ETC is not a neat functional region; it contains, for example, parts of no fewer than four Department of Employment travel to work areas which significantly overlap the putative external boundaries of the Corridor. In addition, some geographical disaggregation of the Corridor was also required in order to assess the amount of internal variation in economic variety. Fig. 6 shows the sub-areas used in the analyses. The ten sub-areas outside London were devised primarily by identifying areas of employment concentration and similar industrial structure (see Church and Frost, 1991 for a more detailed discussion). One advantage of the ETC GIS here was that it allowed the adjustments to be made in an iterative manner and facilitated experimentation with different ward groupings. Because of the complexity of the interactions within the London labour market, the London part of the Corridor was left as a single unit.

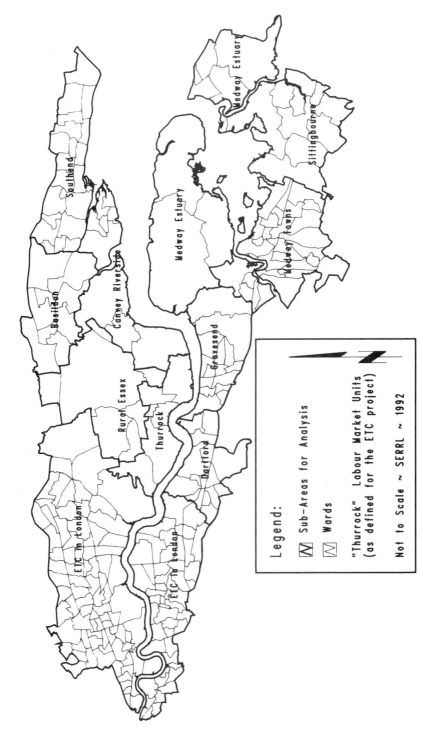

Fig. 6 ETC sub-areas as defined for local economic and labour market analysis

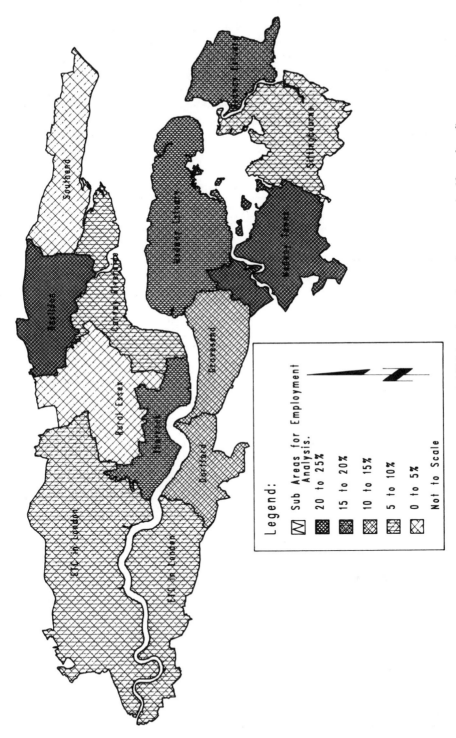

Fig. 7 Employment change in the ETC, 1984–1989 (employees in employment and self-employed)

4.3. Industrial structure and employment in the ETC

Using this geographical framework, Tables 4 and 5 describe the industrial structure of the sub-areas comprising the ETC, while Fig. 7 indicates the nature of employment change within the Corridor. Data from the Censuses of Employment, accessed via 'NOMIS' (Blakemore, 1990) were used for the change analyses and were based on the years 1984 and 1989, the first and last dates in the decade for which reasonably reliable data are available at ward level. In this period the total number of jobs in the Corridor grew from 687 000 to 715 000, a crude increase of just over 4%. A brief comparison of the results in Tables 6 and 7 with the rest of London and the South-East shows clearly the structural context within which change in the Corridor is set.

Table 4 Industrial structure in ETC sub-areas in 1984 (percentage of total employment in each area)

	Standard Industrial Classification divisions							
	Agriculture 0	Energy 1	Manufacturing 2–4	Construction 5	Distribution/catering 6	Transport 7	Banking/finance 8	Other services 9
Medway estuary	3.2	13.2	33.4	3.3	15.4	12.0	2.8	16.8
Sittingbourne	2.8	0.7	32.7	8.3	15.6	5.8	6.8	27.3
Medway Towns	0.9	2.7	30.6	4.6	21.2	6.3	7.1	26.7
Gravesend	1.4	3.6	27.8	5.7	23.0	7.7	7.3	23.6
Dartford	0.1	4.1	32.2	3.9	16.0	8.8	5.1	29.8
Thurrock	0.0	6.8	19.3	4.1	15.8	29.0	4.1	21.0
Rural Essex	3.4	0.0	25.1	2.0	15.7	5.0	3.3	45.6
Basildon	0.0	0.5	45.1	2.7	19.0	3.7	7.0	22.1
Canvey Riverside	0.1	17.1	26.4	6.0	20.2	3.6	4.5	21.9
Southend	0.1	1.3	14.6	4.6	25.7	6.3	16.2	31.3
ETC in London	0.0	1.5	23.5	4.7	19.4	9.6	11.4	29.7
Rest of London[1]	0.1	1.2	18.6	5.5	21.4	9.6	11.2	32.3
Rest of South-East	2.1	1.4	24.2	4.6	21.3	6.0	9.5	30.8

[1]The Department of Employment's definition of London minus the ETC area and Central London, defined by the five central employment offices of the City of London, Borough, King's Cross, Westminster and St Marylebone.

Table 5 Industrial structure in ETC sub-areas in 1989 (percentage of total employment in each area)

	Standard Industrial Classification divisions							
	Agriculture 0	Energy 1	Manufacturing 2–4	Construction 5	Distribution/catering 6	Transport 7	Banking/finance 8	Other services 9
Medway estuary	3.0	10.1	29.8	7.8	15.9	15.3	3.8	14.3
Sittingbourne	2.6	1.7	32.6	5.4	19.6	7.5	6.9	23.6
Medway Towns	0.6	2.1	26.5	4.5	22.3	5.6	10.5	27.9
Gravesend	1.6	2.5	23.7	6.2	24.2	6.7	7.7	27.3
Dartford	0.2	4.2	28.7	3.9	16.4	11.0	6.1	29.5
Thurrock	0.0	5.9	20.5	4.6	21.2	19.1	6.9	21.7
Rural Essex	2.8	0.1	18.1	4.5	20.8	7.1	5.0	41.7
Basildon	0.0	0.6	40.4	3.5	20.9	3.7	7.8	23.1
Canvey Riverside	0.0	12.0	22.0	8.3	24.1	5.0	7.9	20.7
Southend	0.1	0.1	11.9	3.6	25.2	6.8	21.3	30.9
ETC in London	0.0	0.8	19.6	5.5	18.5	8.7	16.3	30.7
Rest of London[1]	0.1	0.9	15.2	4.9	21.8	9.4	14.0	33.8
Rest of South-East	1.6	1.3	20.7	4.6	22.8	6.1	12.9	30.2

[1]The Department of Employment's definition of London minus the ETC area and Central London, defined by the five central employment offices of the City of London, Borough, King's Cross, Westminster and St Marylebone.

To summarize the patterns of total employment growth or decline, including both males and females and the self-employed, Table 6 shows rates of change for each subdivision for all forms of employment. These data suggest that all subdivisions of the Corridor enjoyed some growth during the 1980s and that in some cases this was growth at a very rapid pace. Both Medway and Basildon showed rates of growth in excess of 20% during the five-year period of the study, while even the London area shows a clearly positive performance. In many ways, these generally positive results accentuate the poor performance of Southend which, even with the addition of self-employment, only just registers growth over the period.

Table 6 Total estimated rates of employment change in the ETC sub-areas, 1984–89

Area	Rate of change (%)
Medway estuary	19.4
Sittingbourne	7.5
Medway Towns	21.7
Gravesend	12.2
Dartford	10.4
Thurrock	16.0
Rural Essex	3.8
Basildon	24.3
Canvey Riverside	6.5
Southend	0.9
ETC in London	5.1

Among the conclusions that can be drawn from these data regarding the economic structure of the ETC and economic change in the period 1984–1989 are:

1. There is only limited evidence of a distinctive industrial structure in the Corridor, and although there is a slight tendency for more manufacturing and less service related employment, this is essentially a marginal feature rather than a fundamental divide between the ETC and the rest of London and the South-East.

2. Within the Corridor there are significant variations in employment structure that reflect the historical background, position and other characteristics of individual localities. Hence, there appears to be no such thing as a 'typical' ETC structure.

3. There is some evidence that during the period 1984–1989 parts of the Corridor were 'catching up' with the rest of the South-East in terms of rates of service industry employment growth, but (particularly in the case of business services) these growth rates were generally (with the possible exception of Medway) applied to small initial bases so that the resulting expansion of jobs was limited.

4. There is no general evidence that investment has avoided the Corridor area. Growth rates are highly variable, but the faster growing areas show rates well above those of the context regions. The performance of the ETC in London, however, is relatively weak and job growth is mainly the result of expansion in the City of London and Docklands. The removal of three Docklands wards from the analysis produces a negative figure for the ETC in London.

4.4. The ETC and South-East England: an economic comparison

A report on the above work to the SERPLAN Labour Market Group (Church and Frost, 1991) generated considerable discussion regarding the London section of the ETC compared to the rest of the Corridor. The performance of this part of the ETC had been shown to be poor and it would undoubtedly be included in any policy initiative area that might be defined by central government. After discussion with the SERPLAN group, therefore, it was decided to concentrate comparative analysis on the ten ETC sub-areas outside London.

This research, more explicitly comparative in character, examined change over the period from 1984 to 1989 in two ways:

1. By conducting a 'shift–share' analysis of employment change within the Corridor with the controlling employment totals represented by regional figures for the South-East outside Greater London. Using this approach, positive 'shifts' in employment indicate growth at rates above the average for the rest of the South-East, while negative 'shifts' indicate the reverse.
2. By isolating the differential shift components for all Department of Employment Job Centre Office Areas outside the Corridor, but within the rest of the South-East (chosen because of their similar size to the ETC sub-areas), in order to consider the range of differential shifts seen in the Corridor in comparison to the broader distribution generated by the rest of the region.

The rationale for these approaches was that the differential shifts go some way towards providing a control for the widely different industrial structures found both within the Corridor and within the towns and cities of the region as a whole. They represent a measure of performance for the industries of each area using, in this case, regional changes for each industry as the basis for comparison. The industrial breakdown used was the division level of the Standard Industrial Classification. The analysis did not include the self-employed and was only for employees in employment.

The results of applying a shift–share approach to employment change in the Corridor can be seen in Table 7. They show that, in general, the Corridor started the change period with a mildly unfavourable industrial structure in comparison with the rest of the region. Nine out of the ten subdivisions show negative structural components, the only exception being the clearly positive expectation of growth seen in the case of Southend. These figures do not conflict with the general image of the Corridor as one of the less attractive and less fashionable segments of the South-East region.

On the other hand, in many parts of the Corridor it is clear that the employment performance of economic activity throughout the 1980s diverged significantly from initial expectations based on industrial structures. The two growth centres of the Medway Towns and Basildon are clearly supported by the comparatively strong performance of their industries rather than by favourable initial structures. This is particularly noticeable in the case of Basildon, where the town gained more than 5000 jobs in relation to average regional rates of growth for its industries. A similar gain of jobs, albeit on a larger employment base, can be seen in the case of the Medway Towns. Conversely, the performance of Southend diverges in the opposite fashion. An initially favourable structure of activity in comparison with the rest of the region is totally overshadowed by poor employment performance which, by the end of the period, meant that the town had lost nearly 12 000 jobs in relation to regional rates of change.

Table 7 Shift–share results for the subdivisions of the ETC

Area	Total shift (number of jobs)	Proportionality shift (structural component)	Differential shift (performance component)	Total shift (% of total base year employment)	Proportionality shift (% of total base year employment)	Differential shift (% of total base year employment)
Medway estuary	421.0	−697.3	1118.3	3.3	−5.5	8.7
Sittingbourne	−1268.1	−304.6	−963.6	−7.5	−1.8	−5.7
Medway Towns	4100.0	−1212.9	5312.9	6.7	−2.0	8.7
Gravesend	−436.7	−122.7	−314.0	−2.0	−0.6	−1.5
Dartford	−1273.4	−695.0	−578.4	−4.5	−2.4	−2.0
Thurrock	127.1	−195.8	323.0	0.5	−0.8	1.3
Rural Essex	−924.8	−349.8	−575.1	−10.8	−4.1	−6.7
Basildon	3597.3	−1721.7	5319.0	8.9	−4.3	13.2
Canvey Riverside	−901.5	−392.8	−508.8	−7.6	−3.3	−4.3
Southend	−8438.4	3439.8	−11 878.3	−13.0	5.3	−18.3

Analysis based on employees in employment only.

The problem with all standard shift–share results presented in this manner is that they do not show the range of performances in individual areas that lie behind the regional averages and that determine the significance of the negative values seen in the Corridor in a full regional context. The purpose of comparing the differential shifts in the ETC to those for the 143 Job Centre Office Areas in the rest of the South-East was to establish such a context.

While the bulk of the Corridor's results were encompassed by the range of performance seen in the majority of Job Centre Office Areas in the rest of the South-East, the Corridor still has one representative (Basildon) within the 15 areas of strongest relative growth and another (Southend) that falls within the 15 areas with the greatest relative employment loss. Furthermore, the results for all the Job Centre Office Areas in the South-East do not demonstrate the existence within the region of well-developed corridors or coherent areas of strong relative growth. Rather, the strong growth throughout the period of rapid national and regional expansion was, in spatial terms, a somewhat haphazard process, with the particular mix of circumstances in individual localities like Basildon being of prime importance.

The strong employment loss recorded in many other areas with concentrations of manufacturing jobs suggests that the performance of Basildon in the ETC, a clear centre of manufacturing activity, is an impressive accomplishment. A further finding that may be drawn from the perspective of the ETC is that the problems of Southend were not wholly the result of a poor performance in producer services in comparison with the rest of the region, which might suggest some fundamental lack of locational attractiveness.

The principal conclusion of these results is that there is no evidence to suggest that in a period of strong regional growth firms were deterred from locating new investment or expanding existing operations within the Corridor in favour of alternative locations elsewhere in the region. The range of employment performances seen in the subdivisions of the Corridor suggest that, overall, the area just about maintained its share of activities at a time when the rest of the South-East region was among the fastest growing regions in Europe. This is a considerable, positive achievement for an area whose industrial structure has been somewhat unfavourable as a result of riverside locations losing some of their traditional attractions.

4.5. Transport and access to employment in the ETC

A key consideration given the stated disadvantages of the ETC in terms of strategic and local accessibility is the question of the degree to which the Corridor is linked not only to employment opportunities generated locally but also to those found within the much larger London labour market. All parts of the South-East of England are more or less linked to London within the broad city region comprising London and the South-East, but the strength and nature of these links has been poorly explored in recent research.

In order to examine these issues work-travel tables (Special Workplace Statistics, Table C) from the 1981 Census of Population were used to analyse movement patterns in the ETC at ward level. This section of the project examined the proportions of employed residents within the Corridor who travel to work within the London area. This also provided a measure of local dependence. Residents were identified at the level of Census wards, while workplaces in London were divided into two primary area, Inner London and Outer London, using the Census definition. A further distinction was made between the City of London and the City of Westminster. Work-travel flows were classified on a male/female basis to highlight one of the most important discriminators in work-travel behaviour, the difference between men and women. A selection of the results are mapped in Figs 8–11.

Some of the key results to emerge from this analysis were (more detail is given in Church and Frost, 1992):

- A simple concept of short distance female work-travel in contrast to longer distance male commuting is not sustainable within the Corridor, and some of the longer distance journeys arise from the attractions of the City of London as a source of employment for women, particularly in Essex.
- The pattern of male journeys suggests that Outer London is significantly more important as a source of employment for men in the Essex and Kent sections of the Corridor than for women.
- For both men and women the western part of Central London in the borough of Westminster is of far less importance than the City in the east, a result which tends to confirm the strong 'eastern side' orientation of the Corridor.
- The results for residents of the Corridor within inner London demonstrate the degree to which, while there is a high level of dependence on local jobs with little outward commuting, there are only weak links with both the City and Westminster as sources of employment.
- Levels of dependence on the London area as a source of jobs varies substantially over the Essex and Kent segments of the Corridor (e.g. the areas around the Medway estuary show generally lower levels, especially for working women), a result which demonstrates that any consideration of workplace-based economic change within the Corridor will give only a partial measure of the impact of economic development on its residents.

Finally, given that these analyses are based on 1981 data, it can be confidently expected that substantial changes will have taken place during the 1980s in some of the structures discussed above. The attractions of the City may well have strengthened, while the availability of manufacturing employment in Outer London will almost certainly have diminished. Only the full residence and workplace results of the 1991 Census will establish the impact of these changes on the Corridor.

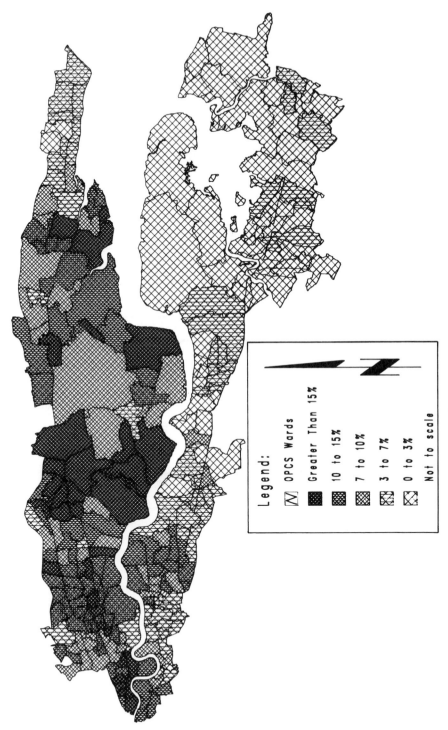

Fig. 8 Travel to work: ETC wards to City, female

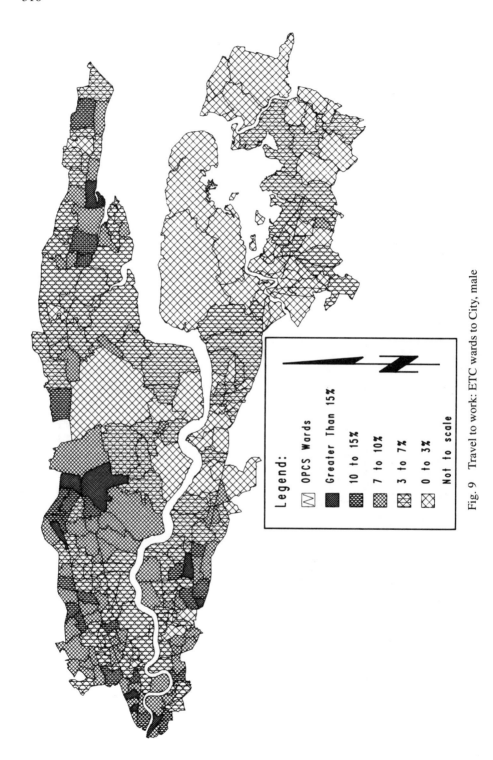

Fig. 9 Travel to work: ETC wards to City, male

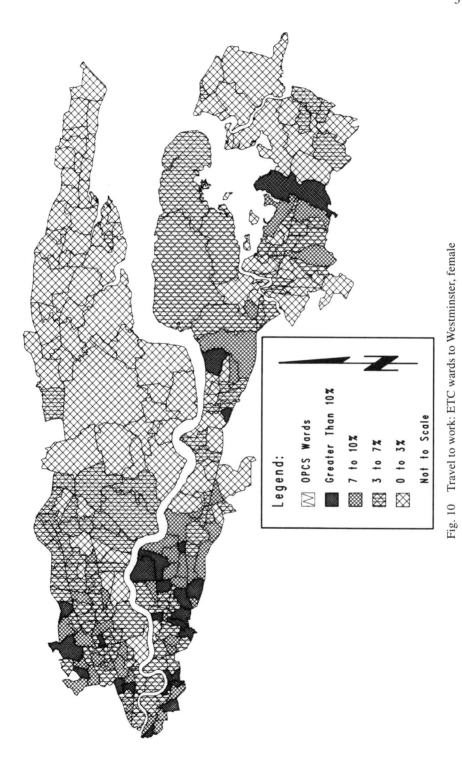

Fig. 10 Travel to work: ETC wards to Westminster, female

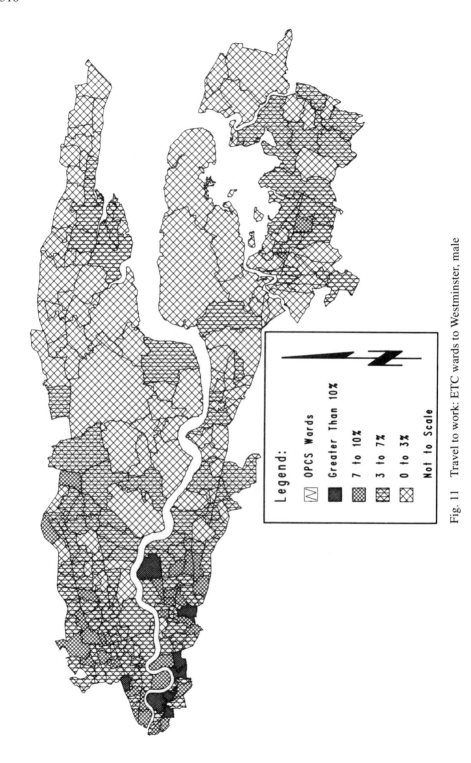

Fig. 11 Travel to work: ETC wards to Westminster, male

5. SOME GENERAL CONCLUSIONS

In making an assessment of the achievements of the ETC GIS project it is useful to understand the background and recent evolution of planning policies for the Corridor itself. Without such an understanding, it is impossible to decide what new data might be obtained and added to the ETC GIS in the future, how they should be added and what types of analyses will be required of them. In short, the information system requirements cannot be understood without reference to the specific tasks they must support. This also entails an understanding of the relationships between (and within) the different public agencies involved. Implicit here is the need for an understanding of the relationship between perceptions of policy at different levels in the planning hierarchy, the role of geographic information in the interpretation of the 'problems' of the Corridor and the nature of the powers available to deal with them. These are obviously major topics in themselves. Moreover, they will continue to evolve as future policies and planning frameworks for the Corridor become clearer. In these circumstances, our final remarks are provisional and prospective, rather than definitive.

5.1. The ETC: a complex mosaic

The idea of an ETC 'subregion' emerged in the mid-1980s as the result of concern among the member authorities of SERPLAN over the unequal pattern of growth between areas to the west and east of London. Growth along the much publicized M4 corridor appeared to have no counterpart along the Thames estuary from Tower Hamlets to Southend and Sheerness. The problems of the ETC were seen as being rooted in an employment structure characterized by older industries, poor accessibility by road and rail, large amounts of derelict and vacant land, and costly constraints on land development. These concerns were reflected, as far as was possible, in the definition of the data requirement for the ETC GIS. The idea of a focused initiative for the ETC received significant new impetus when it was adopted by the secretary of state for the environment and developed in studies by the Department of the Environment and external consultants (Llewellyn Davies *et al.*, 1993; Department of the Environment, 1993) This too, no doubt, will impact upon the future information requirement.

In the course of these developments, it has been necessary to find appropriate data, an appropriate framework for the data and an appropriate analytic stance to bridge the disjunction in scales between the level at which problems are perceived and responsibility for policy implementation. The problem of economic and social disparity, for example, was articulated by an advisory regional planning body and adopted by central government, whereas responsibility for concrete action, i.e. for intervention, lies with the twenty-four local government bodies operating within the Corridor and with such unelected bodies as central government may determine. Intervention in this sense involves capital spending on particular development sites as identified, for example, by SERPLAN and the East Thames Corridor Action Group. A suitable database for such an approach is thus required both to hold data at the site level and, on a broader basis, to permit an investigation into the potential contribution of 'micro-level' changes to shifting the broader scale disparities that prompted the policy.

A central and recurring problem in building the ETC database has therefore been that of tackling the problem of conceptual and administrative fragmentation. At least three dimensions to this problem were experienced in the course of the project:

1. Thematic heterogeneity, in the sense of which data are pertinent to the aims of the GIS.

2. Geographic heterogeneity, which implies difficulty in determining the appropriate scale at which data should be incorporated and analysed.

3. A division between (and within) public authorities with regard to statutory responsibilities and geographic remit.

These broad themes inform our concluding remarks, which are grouped as follows: database definition and data assembly; research on problems and opportunities within the Corridor using the database; analyses of the use made of GIS by local authorities within the Corridor; and the nature of the discourse between the GIS research team and those responsible for devising and implementing policies for economic and land development.

5.2. Database definition and data assembly

An important aim of the project, stated simply, was to examine to what extent a practically useful subregional planning database could be assembled from existing sources, what the problems associated with this objective were and what was the nature of the gaps in information revealed by this process. It is immediately recognized, of course, that the processes that influence the aggregate level of economic activity and employment in the labour market areas (sensu strictu) that impinge upon the Corridor operate at quite a different geographic scale from those that 'shape' housing opportunities and outcomes. These processes are mediated, or even managed, by local government agencies whose geographic domains are defined at yet other scales. While these processes imply that pertinent areas should be described at different scales, or even with differing degrees of 'fuzziness', a GIS framework obliges us to treat all geographical units with uniform (and usually quite unreasonable) precision. There is an obvious point of contact with other projects in the ESRC/NERC geographic information handling programme here which could be usefully pursued.

Compounding the problems of compatibility of map scales are a number of technical issues of finding appropriate links between data sets. GIS functionality provides means of associating sets of data via a common 'key', or via more distant data sets which contain attributes present in each of those to be joined. Where unambiguous numeric codes are present, such linkage may be exact (even if of no substantive meaning). Our work, however, highlighted the importance and difficulty of linking textual data together, particularly addresses. Such problems proved most intense where property professionals eschew names such as 'Stockley Road Tip' in favour of more attractive nomenclature!

These difficulties notwithstanding, the project succeeded in drawing together a diverse range of data into a common, flexible and analytically meaningful framework. The emphasis of the work was thus on 'harvesting' and 'recycling' of different data sets. The former relates to the exploitation of available digital data for purposes which 'mesh' closely with those intended by the data provider and where there is at least an implicit 'guarantee' of appropriateness for the task (e.g. use of DoE/OPCS boundaries as an indicator of urban extent). The latter relates to the use of digital data that were originally produced for quite different purposes, such as the use of address lists (from, say, PAF on CD) or business directories to proxy urban areas or economic activity.

The disadvantage of an approach where the content of the knowledge base is driven, at least in part, by availability, lies in unbalanced coverage of issues of interest. A long tradition of state intervention in a particular domain, and hence a large bureaucracy associated with gathering and using information, tends to imply relatively rich sources of data. Elsewhere, despite significant levels of current interest or topicality, sources of data are more limited. This was most marked in the present study by the paucity of environmental data, especially geo-coded data on habitats and wildlife. Where data do purport to bear upon such issues, they may rely on extrapolation from small samples and hence raise a further set of data issues. Despite these significant problems, a key achievement of the project was the production, at modest cost, of a database with real value at the subregional level.

5.3. Problems and opportunities in the Corridor

A good example of the value for money derived from the database was the series of substantive research studies carried out by experts in specific fields. This process further enhanced the discourse between the research team and the ETC planners.

Nowhere was the significance of the issue of analytic scale greater than in the examination of the economic and employment base of the Corridor. Our work indicates that there is no general evidence of a uniformly unfavourable industrial structure or that investment has avoided the ETC. Instead, it highlights intense variations in growth rates within the area. While the performance of the ETC in London, for example, is relatively weak, with job growth confined to the City and Docklands, the faster growing localities show growth rates well above those of the comparator regions. There is, moreover, no such thing as a 'typical ETC employment structure': the pertinent scale of analysis is the individual locality. This implies a need to focus future work on investigating 'the local dynamics of change to attempt to identify key features of poorly performing areas'. In particular, we would seek to explore the role and scope of local intervention in land and property development in influencing local economic outcomes.

An important component of such an examination would be the development of the analytic perspective on land use change introduced in this study. The investigations of land use change undertaken in the study (though not reported here) are more tentative than the employment studies, reflecting the relative novelty of the data set employed (SERRL, 1993). Nevertheless, the analyses of the Department of the Environment's Land Use Change Statistics (LUCS) on which this part of the work is based appear to allow an approach to land use studies that embodies both a clearer integration of information about economic and demographic underpinnings and a more flexible approach to the building of geographic units (Bibby and Coppin, 1993).

Studies of land use change in the ETC highlight the marked variation in rates of land development in different localities relative to population. They also show marked variations in the sources of land for urban development. Generally, development in ETC localities has been more conservative with respect to rural land than is the case in England as a whole. Nevertheless, the analyses point to relatively extensive development in Thurrock and Basildon but with intense urban land recycling not only in London but also in Gravesend, for example. In its most extreme form, this process of urban land recycling has involved the redevelopment of large areas of industrial and commercial land for housing, reflecting and reinforcing fundamental change in the economic role of certain urban areas. There is clearly scope for further research on the more precise specification of the relationship between such land use changes and their socio-economic determinants. Moreover, such work should be augmented by a further programme of work carrying forward the investigation of the particular nature of the LUCS data set. This research might play an important role in extending the range of users of this potentially important source and enhance their value to planning practitioners in local government.

5.4. GIS and local government

The achievements of the ETC project have been realized in an environment free from the bureaucratic constraints and pressure for operational decisions that characterize local government. Moreover, the use of 'recycled' data using GIS has allowed progress to be made without recourse to data through administrative procedures. The results presented here thus demonstrate what can be achieved with 'off the shelf' data, an appropriate technological resource and the freedom of disinterested inquiry. To make this point begs the question of the potential of GIS as an aid to policy making within the ETC by the public agencies themselves.

Local government officers clearly show a high level of awareness of GIS, though the work reported in this study shows that all too frequently the potential is not easily realized. Much of the power of GIS lies in the possibility of data integration, implying a need for data sharing. Such a notion, however, may be antithetical to the objectives of key officers in bureaucracies where departmentalism is rife and only intensified in the more stringent financial regime of the late 1980s. In such circumstances, the results of our work, while salutary, come as little surprise. The emphasis upon the acquisition of hardware and software identified by our research reflects an imperative to spend. What is lacking is a will to share data, and real progress in the establishment of data standards, and there is frequently uncertainty over how best to render data into a usable form. Overlain on all this there is frequently a tension between aspirations for GIS as a means of informing policy and a reality which ties it firmly and often exclusively to operational management concerns.

5.5. A GIS discourse

An important objective of the project was to engage regional and local government planners directly in the process of establishing and using the ETC GIS. A precondition of such a discourse is shared understanding, on the one hand, of the analytical and technical possibilities of GIS espoused by researchers and, on the other, of the particular pressures and constraints to which local government officers are subject. The ETC GIS project, set up as a 'near joint' endeavour between academia and a regional planning body, went a significant way towards building a bridge of understanding between the two groups. As such, the project could be seen as a prototype upon which to develop a formally structured and monitored programme of action research in GIS applications. Such a programme would contribute, among other things, to a better understanding of the way GIS might contribute in institutional situations characterized by particular, subtle relationships between information and action.

5.6. Postscript

It will be evident from the wide ranging nature of this report that the ETC GIS project is not a straightforward piece of research to summarize. It has resulted in a number of clearly defined 'products', including the ETC database itself, a 'planner friendly' interface to the GIS, several reports and research papers, participation by research staff in conferences and seminars and the presentation of numerous demonstrations of a subregional GIS at work. In substantive terms, the project has also demonstrated the still relatively unprepared state of local government to take on board GIS technology and the nature of the data problems inherent in establishing a GIS at the subregional scale. Research carried out *on* the database has demonstrated clearly the heterogeneous nature of the Corridor and hence the need for an analytical perspective that is both synoptic and particular.

Many of the 'outputs' of the project are less tangible and less easy to assess. A key characteristic has been the collaboration with SERPLAN and, through that body, direct and close links with local government officers and the 'real' policy setting in which a GIS might operate. These links, developed as they have been in an evolving situation, have given us a deeper understanding of the way in which planning institutions in a subregional setting work, hence allowing us to devote further attention to the research 'tactics' required to understand the role of GIS in such situations. Moreover, a lasting product of the research is the ETC database itself. So far as resources permit, this will continue to be developed jointly by SERRL and SERPLAN and will be used to take substantive research on the Corridor further. As policies for the Corridor evolve, we hope to use GIS technology to participate in and inform the debate that is likely to follow.

ACKNOWLEDGEMENTS

The ETC GIS project was a cooperative and collaborative effort. At SERRL Sian John, Simon Lewis and Peter Bujwid dedicated significant periods of time to the project, and Graham Sadler, Duan Ming and Ann Low made other important contributions. In the later stages, the advice of Peter Bibby, SERRL senior research fellow, was invaluable. John Barber, a regional planner, acted as the key day-to-day liaison officer for the project at SERPLAN. His commitment, and the contributions of other SERPLAN staff, represented a considerable additional manpower contribution to the project. The ETC GIS research team also wishes to acknowledge the support of ESRC and NERC and the encouragement that the geographic information handling programme has given to collaboration with the public and private sector in research both on and with geographic information. At all times, Professor Paul Mather, the programme coordinator, gave essential support and encouragement. Finally, the usual disclaimer applies to all the work reported in this document: none of the public or private bodies supporting or involved in the research is responsible for the opinions expressed.

REFERENCES

Aronoff, S. (1989) *Geographic Information Systems, A Management Perspective*. WDL, Ottawa.
Bibby, P. (1992) Postcodes in town planning/Postcodes in local economic analysis. In Raper, J., Rhind, D.W. and Shepherd, J. (eds.), *Postcodes – The New Geography*. Longman, Harlow.
Bibby, P. and Coppin, P. (1993) *Analyses of Land Use Change Statistics*. SERRL Occasional Monographs no. 1, Birkbeck College, London.
Bibby, P. and Shepherd, J. (1992) GIS: strategic choice or laundry lists? *Proc. Ass. Geogr. Inf. Conf. (AGI'92)*, Birmingham, November. Association for Geographic Information, London, 2.31.1–2.31.7.
Blakemore, M. (1990) The UK networked GIS (NOMIS) for monitoring and strategic planning, *Proc. 1st European GIS Conf. (EGIS'90)*, Amsterdam, 10–13 April, 77–86.
Bromley, R. and Coulson, M. (1989) The value of corporate GIS to local authorities. *Mapping Awareness*, 3(5), 32–35.
Campbell, H. and Masser, I. (1991) The impact of GIS on local government in Great Britain. *Proc. Ass. Geogr. Inf. Conf. (AGI'91)*, Birmingham, November. Association for Geographic Information, London, 2.5.1–2.5.6.

Charlton, M. and Ellis, S. (1991) GIS in Planning. *Plann. Outlook*, **34**(1), 20–26.
Church, A. and Frost, M. (1991) *Employment change and structure in the East Thames Corridor*. SERPLAN Regional Planning Circular no. 2150, London.
Church, A. and Frost, M. (1992) *Employment and Accessibility in the East Thames Corridor*. SERRL Working Report no. 35, Birkbeck College, London.
Clark, A.M. and Thomas, F.G. (1990) The geography of the 1991 Census, *Popul. Trends*, **60**, 9–15.
Cowling, T.M. and Steeley, G.C. (1973) *Sub-Regional Planning Studies: An Evaluation*. Pergamon Press, Oxford.
Department of the Environment (1988) *Regional Planning Guidance for the South East*, PPG9. HMSO, London.
Department of the Environment (1993) *Regional Planning Guidance for the South East*, Consultation Draft. HMSO, London.
Deloitte, Haskins and Sells (1989) *Action Plan for the East Thames Corridor*. SERPLAN, London.
ESRC/NERC (1989) *Collaborative Programme on Geographic Information Handling*. ESRC, Swindon, January.
Huxhold, W. (1991) *An Introduction to Urban Geographic Information Systems*. Oxford University Press, Oxford.
John, S. and Lewis, S. (1992) Compiling a GIS database jigsaw: the question of data quality. *Proc. 3rd European GIS Conf. (EGIS'92)*, Munich, 23–26 March, 521–531.
John, S. and Lopez, X. (1992) Integrating data derived from within the British local government institutional framework. *Proc. Ass. Geogr. Inf. Conf. (AGI'92)*, Birmingham, November. Association for Geographic Information, London, 1.18.1–1.18.8.
Lewis, S. (1991) Hypermedia geographical information systems. *Proc. 2nd European GIS Conf. (EGIS'91)*, Brussels, 2–5 April, 637–645.
Lewis, S. and Rhind, D.W. (1991) Multimedia geographical information systems. *Mapping Awareness*, **5**(6), 43–49.
Llewellyn Davies, Roger Tym and Partners, TechnEcon and Environmental Resources Ltd (1993) *The East Thames Corridor. A Study of Development Capacity and Potential*. HMSO, London.
Lopez, X. and John, S. (1993) Data exchange and integration in UK local government. *Mapping Awareness*, **7**(5), 37–40.
Raper, J. (1993) Multi-media and GIS, an interactive future? *Mapping Awareness*, **7**(1), 8–10.
Raper, J., Rhind, D.W. and Shepherd, J. (1992) *Postcodes – the New Geography*. Longman, Harlow.
SERPLAN (1987) *Development Potential in the Eastern Thames Corridor*. SERPLAN Regional Planning Circular no. 700, London.
SERPLAN (1988) *Increasing Activity in the Eastern Thames Corridor*. SERPLAN Regional Planning Circular no. 1000, London.
SERPLAN (1989) *Into the Next Century: Review of the Regional Strategy*. SERPLAN Regional Planning Circular no. 1500, London.
SERPLAN (1990a) *A New Strategy for the South East*. SERPLAN Regional Planning Circular no. 1789, London.
SERPLAN (1990b) *Action in the East Thames Corridor*. SERPLAN Regional Planning Circular no. 1800, London.
SERPLAN (1991) Communication to Central Government.
SERRL (1992) *A Land and Economic Development Information System for the East Thames Corridor*. Final Report to ESRC, Grant no. A505255014.
SERRL (1993) *A Land and Economic Development Information System for the East Thames Corridor: Summary Report*. SERRL Occasional Monograph no. 2, Birkbeck College, London.
Shepherd, J. and Brewer, A. (1992) Postcodes as a geographic database, added value and added analysis. *Proc. Ass. Geogr. Inf. Conf. (AGI'92)*, Birmingham, November. Association for Geographic Information, London, 1.2.1–1.2.4.
Townshend, J.R.G. (1991) Environmental databases and GIS. In Maguire, D.J., Goodchild, M.F. and Rhind, D.W. (eds.), *Geographical Information Systems: Principles and Applications*. Longman, Harlow, 201–216.

Chapter Twenty-Two

Intelligent, interactive and analysis-based GIS: principles and applications

M. CLARKE
Gmap Ltd, Leeds

AND

J. CHESWORTH, J. HARMER, A. McDONALD, Y. L. SUI AND A. WILSON
School of Geography, University of Leeds

1. INTRODUCTION

1.1. The objectives of the research project

In our original research proposal a number of objectives for a programme of research were identified. These can be summarized as follows:

- The development of a set of design principles for the implementation of intelligent, interactive and model-based geographical information systems (GIS).
- Illustrations of these design principles through three main applications, in the water, energy and health care sectors.

In this chapter the context in which the research work was undertaken is described; this is followed by an outline of the main design principles and design issues identified. The subsequent sections provide an overview of each of the three applications that were developed during the project. The final section summarizes the main outstanding research issues in the development and implementation of intelligent spatial decision support systems.

1.2. The context of the research project

This research project commenced in late 1989 as part of the ESRC/NERC Joint Programme on Geographical Information Handling. At that time the second phase of the ESRC Regional Research Laboratory (RRL) initiative was well under way. During the subsequent three years the GIS industry has expanded and diversified around the world and can now be seen to be an academic sub-discipline as well as a major business sector. There tends to be a consensus that GIS has been a major success story, in both an academic research context as well as in applied business practice. We disagree with this view for a number of reasons, including the following:

- Most of the advances in GIS have been related to technology rather than to understanding the ways in which geographical analysis can help solve problems, especially in business and commerce.
- There has been a strong bias within GIS towards digital mapping. While this area is of interest to many, it does not offer a great deal more challenging intellectual problems than, for example, CAD/CAM, where vector GIS may arguably have originated. We believe that a much more challenging set of problems is concerned with the design and use of systems that address strategic planning issues.
- An overt reliance upon proprietary GIS software in the academic community has resulted in a concern with the capabilities and application of this software as opposed to more broad-ranging issues. This does not mean that proprietary GIS have provided the researcher with a set of questions rather than the reverse. It has limited the nature of the questions and restricted the way in which the questions could be posed. In the UK at least, the supply of proprietary software to the RRLs has exaggerated a trend encouraged by education software.
- GIS applications are largely restricted to utility companies and local government, where they are mainly used in AM/FM-type applications as opposed to more substantive areas such as strategic planning. In most large businesses, GIS currently contributes virtually nothing to the central business functions, including strategic planning (Clarke, 1993). If GIS applications exist they do so as a relatively marginal, technical adjunct to management, usually at the operational level performed by junior staff.

2. CONSIDERATIONS OF PRINCIPLES

2.1. Introduction

One purpose of this project is to demonstrate that geographical analysis, suitably embellished with a range of information systems technology, can contribute effectively to business planning. We have specifically aimed to move a GIS-type capability from the top left-hand corner of Fig. 1 towards the bottom right-hand corner. It is interesting to note that during the course of the project we have been joined in our views by a number of observers who have also expressed concern about the current focus of much current GIS research.

From Fig. 1 it is apparent that the majority of applications in the technical/operational element of the matrix are characterized by problems that are relatively easy to define, the use of proprietary GIS software, and a digital mapping/facilities management focus which is of marginal concern to the majority of business at a strategic level.

As a consequence the major research themes that have dominated the GIS community have tended to be largely technical, such as:

- Data, including data capture, data structures, data transfer formats and copyright issues.
- Software – databases, object-orientated languages, raster versus vector representations.
- Choice and attributes of hardware platforms.

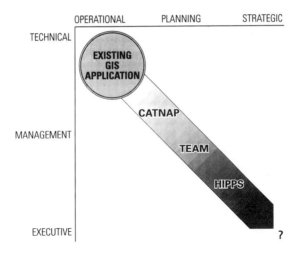

Fig. 1   Extension of GIS-type capability, to contribute to business planning

During this research project the issues surrounding the development of executive/managerial applications of GIS in a more strategic role have been addressed. Here the application issues surround a different set of characteristics.

At a senior management level, problems are often poorly defined. In part, this occurs because the processes of concern relate to many different business functions – e.g. marketing, planning, finance, property, and so on. The resultant decision support systems rely more on a customized than a proprietary approach. This is not to suggest that the problems are unique to each business for we recognize generic sets of questions, issues and decision needs. However, since these are potentially a central core of the business they often operate with different data sets, not normally evaluated in an integrated manner, and require a decision output more flexible than that easily available in proprietary systems.

Given the nature of the problems outlined above, a different set of research themes emerge, which have been given much less attention than the technical issues discussed earlier:

- Relevance: how can the system be designed to address the real strategic problems faced by an organization?
- Intelligence: how can a system be designed to enable a user to generate useful insights from large databases and modelling?
- Culture: how does a new technology become adopted within an organization – especially when it might challenge the existing status quo and threaten conventional methods of working?

Some of the design principles and issues that have been distilled from experience gained in developing decision support systems are discussed below.

2.2. Design principles and design issues

In this section the design principles developed for the construction of business-orientated spatial decision support systems are reviewed. Adoption of these principles will increase the probability of successful implementation by an organization, but will by no means guarantee it. Following on from these principles are a number of issues that also have to be resolved in the implementation phase. A more detailed discussion of the themes described in this section is contained in Clarke, Chesworth and Harrison (1993).

The following principles have been identified:

1. Built-in usability, through business process focus and comprehension. Understanding of the way in which a system will contribute to business planning is essential.

2. Objectively defined development. Here it is recognized that, because of the existing technical basis of GIS interests, the prime consideration has not been the needs of decision takers or policy makers. In our applications we have sought to develop systems that provide direct support to decision making. This will usually require a considerable amount of time in the predesign stage, conducting unstructured interviews with managers and programmers.

3. Transparency in operation. If strategic decisions that might affect company viability are to be aided by GIS systems then decision makers must have confidence in the mode of operation of the system. Assumptions must be made explicit and data quality measures must be provided.

4. Parameter audit and analysis adaptation, which implies that the results of an analysis are completely dependent on the selection of parameters and the ability of the analysis to run any of a wide selection of scenarios. Failure to be capable of running a chosen selection will reduce the strategic planner's confidence in the system, leading to a belief that the full range of possible outcomes has not been explored.

5. High-level system endorsement. To succeed, the design of an intelligent GIS must be agreed and endorsed at a high level within the client organization. Sponsorship by a senior individual is almost a prerequisite for success.

Following on from these principles is a set of design issues relating to system implementation. At a fundamental level these issues involve definition of the system container, system entitation and system process. The system container is the geographical extent of the system being studied. This may be a river basin, a market area, a region or a whole country. Markets and river basins may be larger or smaller than countries, and subsets of operational areas are important issues. System entitation requires the identification of those elements within the system that are to be represented. The system processes element requires an understanding of and representation of the main relationships between the system elements. This latter element moves away from the traditional concerns of GIS, but we note its growing importance in the research reported in this volume.

It can be noted that most GIS applications address issues of container and entitation (and this represents the limit of digital mapping). Of particular interest is how process modelling is represented within a GIS framework. It is clear that this area has been largely ignored to date in GIS research. However, it is gradually emerging as an important theme as a more critical view of GIS develops (Openshaw, 1992).

In the three applications described in Section 3 we have attempted to address these design principles and issues explicitly. We also recognize that the software design dimension to system implementation is an important one, and consideration of graphical user interfaces and related issues are inevitable.

Intelligent, Interactive and Analysis-based GIS 329

3. APPLICATIONS

3.1. Introduction

One aim of the research project was to illustrate how the design principles for spatial decision support systems could be translated into practical applications. This has been achieved through three distinct prototype spatial design support systems in health care planning, water resource planning and energy planning. It was suggested in our grant application that we would seek external collateral to support the development of these prototypes. This was achieved, and the support of the Nuffield Provincial Hospitals Trust and Yorkshire Water PLC has enabled the applications in health care and water resource planning to be developed to the stage of working systems.

3.2. Health Information for Purchaser Planning System

The reform of the National Health Service following the White Paper 'Working for Patients' has led to an internal market for health care in the UK. Purchasers (District Health Authorities, fundholding general practices) determine the health care requirements of their resident populations and negotiate contracts with providers (Hospital Trusts, Direct Managed Units) for the supply of services. Consequently, purchasers have a number of key requirements:

1. Understanding the current and future health care needs of their residents by person type, condition and small area.
2. Understanding where this need is currently being met and how need and utilization vary due to external factors.
3. An ability to predict the impacts of changes at both the purchaser and provider end, and to evaluate this on both an efficiency and effectiveness basis.

The main processes that operate in a purchaser–provider environment are shown in Fig. 2.

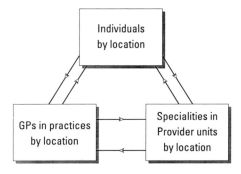

Fig. 2 Key relationships in the Health Information for Purchaser Planning System (HIPPS)

The objective was to build a demonstration system that would help a purchaser to address these three requirements by focusing on the main processes identified in Fig. 2. Bradford Health Authority and Bradford FHSA cooperated to provide information and critical design input and evaluation of the prototype system.

The Health Information for Purchaser Planning System (HIPPS) has two main components, an information system and a planning system. The information system contains detailed information on characteristics of residents by enumeration district, the population registered with each GP (and GP practice) in Bradford, the referrals by each GP to a provider unit, and the activity of each provider by speciality, condition, and outcome, including cost. The planning system allows the user to change any component of the supply–demand equation.

HIPPS provides a broad range of capabilities in an extremely user friendly but flexible system. In terms of design principles, an attempt was made to understand the key business processes and business objectives that the system would address. An initial user requirements document was generated following discussions with the potential users, in this case the local providers. The resultant system that was developed has been installed in a wide variety of locations within Bradford Health Authority, and critical feedback has been obtained (Clarke, Chesworth and Harrison, 1993). The system was developed in Windows 3.1 with a SQL database.

While we have attempted in this chapter to provide a more philosophical discussion of the direction and rationale for the development of new generations of GIS, it may be appropriate here to summarize the value of the HIPPS system for international readers unfamiliar with the UK system. UK health services are in flux. In essence, there is now a group of purchasers who have government money to provide healthcare. They are involved in diagnosis and the purchase of further specialized diagnostic information and treatment. These services are furnished by providers. Both need spatial information on performance as an aid to the selection of providers and to the monitoring of the wise use of funds by the purchasers. Providers also need to be able to judge the type of service that may be in demand in particular areas in the future and to evaluate whether they or nearby competitors are best placed to provide the service. HIPPS provides this information at a variety of scales, mapping, where appropriate, from both provider and purchaser viewpoints in relation to current performance and future scenarios.

3.3. TEAM: Energy modelling

In many respects, energy is the driving force of the economy and both a direct cause of and surrogate for environmental degradation. The environmental influence of energy is manifested both before use, in the form of opencast land despoliation, and after use, as enhancement of atmospheric acid levels. TEAM is a toolkit of software capable of modelling the economics of energy in a spatial context. TEAM has been developed expressly for the use of energy modellers and analysts whose primary interest lies in the energy system, not in the abstraction of packages or programming languages.

The TEAM package models energy demand and supply at both the national and the regional levels. This is the appropriate scale for the needs of the Regional Electricity Companies (RECs) and for national energy planning. In addition the system integrates socio-economic and energy data, and combines inputs from both the natural and the social sciences with integrated mapping and graphical capabilities tailored to meet the needs of a

Intelligent, Interactive and Analysis-based GIS 331

wide range of users. The package has a strong scenario modelling capability able to simulate and provide answers to many types of 'what if' questions, such as:

- The energy implications of different patterns of economic activity. This would yield information concerning demand for fuel types and thus link to the exploitation and depletion of energy resources. The Cambridge Economic model, among many, has a regional modelling component and this component is currently being developed to give subregional predictive economic activity.
- The environmental implications of these energy requirements in terms of heat, and 'greenhouse' gases and other pollutant outputs.
- The resource implications, specifically scarcity, of the predicted patterns of energy usage.
- Conversely, the economic implications of energy, resource or environmental constraints operating nationally or regionally.

The key features of the TEAM system are that it is a self-contained PC- or workstation-based analytical tool. It is designed to be user friendly, multifunctional and fully integrated. For the purposes of this chapter, the features can be conveniently described as a full relational database capability allowing the storage, retrieval and display of energy and energy-related data. This includes extensive data validation, and data import and export facilities; for example, to and from dBase formats; a wide range of statistical tests and analysis; a versatile modelling capability, incorporating a wide variety of energy modelling approaches such as input–output, economic and system dynamic models capable of different degrees of modification including in-built energy models, user-adapted and customized energy models and user-defined energy models; and graphical and mapping displays of all modelling and changes in spatial modelling output. In addition the package provides full help and support at all levels, in-built system management and configuration, and in-built easy to use multi-window text editor. The system is highly portable and can be readily customized to provide analysis of regional or country-wide energy profiles tailored to meet the exact needs of specific users.

The model has been tested in applications to two widely different data sets. The first is UK energy data for the twenty-year period from 1970 to 1989. The second data set is derived from the International Institute for Applied Systems Analysis (IIASA) Module d'Evolution de la Demand d'Energie (MEDEE-2) programme for the Pacific and Far East, which is the energy demand forecasting model used by IIASA in the Energy System Programme (ESP) for global systems.

3.4. CATNAP: Water quality modelling

The quality of the environment is an issue of great concern to both the public and decision makers. The National Rivers Authority (NRA) has, as a significant element of its mandate, a responsibility for the improvement and control of the water environment. To control river water quality, the NRA uses a system of consents through which the discharge of pollutants into, and the abstraction of water from, the river system are controlled. These two strategies are interconnected because any reduction in the volume of water will reduce the amount available for dilution of pollutants, and thus the consent limit on a discharge must be reduced if river water quality is not to decline.

National discharge consent standards do not exist. The strategy evolving within the NRA is based on the recognition that the consents to individual companies will depend on the number and size of pollutant inflows to a particular river, the targets set for water quality on that river and the ability of the particular river to absorb (in a non-technical sense) the pollutants. All of the UK's flowing surface waters are divided into defined reaches. There are some 40 000 reaches, and target levels of water quality, for all significant parameters, have been defined for each of these. In addition, and of great significance, is that a time of attainment of target has been set for every reach. Therefore, the NRA has set itself clear attainment targets and its performance in attaining the targets can be monitored.

While there are a number of catchment water quality models, few focus on the detailed prediction of in-channel water quality for a series of reaches. Only two models in use by the UK water industry attempt this. These are Simcat (simulation of catchments) and TOMCAT (temporal/overall model of catchments). Each has advantages and drawbacks. Simcat is relatively simple in scientific terms, therefore any results are open to debate and differing interpretation. Its simplicity does, however, make it relatively easy to apply. TOMCAT, on the other hand, is much more complex, attempting to model the river system at the leading edge of modelling expertise available to the industry. It is, unfortunately, extremely user- unfriendly and the number of capable operators in a water plc or NRA region will typically be one or two. The much more serious result is, however, that the results must, to some extent, be taken on trust since very few people understand the coding, and changes in 'runs' are made by the direct alteration of the data files and, sometimes, code, neither of which is annotated.

In this application we sought to provide the scientific integrity of TOMCAT but with the user utility and transparency of simpler models. This being a spatial problem, mapping facilities would be provided to display the modelled results and, more unusually, as a means to identify and automatically select the input data files.

CATNAP (catchment numerical analysis and presentation) is a GIS inclusive water quality and management system developed from a customization of the existing TEAM package outlined earlier. Existing packages had a number of basic drawbacks in addition to the limited user interface. These include the lack of portability because of mounting on mini-computers and dependence on external routines, host system data editing, poor reporting facilities, manual input data organization, and no mapping or graphics facilities.

To alleviate these problems new independent routines were developed to replace external calculation. The original code was reviewed and rewritten to modern standards. Errors in the original code were detected and corrected. Simplifications of the science were evaluated and algorithms were developed to simulate more accurately the response of the river to pollutant inputs. The code was developed to allow a variety of modelling strategies to be applied. For example, estimates of water velocity can be made in a variety of ways. Since this controls the time of degradation of pollutants in a reach it is important that users can explore the effects of different assumptions on water velocity. To be more specific for a moment, the original reach time was determined by assuming a constant cross-section through which a variable discharge was routed, enabling velocity to be estimated. This is converted into reach time by multiplying by reach length. CATNAP, at the suggestion of both the water industry and the water regulator, allows travel times to be calculated through Manning's equations, as $1/Q$ or as $\log Q$. This was one of 26 specific suggestions made through a review process of the penultimate version of the model-based GIS. It reflects perhaps the degree of tailoring of software needed in order to promote the technology to the policy arena.

All code was fully annotated. The result is a package with the following features: a powerful, user friendly data entry, editing, storage, updating and display module; a flexible water quality and hydrological modelling capability; interfaces to permit preprogrammed, user-adapted or user-defined modes of model implementation; and fast, streamlined program execution with reporting of the stages of execution to provide confidence in the model operation to non-technical policy makers. The speed of execution allows a realistic 'what if' capability; a powerful visual presentation module integrates digitized river catchment networks with model results and provides graphical and summary output. (Legislation demands a statistical basis to the evaluation for consents and this must be provided in addition to the basic GIS display.)

Again, to be more specific about the context and uses of CATNAP, the river water quality in two Yorkshire catchments in 1989, at the start of this project, and the river water quality objectives are given in Plate XVI. These catchments have serious environmental problems. Below Leeds, for example, the dry weather flow of the river is only 30% natural, with 70% of flow derived from effluent. The broad water quality objectives for the river as defined by the NRA are shown in Table 1. That these objectives cannot be attained at present is attributable to many causes, the principal of which are shown in Table 2. Many different strategies and mixes of strategies may be employed to remedy the situation. For example, it is possible to reduce effluent volume, improve effluent quality, alter effluent input sites, create oxidation lakes, introduce oxygen curtains, develop inter-basin transfers for dilution, and so on. CATNAP replicates and demonstrates the current fluctuations in water quality in the basin and simulates the basin response to various remedial strategies as expressed in new consent limits and river conditions.

Table 1 Water quality classification – rivers and canals

River and class		1989		Objective	
		(km)	(%)	(km)	(%)
Aire	1a	110.5	23.3	137.3	28.9
	1b	146.4	30.8	165.0	34.7
	2	72.4	15.2	169.3	35.6
	3	129.2	27.2	3.4	0.7
	4	16.5	3.5		
Calder	1a	102.4	24.1	118.3	27.8
	1b	99.0	23.2	129.0	30.3
	2	106.3	25.0	161.0	37.8
	3	98.3	23.1	17.3	4.1
	4	19.6	4.6		
Canals	1a				
	1b	66.0	33.6	121.7	61.9
	2	77.3	39.3	73.7	37.4
	3	53.4	27.1	1.3	0.7
Total	1a	212.9	19.4	255.6	23.3
	1b	311.4	28.4	415.7	37.9
	2	256.0	23.3	404.0	36.8
	3	280.9	25.6	22.0	2.0
	4	36.1	3.3	22.0	2.0
Grand total		1097.3	100	1097.3	100

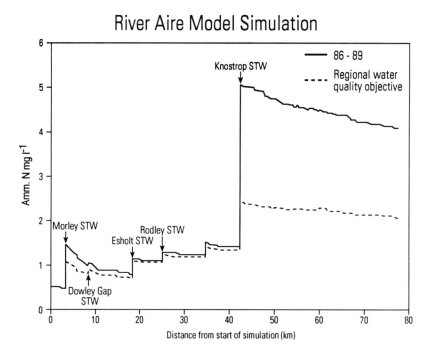
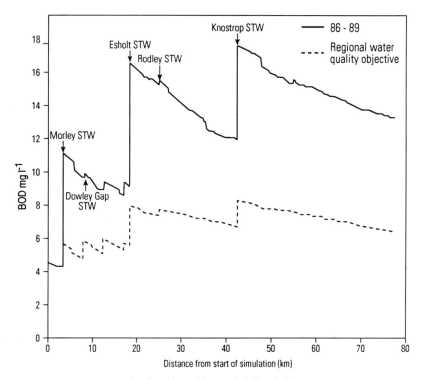

Fig. 3 River Aire model simulation

Table 2 Sources of pollution of the Aire and Calder rivers

Pollution Source	Aire	Calder
Sewage treatment works (STW)	48.3	45.8
Storm sewage overflows (SSO)	21.0	9.7
STW and SSO	11.8	10.0
Urban runoff	5.1	4.2
Industrial discharges to river	1.8	0.6
Landfill leachates	2.3	9.4
Minewaters	1.6	12.1
Agriculture	2.7	1.8
Other	5.4	6.4

The current consents are shown in Table 3, and an example of the existing river situation on the Aire averaged over the period 1986–1989 compared with the situation that will prevail under these consent conditions is given in Fig. 3. CATNAP can operate for different basin sub-catchments and the appropriate data sets can be drawn from a map interface. Output can be map, graph or statistical overview, as appropriate. In any event, the output has always to be a statistic, because CATNAP is based on a Monte-Carlo simulation since the key determinant for legislative purposes is the 95th percentile.

In summary, CATNAP provides a generic practical computer-aided analytical modelling framework to aid in the formulation and application of strategic environmental planning and decision making with the aid of advanced GIS techniques.

At all stages the development was subjected to robust and independent tests by both the Yorkshire region of the NRA and Yorkshire Water PLC.

4. CONCLUSIONS AND RECOMMENDATIONS

In this chapter an attempt has been made to provide a summary of the findings of the research project. Our general conclusion is that, despite views to the contrary, the applied business GIS area is at a very immature stage of development. There are a crucial set of (largely non-technical) issues that need to be addressed if there is to be a continued development path in this area. These largely relate to the design principles that were articulated in Section 2. The main recommendations are that future research in this area is directed towards the following questions:

1. Developing extended enterprise approaches (extended to all aspects of the enterprise) to intelligent GIS. If GIS is to become a general management tool it must be distributed across an organization. How can this be designed and implemented?

2. Integration of GIS within a general business planning approach.

3. Development of analytically-based GIS. A move from digital mapping to more intelligent orientated systems based upon modelling and analysis as needed.

We suggest that ESRC should implement a programme of research between academics and business partners aimed at addressing at least some of these issues. The UK has the opportunity to exploit a major future development opportunity in an area where it has a considerable competitive research advantage.

Table 3 Provisional consent conditions for sewage treatment works in relation to river quality objectives

Sewage treatment works	Current consent BOD	Current consent Solids	Review date	Proposed consent BOD	Proposed consent Solids	Proposed consent Ammonia	Proposed consent Dry weather flow (m³ per day)	Receiving watercourse
Aire catchment								
Earby: treated	35	50	1/2/91	To be determined				Earby Beck
Earby: land treatment	25	35	1/2/91	To be determined				
Oxenhope	70	50		15	25	2	400	Bridge House Beck
Marley	90	100	1/3/94	20	30	10	25 800	Aire
Esholt	60	70	1/9/93	20	30	5	103 600	Aire
Rodley	65	80	30/11/93	60	60	20	9817	Aire
Knostrop	55	70	1.6.93	18	30	7	155 900	Wyke Beck
Smalewell	40	60		15	25	5	1632	Tyersal Beck
Lemonroyd	50	100	31/11/92	Transfer to Aire under consideration				Oulton Beck
Owlwood	30	40		12	20	5	6784	Lin Dyke
Calder catchment								
Eastwood	35	55	1/10/92	25	50	5	7740	Calder
Redacre	50	60		50	60	7	3510	Calder
Ripponden Wood	70	70		20	30	3	1285	Ashfield Beck
Barsey Green	40	60	19/11/90	To be determined				Barsey Clough
Meltham	50	50	19/8/90	To be determined				Mag Brook
Neiley	60	80	30/4/92	To be determined				Holme
High Royd	20	40		20	40	5	2635	Calder
Milner Royd	55	70	31/3/90	20	30	15	5060	Calder
Halifax	40	50		30	45	10	46 070	Calder
Brighouse: final	70	35	1/1/91	20	40	20		
Brighouse: tertiary	30	40	1/1/91	20	30	8	25 300	Calder
Huddersfield	70	130		20	30	8	108 600	Calder
Mitchell Laithes	65	90	30/4/93	40	60	8	57 000	Calder
Calder Vale	20	30		20	30	14	31 321	Calder
North Bierley	65	65	28/2/92	9	15	3	16 000	Spen
Spenborough: old	75	100	31/3/92	12	20	4		
Spenborough: new	60	70	31/3/92	12	20	4	13 800	Spen
Birkenshaw	35	80		To be determined				Oakewell Beck
Crofton: final	35	35	31/3/92	To be determined				Oakenshaw Beck
Crofton: tertiary	35	30	31/3/92	To be determined				
Normanton	35	40	30/11/90	To be determined				Ashfield Beck

ACKNOWLEDGEMENTS

We are grateful for the additional financial support provide by the NRA and Yorkshire Water. Of much more importance was the joint development and testing of the models on full regional catchment and river data sets. We are also grateful to the Nuffield Provincial Hospital Trust for their support in both a financial and intellectual sense.

REFERENCES

Clarke, M. (1993) Going beyond GIS: The next step to getting business the answers it needs. *The Corporate Real Estate Executive*, **8**(2), 36–38.

Clarke, M., Chesworth, J. and Harrison, S. (1993) Information and National Health Service Purchasing Authorities: the HIPPS system. Nuffield Institute, University of Leeds.

Clarke, M. and McDonald, A.T. (1993) Design principles for intelligent GIS – theory and applications, forthcoming.

Openshaw, S. (1992) Some suggestions concerning the development of artificial intelligence tools for spatial modelling and analysis in GIS. *Ann. Regl Sci.*, **26**, 35–51.

Index

Abstract data types   130
Adaptive resonance theory   56
Address-Point   24, 28
Adjacency vector   152-153
ADT, see Abstract data types
Advanced Very High Resolution Radiometer   104, 108, 166
Aerial photography   7, 147, 163, 173, 215, 305
AGI, *see* Association for Geographic Information
Airborne Thematic Mapper   194
Alper Records   278, 279
Altimeter   14, 111
Analytical plotter   213
Antarctic Environmental Data Centre   10
Antarctic   9, 10, 12, 13
Arc   95, 96
ARC/INFO   11, 12, 50, 51, 93, 98, 115, 119, 120, 128, 177, 239, 243, 244, 267, 271, 278, 279, 281, 300
Area   131
Artificial intelligence   72, 77, 183, 337
Association for Geographic Information   129
Autocad   128
AVHRR, *see* Advanced Very High Resolution Radiometer
Axis Amis   279

Basin hydrology   205
Belief functions   190
Benchmarking computer system   35-37
BGS, *see* British Geological Survey
Bintree   105-109
British Oceanographic Data Centre   10, 12
British Antarctic Survey   12
British Geological Survey   8, 10, 11, 12, 104, 240, 244
British National Space Centre   186
Buffers   50

CAD/CAM   128, 129, 326
Calibration problem   198

CASE tools, *see* Computer-Aided Software Engineering
Catchment area   204, 210
Catchment data   198
Catchment length   210
Catchment shape   210
CATNAP   332
Census   297
Chain   131
Channel flow velocity   204
Chorley Report   274
Classification   149
Classification, contextual   149, 150, 184
Classification, unsupervised   153
Clustering   193
Coastal ecosystems   14
Computer-Aided Software Engineering   123, 127–128, 129
Computer vision   186
Conductivity–temperature–depth cast   104, 105, 108
Confusion matrix   155, 158
Contextual classification   149, 150, 184
Contour encoding   164
Convolution kernel   148
Coordinate   279
CORINE   15
Countryside Commission   187
Countryside Stewardship   187
Critical path analysis   37
CTD, *see* Conductivity–temperature–depth cast
Customization   116

Data Centre   10
Data dictionary   297
Data effect   35
Data integration   14
Data model   105
Data quality   25, 270
Data structure, hierarchical   105
Data visualization   10, 14, 104
Data volumes   33, 35
Database   320

Database Management System 92-94, 108
DB2 128
dBase 331
DBMS, *see* Database Management System
Decision support system 248
Delauney triangulation 254
DEM, *see* Digital elevation model
Department of the Environment 168, 169, 186, 261, 265, 297, 319, 321
Digital elevation model 173, 174, 207 *et seq.*
Digital photogrammetric workstation 213
Digital terrain model 199, 203, 252
Digitizer error 56-57
Digitizing 10, 51, 52
Directorate of Military Survey 222
Discriminant analysis 242
Disk storage, effect on performance 34
Disk input–output 39-40
DoE, *see* Department of the Environment
Doppler effect 82
Dorset Heathland Information System 189
Dorset heaths 188, 196
Douglas–Peucker algorithm 69, 70, 77, 78
Drainage density 216
DTM, *see* digital terrain model
Dynamic visualization 148, 173
Dynamic Visualization Toolkit 177

Eastern Thames Corridor, contaminated sites 293
Eastern Thames Corridor, database 295
Eastern Thames Corridor, defined 288, 289
Eastern Thames Corridor, GIS project 290–291
Eastern Thames Corridor, potential data flows 296
Ecological modelling 185
Ecological models 186, 187
EIA, *see* Environmental Impact Assessment
Elapsed time, computer 35
Electronic distance measurement 82
Energy modelling 330–331
English Heritage 264
Environmental impact 254
Environmental Impact Assessment 173
Environmental Information Centre 10, 185, 186
Error, categorized 47
Error, definition 47
Error, sources 48
Error epsilons 50
Error propagation 14, 49
Error simulation 48-49
ERS-1 207, 212, 214
ESPRIT II 118, 119
ESRC/NERC Joint Programme, aims 1

ETC, *see* Eastern Thames Corridor
Ethernet 40
Euclidean operations 97, 98
European Geotraverse 8
European Science Foundation 18
Experimental Cartography Unit 8
Expert systems 14, 77, 183, 186, 187, 239, 243
Extended Region Adjacency Graph 164, 165

File server 40
Filtering, spatial 150
Fine Resolution Antarctic Model (FRAM) 9
Flood frequency 205
Flow network 220
Freeman chain code 164
Function 116
Fuzzy clustering 190, 193

Genamap 115
General Practitioner 329-330
Generalization, automatic 64, 71, 77
Generalization, examples 66-69
Generalization, intelligent 74
Generalization, objective 69
Generalization, paper maps 65-66
Generalization of spatial data sets 63-78
Generalized likelihood uncertainty estimator 197, 203
GENIE, *see* Global Environmental Network for Information Exchange
Geodetic coordinates 80
Geomorphological characteristics 207
GEOVIEW 101
GeoVision 128
GEWEX, *see* Global Energy and Water Cycle Experiment
GFIS 278, 279
GIS design principles and issues 327-328
GIS customization 126-127
GIS disasters 48
Global Change Database 15
Global Positioning System 25, 30, 79 *et seq.*, 212, 231, 232
Global Energy and Water Cycle Experiment 9
GLUE, *see* Generalized likelihood uncertainty estimator
Goal 116
GP, *see* General Practitioner
GPS, *see* Global Positioning System
Graphical User Interface 116, 122-123, 135, 148, 173
Greater London Council 278
Greenhouse gases 331
GUI, *see* Graphical User Interface

Habitats   187
Hadrian's Wall   264
HCI, *see* Human–Computer Interaction
HM Land Registry   262
Hoskyns G-Gp   278, 279
Human–Computer Interaction   118
Hydrological model   197
HyperArc   119, 120
HyperCard   305

IACGEC, *see* Inter-Agency Committee on Global Environmental Change
IDRISI   93
IGBP, *see* International Geosphere–Biosphere Programme
Image analysis software   190-195
Image segmentation   186, 196
Industrial structure   310
INGRES   88, 94, 128, 141
Input–output   35, 38-39, 41, 42, 43
Institute of Hydrology   10, 11, 198
Institute of Oceanographic Sciences   104
Institute of Terrestrial Ecology   10, 11, 185, 189
Inter-Agency Committee on Global Environmental Change   18
Inter-visibility   173
Interferometry   212, 214
Intergovernmental Oceanographic Commission   10
Intergraph   11, 128
International Geosphere–Biosphere Programme   15
Interpolation of contours   215
Interpolation   109-110
ITE, *see* Institute of Terrestrial Ecology

Kappa coefficient   155, 158
Knowledge based segmentation   187
Kriging   109-110

Land parcel   261
Land use change   11, 147, 168
Land Use Change Statistics   297, 321
Land use map   157, 160
Land cover   185-186, 195
Land cover classes   153
Land cover classification   153, 155, 182
Land use   152
Land use planning   251
Land–Ocean Interaction Study, *see* LOIS
Landsat   12, 182, 189, 215, 232
Landscape simulation   251
Landsliding   244-246
Large data sets, handling   219
LaserScan   11, 65, 88, 253, 279

Line complexity   52-54
Line-following   214
Linear regression   242
Link   131
LITES2   65, 88
Local authority   273
Local government   298-300, 322
Logistic regression   242
LOIS   14-15
London Planning Advisory Committee   298

Magnetometry   7
Majority filter   189-190
Map scale   63 *et seq.*
Maximum likelihood decision rule   149, 153, 155
McDonnell-Douglas   279
Measurement error   50, 51
Memory management   107
Memory (computer), effect on performance   34
Meteorology   9, 104, 108, 109
Metric operations   97, 98
Modelling   14, 328
Modelling, energy   330-331
Modelling, water quality   331-335
Monte-Carlo simulation   49-50, 78, 197, 203
Morphometry   209
Multi-temporal remote sensed data   149, 195
Multimedia   305
Multiprocessing   41
Multispectral data   147
Multispectral Scanner   215
MuSIP project   186
National Atmospheric and Oceanographic Administration   166
National Grid   84
National Water Archive   10
National Center for Geographic Information and Analysis   49, 118
National Topographic Database   23
National Rivers Authority   331
National Health Service   329
National Street Gazetteer   265
National Geosciences Information Service   10
NATO Advanced Study Institutes   118
Natural hazard   239 *et seq.*
Natural hazard, definition   239
Natural hazard, estimating probabilities   240–241
NAVSTAR, *see* Global Positioning System
NCGIA, *see* National Center for Geographic Information and Analysis
Network   40, 44
Neural computing   52, 60

Neural net   52, 55–56, 57-58, 61
Nexpert   239, 243, 244
NOAA, *see* National Atmospheric and Oceanographic Administration
Node   131
NOMIS   323
North Sea Project   9, 12
NRA, *see* National Rivers Authority
NUTIS   14, 103, 185, 186, 239

Object classes, spatial   96
Object data   108
Object identification   163
Object labelling   166
Object oriented data modelling   94-95
Object oriented database management   94
Object search map   164
Object-based data structures   163
Object-based paradigm   91 *et seq.*
Object-based spatial re-classification   162–172
Ocean circulation   9
Ocean temperature   110
Octree   112
Office of Population Censuses and Statistics   169
OPCS, *see* Office of Population Censuses and Statistics
ORACLE   11, 98, 99, 100, 128
Ordnance Survey   23 *et seq.*, 79, 90, 169, 171, 222, 262, 267
OSCAR   24, 28
OSGB36 datum   79 *et seq.*
OSM, *see* Object search map
OSSN80 datum   81 *et seq.*
Ozone depletion   8

Pafec DOGS   279
Parallel computing   41-43, 44, 60, 61
Performance assessment   37
Performance (of system)   34
Planes Spatial   279
Point   131
Poisson probabilities   241
Polygon   131
Polygon overlay   35, 50
Population change   305-307
Post-classification spatial processing   150
Postcode   267, 297
Postgres   94
Pre-classification image transforms   149
Prime   128
Processor time   35, 38
Property terrier   265
Proudman Oceanographic Laboratory   10

Quadtree   107, 108, 111, 141
QUEL   129
Query-by-Example   129, 141

RAID   43
Raster–vector debate   91, 101
Recession parameter   204
Region   97
Region editing   190, 192-193
Regional Research Laboratories   17, 18, 325, 326
Regionalized variables   109, 111
Register of Contaminated Land   265
Relational database   115, 116, 141
Relational database management system   92, 93, 101, 127, 129
Remote sensing   11, 14, 147, 168, 186, 216, 232
Remotely-sensed images   148
Rhind's rule   69
River catchment characteristics   209-211
Root zone storage   199
Rotational invariance   53
Royal Society for the Protection of Birds   187
RRL, *see* Regional Research Laboratories
Runoff   208
Rural environmental planning   252

SAP, *see* Spatially-Aware Professional
Saturated zone   201
Scanning, raster   214
Segmentation   111, 182, 191, 192
Semi-variogram   110
Set-oriented operations   97
Sheet edge mismatching   267
SIA Datamap   279
SLIMPAC   265
Smallworld GIS   127, 271
Soil moisture   199, 203
SPANS   115, 278, 279
SPARK, *see* Spatial Reclassification Kernel
Spatial data types   133-135
Spatial classification   196
Spatial filtering   150
Spatial modelling   241
Spatial operations   97
Spatial re-classification   150-172
Spatial re-classification, object-based   162–172
Spatial Reclassification Kernel   151
Spatial relationships   95
Spatial resolution   149, 150, 163
Spatial scale   147, 148
Spatially-Aware Professional   114
Spatially-extended SQL   141

Spatiotemporal GIS   103-112
Spectral reflectance characteristics   149
SPOT   147, 154, 168, 182, 207, 212, 213
SQL   94, 98, 101, 113, 116, 118, 128, 131, 141, 142, 198
SQL, spatial extensions   123, 128-135, 141
SQL3   130 *et seq.*
Starburst project   94, 101
Stereo digitizing system   148, 175, 176
Stereo matching   174
Stereo matching, algorithm   174
Stereoscopic   147
Strand   97
Stream channel network   207 *et seq.*
Stream ordering methods   211, 227
Streamflow   208
Subsidence   246-248
Sybase   128
Synthetic aperture radar   207, 212, 214

Task   116
Thematic Mapper   182, 184, 189, 215
Three-dimensional GIS   178
TIN, *see* Triangulated Irregular network   216
TOMCAT   332
Topographic characteristics   208
Topographic data extraction   215
Topological operators   97, 98
Topological relationships   108
Topology   35, 119
Topology building   100
Transmissivity   201
Transputer   44, 105, 108, 203
Triangulated Irregular network   216, 217-218

UGIX   113, 119-126, 136, 141
UK Digital Marine Atlas   12
Uncertainty   47
Unsaturated zone   201
Unsupervised classification   153
Upper Atmosphere Research Satellite   8
User interface   114

VAX   119
Viewmap   279
Virtual memory   219
Visual impact   251, 254
Visualization   108-109, 115
Volume data   105
VTRAK   65, 70

Water Information System   12, 197, 198, 199, 202, 203
Water quality modelling   331-335
WGS84, *see* World Geodetic System
Widgets   124
Wildlife and Countryside Act   263
WIMP interface   114, 116, 125
Wings   279
WIS, *see* Water Information System
WOCE, *see* World Ocean Circulation Experiment
Workstation   40, 105
World Climate Research Programme   9
World Geodetic System   79, 81, 82, 83
World Ocean Circulation Experiment   9, 103

X-Windows   115
XRAG, *see* Extended Region Adjacency Graph